Engineering Optimization: Methods and Applications

Optimization carries great significance in both human affairs and the laws of nature. It refers to a positive and intrinsically human concept of minimization or maximization to achieve the best or most favorable outcome from a given situation. Besides, as the resources are becoming scarce there is a need to develop methods and techniques which will make the systems extract maximum from minimum use of these resources, i.e. maximum utilization of available resources with minimum investment or cost of any kind. The resources could be any, such as land, materials, machines, personnel, skills, time, etc. The disciplines such as mechanical, civil, electrical, chemical, computer engineering as well as the interdisciplinary streams such as automobile, structural, biomedical, industrial, environmental engineering, etc. involve in applying scientific approaches and techniques in designing and developing efficient systems to get the optimum and desired output. The multifaceted processes involved are designing, manufacturing, operations, inspection and testing, forecasting, scheduling, costing, networking, reliability enhancement, etc. There are several deterministic and approximation-based optimization methods that have been developed by the researchers, such as branch-and-bound techniques, simplex methods, approximation and Artificial Intelligence-based methods such as evolutionary methods, Swarm-based methods, physics-based methods, socio-inspired methods, etc. The associated examples are Genetic Algorithms, Differential Evolution, Ant Colony Optimization, Particle Swarm Optimization, Artificial Bee Colony, Grey Wolf Optimizer, Political Optimizer, Cohort Intelligence, League Championship Algorithm, etc. These techniques have certain advantages and limitations and their performance significantly varies when dealing with a certain class of problems including continuous, discrete, and combinatorial domains, hard and soft constrained problems, problems with static and dynamic in nature, optimal control, and different types of linear and nonlinear problems, etc. There are several problem-specific heuristic methods are also existing in the literature.

This series aims to provide a platform for a broad discussion on the development of novel optimization methods, modifications over the existing methods including hybridization of the existing methods as well as applying existing optimization methods for solving a variety of problems from engineering streams. This series publishes authored and edited books, monographs, and textbooks. The series will serve as an authoritative source for a broad audience of individuals involved in research and product development and will be of value to researchers and advanced undergraduate and graduate students in engineering optimization methods and associated applications.

Kishalay Mitra · Richard Everson ·
Jonathan Fieldsend
Editors

Optimization, Uncertainty and Machine Learning in Wind Energy Conversion Systems

Springer

Editors
Kishalay Mitra
Department of Chemical Engineering
Indian Institute of Technology Hyderabad
Hyderabad, Telangana, India

Richard Everson
Department of Computer Science
University of Exeter
Exeter, UK

Jonathan Fieldsend
Department of Computer Science
University of Exeter
Exeter, UK

ISSN 2731-4049 ISSN 2731-4057 (electronic)
Engineering Optimization: Methods and Applications
ISBN 978-981-97-7908-6 ISBN 978-981-97-7909-3 (eBook)
https://doi.org/10.1007/978-981-97-7909-3

This Springer imprint is published by the registered company Springer Nature Singapore Pte Ltd.
The registered company address is: 152 Beach Road, #21-01/04 Gateway East, Singapore 189721, Singapore

Foreword

The current surge in research on optimization and machine learning is opening up tremendous opportunities in applying them to practical and societal problems. While most studies focus on developing efficient computational algorithms for solving various practical problem-solving tasks, such as, prediction, classification, clustering, regression, modeling and optimization, they are often demonstrated to work on synthetic data and artificial problems. Another practicality which is common to most problems is the uncertainty in data, environmental conditions, models, and implementational parameters. A problem-solving algorithm which does not take into account uncertainties is incapable of producing any meaningful result.

Among various alternative energy sources, wind energy is the most extensively utilized, emerging, and economically profitable energy source. However, despite its ongoing implementation, there is a huge scope for improving the current practice of wind farm layout design. Wind energy harvesting problems must consider uncertainties in wind speed, making the optimization and use of machine learning tasks challenging.

This book is unique in the sense of its focus on using artificial intelligence, machine learning, and efficient single and multi-objective optimization techniques in achieving various tasks of designing, predicting, forecasting and modeling of wind energy systems. This book is a culmination of authors' many years of studies in this area. Each chapter addresses a single aspect of computational methods in wind energy systems, making the presentation of the book lucid, focused, and fundamental.

Although this book is focused on wind energy conversion systems, the methodologies and algorithms described in the book can be extended to address other similar energy systems, or in other engineering and societal problem-solving tasks.

The book advances the knowledge of applying artificial intelligence, machine learning and advanced optimization techniques to an important engineering and societal problem. The book does so, not for the sake of being up-to-date with current technologies, but in demonstrating the usefulness of advanced computational methods in addressing challenging practical problems.

I strongly encourage you to read the book, share its contents with your colleagues, and replicate the concepts to problems of your interest.

Kalyanmoy Deb
University Distinguished Professor
Michigan State University
East Lansing, USA

Preface

Wind energy is second only to Hydropower in terms of renewable energy generation worldwide. This is in no small part due to a number of beneficial characteristics. It is freely available, abundant, green and cost-effective to operate. It also is able to function flexibly under a range of different wind conditions and across a range of locations. The popularity it has earned since the beginning of the third millennium has helped the total cumulative wind energy electricity generation capacity worldwide to reach 906 GW at the end of 2022. India is showing tremendous interest and has become a great contributor in this green energy generation process, standing after China, the United States and Germany, providing impetus for researchers working in this field to advance the international body of knowledge in this area through collaborative research. Responding to the unique call for proposals under the Scheme for Promotion of Academic and Research Collaboration (SPARC) run by the Ministry of Education, Government of India, a project titled "Wind farm layout optimization under uncertainty using wind speed forecasting through probabilistic models and comparison with machine learning algorithms" was developed by a few of us from the Indian Institute of Technology Hyderabad and University of Exeter, UK as a step towards this knowledge enhancement effort. This was originally envisioned as a mobility support programme with grand vision of an academic and cultural exchange and collaboration between research scholars and professors from both locations. Despite the unprecedented situation of the global COVID- 19 pandemic beginning in the early stages of the programme, the team was fully committed to deliver more than was initially planned. All commitments, including the international workshop as well as teaching a course, were fulfilled in an online mode, the novel way of communication and collaboration that the COVID-19 era taught us. Additionally, research scholars' visits were enabled upon relaxation of restrictions towards the end of the period. This book is a humble effort towards the team's commitment to this collaborative research.

Events such as irregularities on the earth surface, the uneven heating of the atmosphere by the sun and the earth's rotation lead to solar energy being converted into wind energy. Wind mills, familiarity with which we developed since our childhood through our kindergarten or nursery books, have been used for applications such as grain grinding, pumping of water, and oil extraction from seeds etc. for considerable periods of time. This device has undergone technological transformations (currently known as a wind turbine) over the last 2-–3 decades and has been increasingly utilized to gather and transfer the kinetic energy that wind produces into electricity to power the grid. The job of electricity generation is performed through a wind farm, where a number of wind turbines is organized optimally to harness maximum energy from this renewable resource. Several practical challenges such as handling wake interactions, noise control, investment cost, longevity of turbine, and visual impacts, etc. might affect the assessment of a particular farm configuration. Additionally, the supply of wind as a source of energy has its own uncertainty associated with it. When wind farms are designed and operated not considering without sufficient consideration of such sources of uncertainty, the outcome is far below expectation. This in turn means the task of optimal placement of wind turbines, also known as micro-siting, is not as trivial as it may at first appear. As wind farm design and control significantly affects power generation and thereby return on investment, this area of research has attracted the attention of wind farm designers, addressing many challenges it faces. With the recently acclaimed success over diverse fields, researchers from the wind energy domain are also extremely keen to see the application of machine learning techniques in several aspects of wind farm layout optimization and control. Addressing many such advanced aspects, the objective of the book is to present the state-of-the art technologies in the fields of wind farm layout optimization and control, which can improve the current industry/research practices. The approach adopted in this book is, therefore, to articulate several aspects of wind farm micro-siting through detailed discussions in different chapters and explore novel ways to provide solutions to these problems utilizing recent techniques. First, an introduction to the wind farm micro-siting problem has been presented in Chap. 1, where challenges involved in it, different approaches adopted by researchers to solve such issues and the avenues for further research are discussed. As a basic building block, the fundamentals of deterministic multi-objective optimization and an overview of several methodologies for solving such problems are outlined in Chap. 2. As several parameters during optimization should ideally be treated as uncertain to mimic micro-siting under uncertain environments (e.g. wind state uncertainty), the most realistic case, handling optimization problems under uncertainty and methodologies to solve them are also covered in this chapter. Despite having the capability to identify global optima in the presence of several local optima and having an edge to solve multiobjective optimization problems over classical optimization methods, the population-based evolutionary techniques, though popularly adopted by researchers, are generally computationally expensive. This case becomes more intensive if the mathematical model connecting the decision variables and objective functions and constraints also becomes expensive to evaluate. Handling such computationally expensive models inside an optimization framework becomes a challenging proposition in a given time frame as optimization techniques are iterative in nature. Moreover, in some cases, optimization problems rely on existing data sets and there might be no direct functional mapping available between decision variables and the objective functions or constraints. Though surrogates or metamodels become useful in such cases to find solutions, they bring uncertainty in their predictions leading to lack of reliability. Chapter 3 attempts to address this challenge by quantifying the uncertainty while using surrogates and further utilizing them in optimization. Next, the focus is shifted towards accurate evaluation of wind speed inside wind farms. As accurate wind speed prediction helps in realistic wind farm power calculations and other constraints, a concise description of predominant wake models covering the approaches of computational fluid dynamic (CFD), Large Eddy Simulations (LES) and analytical models is covered in Chap. 4. The role of key factors such as inter-turbine distance, yaw misalignment, atmospheric thermal stability and wind direction in influencing wake dynamics within wind farms is discussed in Chap. 5. Further, it covers a laconic introduction to the Simulator for Wind Farm Applications (SOWFA), an opensource, finite volume, LES code based on OpenFOAM, which can help to model wake interactions efficiently inside a wind farm. The discourse also encompasses a case study using SOWFA demonstrating the impact of yaw misalignment on the performance metrics of a wind farm. On the one hand, CFD-based high-fidelity simulations are reported to be highly accurate; however, attaching such models in an iterative environment of optimization and looking for quick results in a realistic time frame using manageable computational resources is a utopian idea. One of the approaches used under such a scenario is to build hybrid models, if not a full surrogate, where a part of the complicated phenomena can be captured using supervised machine learning (ML) techniques while keeping physics-based understanding to predict the rest of the phenomena. Surrogates can be validated with experimental data, if available, or the data generated by highfidelity CFD models. One such hybrid approach, where machine learning-based artificial neural networks are used to approximate certain non-linear phenomena, keeping physics involved in the process as much as possible, is covered in Chap. 6. Wind data availability for the intended site location of a wind farm is not typically available beyond a decade or two historically due to the expense and challenge in placing and maintaining sensors and data acquisition systems in the field. Hence, wind frequency maps (WFMs), which are joint probability mass functions of wind speed and direction and are used quite often by the power calculation models, are generated using short-term wind data. Such WFMs, sitting at the heart of power calculations, can be made more accurate with the help of forecasting techniques, which forecast wind behaviour by studying wind data from the past. ML techniques, specifically long short-term memory network (LSTM) and least square support vector regression (LSSVR), which have shown accurate and improved forecasting results compared to the existing statistical models, can be utilized to generate accurate forecasting results and thereby improve the quality of WFMs by predicting unseen wind scenarios of the future. Using live data from a realistic wind farm, the potential of such approaches is demonstrated in Chaps. 7 and 8, respectively. Such accurate forecasting can help generate several unseen wind scenarios and help the robust design of wind farms against the variation of wind state parameters. However, forecasting over a very long-term using a limited amount of past data might be erroneous. One way to capture the nature of WFMs is to identify the regions of interest using machine learning- based clustering techniques and sampling in these focused regions to generate new scenarios which might appear in future but not present in the past data. Another approach of scenario generation is to capture the embedded distribution in the given data and use that for more scenario generation for future. ML-based generative adversarial network (GAN) has shown remarkable performance in capturing embedded distribution from images in the recent past. Utilizing the high-resolution WFMs of the past data, GANs can be trained to capture the embedded distribution in the data to generate more unseen scenarios for the future, which can, in turn, be used for robust wind farm designs. Such demonstrations are shown in Chaps. 9 (Clustering) and 10 (GAN), respectively. Once the design of the wind farm is optimized, operational optimization of wind turbines can be performed using several turbine parameters, such as yaw angle and blade pitch angle control actions, to further enhance their performance towards generating more power. Utilizing emerging techniques, a controlled study on an optimally designed wind farm has been showcased in Chap. 11 using reinforcement learning (RL). Model predictive control (MPC), one of the most successful techniques in the control domain, has also been used next to see its applicability to the wind farm control problem, and the results are discussed in Chap. 12. Most literature reports wind speed and direction as the two most significant contributing factors to wind farm design and control. To examine the sensitivity of various other wind farm parameters such as pitch angle, vane position and, operating temperatures etc. in wind power prediction, Exploratory Data Analysis (EDA) and feature selection techniques have been performed on realistic wind data and reported in Chap. 13. Many Artificial intelligence (AI) models, especially deep learning models such as long shortterm memory models, are often considered black boxes because they lack transparency owing to their inherent opacity regarding the decision-making process. Explainable AI (XAI) addresses the challenge of model opacity by providing insights into how AI models arrive at their decisions. XAI, especially TimeSHAP, has been utilized for interpreting the LSTM model that predicts the active power generated in a realistic wind farm data and the results are reported in Chap. 14. In this way, several aspects of wind farm design, e.g. state-of-the-art research performed so far—Chap. 1, optimization fundamentals—Chap. 2, surrogate optimization—Chap. 3, high-fidelity wake calculation and power estimation—Chaps. 4 and 5, realistic wake calculations and power estimations under resource constraints—Chap. 6, wind forecasting—Chaps. 7 and 8, optimization placement of turbines under uncertainty—Chaps. 9 and 10, wind farm control—Chaps. 11 and 12 and wind data analysis—Chaps. 13 and 14 are covered in this book which can be made useful as a reference for research for the community involved in improving state of the art in wind energy.

The research team involved in this project at Indian Institute of Technology Hyderabad and University of Exeter is extremely grateful to the SPARC programme (SPARC/20182019/P1084/SL) of the Ministry of Education, Government of India, which catalysed this collaboration. The seeding of this collaborative effort dated back in 2018 when two teams were involved in preparing the SPARC project proposal and a common thread in this whole effort was Dr. Tinkle Chugh, an alumnus of Indian Institute of Technology Hyderabad doing his post-doctoral research at the Computer Science department, University of Exeter. The team at Indian Institute of

Technology Hyderabad is also grateful to the funding provided by the National Supercomputing Mission (DST/NSM/R&D_HPC_Applications/2021/23), Department of Science and Technology, Government of India for carrying out part of the research presented in this book. Discussions with Prof. Raja Banerjee and Prof. Soumya Jana on project proposal preparation and submission were highly appreciated in this regard. During the entire project execution, several research scholars were extremely active to take this project forward, e.g., Dr. Srinivas Soumitri Miriyala, Dr. Priyanka D. Pantula, Dr. Prateek Mittal, Dr. Inapakurthi Ravi Kiran, Ms. Keerthi N. Pujari—the entire project team is grateful to them without which the progress in this project would have not been possible especially with the significant impact of COVID-19 during the project. Fuelling additional impetus post- COVID-19 era by the research scholars, Mr. Utkarsh Deep Tiwari, Ms. Aswitha Tadepalli, Mr. Vishnu Swaroopji Masampalli and Ms. Rashi Verma, towards wind research in Global Optimization and Knowledge Unearthing Lab (GOKUL) is also gratefully acknowledged. Arranging the international online workshop during the COVID-19 crisis was an experience in itself and the team here conveys gratitude to the coveted speakers, Prof. Kalyanmoy Deb, Dr. Eric Tromeur, Dr. Niranjan S. Ghaisas apart from the editors of this book as well as research scholars who agreed to present their research and spend time with the attendees of the workshop. Finally, a special thanks to the many health professionals and key workers worldwide who helped us all through the COVID-19 pandemic and gave so much.

Hyderabad, India — Kishalay Mitra
Exeter, UK — Richard Everson
Exeter, UK — Jonathan Fieldsend

Contents

Editors and Contributors

About the Editors

Prof. Kishalay Mitra received his B.E. from the National Institute of Technology (NIT) Durgapur, India, in 1995, M. Tech. from Indian Institute of Technology (IIT) Kanpur, India, in 1997, and Ph.D. from the Indian Institute of Technology (IIT) Bombay, India, in 2009 (all in Chemical Engineering). He is a Professor in the Department of Chemical Engineering and associated faculty in the Departments of Artificial Intelligence and Climate Change at IIT Hyderabad, India. His work interests lie in the interface of data analysis and process optimization such as machine learning, evolutionary optimization, optimization under uncertainty, planning and scheduling of supply chain, and analysis of systems involving biology, climate change, and renewable energy. He worked in several engineering leadership positions at General Electric Global Research, Bangalore, and Tata Research Development and Design Centre, Pune for close to 15 years. His research has been supported by several leading agencies in India like the Department of Science and Technology, Department of Bio-Technology, Ministry of Education, Defence Research and Development Organization, and Tata Steel through several nationally important projects (~ INR 50 million). He has over 200 international journal and conference publications to his credit. Apart from serving on the editorial board of several reputed journals and conferences, he has been visiting Washington University in St. Louis, USA, and the University of Washington, Seattle as a visiting professor on several occasions. He has been named among the World's top 2% scientists according to the latest profile review conducted by a group from Stanford University since 2021.

Prof. Richard Everson is a Professor of Machine Learning and Director of the Institute for Data Science and Artificial Intelligence at the University of Exeter. He is a Fellow of The Alan Turing Institute and the Turing University Lead for Exeter. His research interests focus on statistical machine learning, multi-objective optimization, and the interactions between them. Particularly relevant is his work on measuring and accounting for uncertainty in evolutionary and robust Bayesian optimization. He was the principal investigator for the EPSRC 'Data-Driven Surrogate-Assisted Evolutionary Fluid Dynamic Optimisation' which pioneered Bayesian optimization methods for Computational Fluid Dynamics problems.

Prof. Jonathan Fieldsend received a BA in Economics from Durham University in 1998, an M.Sc. in Computational Intelligence from the University of Plymouth in 1999, and a Ph.D. in Computer Science from the University of Exeter in 2003. He is a Professor of Computational Intelligence at the University of Exeter. He primarily works on the interface of optimization heuristics and machine learning. This includes both fundamental advances, as well as solving immediate and near-term problems with partners from industry and the public sector. He has particular interests in multi-objective, expensive, and uncertain design optimization problems, along with search landscape characterization. He has contributed to several professional activities, including acting as Editor-in-Chief for the ACM Genetic and Evolutionary Computation Conference in 2022.

Contributors

Tinkle Chugh Department of Computer Science, University of Exeter, Exeter, UK

Jonathan Fieldsend Department of Computer Science, University of Exeter, Exeter, UK

Niranjan S. Ghaisas Department of Mechanical and Aerospace Engineering, Indian Institute of Technology Hyderabad, Hyderabad, Telangana, India; Department of Climate Change, Indian Institute of Technology Hyderabad, Hyderabad, Telangana, India

Ravi Kiran Inapakurthi Department of Chemical Engineering, Indian Institute of Technology Hyderabad, Hyderabad, Telangana, India

Naveen N. Kethavath Department of Mechanical and Aerospace Engineering, Indian Institute of Technology Hyderabad, Hyderabad, Telangana, India

Vishnu Swaroopji Masampally Department of Chemical Engineering, Indian Institute of Technology Hyderabad, Hyderabad, Telangana, India; TCS Research, Tata Consultancy Services, Tata Research Development and Design Centre, Pune, India

Atanu Mazumdar Department of Electrical Engineering and Automation, Aalto University, Espoo, Finland

Srinivas Soumitri Miriyala Department of Chemical Engineering, Indian Institute of Technology Hyderabad, Hyderabad, Telangana, India

Kishalay Mitra Department of Chemical Engineering, Indian Institute of Technology Hyderabad, Hyderabad, Telangana, India;
Affiliated Faculty, Department of Climate Change, Indian Institute of Technology Hyderabad, Hyderabad, Telangana, India;
Affiliated Faculty, Department of Artificial Intelligence, Indian Institute of Technology Hyderabad, Hyderabad, Telangana, India

Prateek Mittal Department of Computer Science, University College London, London, UK

Kingshuk Mondal Department of Mechanical and Aerospace Engineering, Indian Institute of Technology Hyderabad, Hyderabad, Telangana, India

Priyanka Devi Pantula Department of Chemical Engineering, Indian Institute of Technology, Hyderabad, Kandi, Telangana, India

Jay A. Patel Department of Mechanical and Aerospace Engineering, Indian Institute of Technology Hyderabad, Hyderabad, Telangana, India

NagaSree Keerthi Pujari Department of Chemical Engineering, Indian Institute of Technology Hyderabad, Hyderabad, Telangana, India

Aswitha Tadepalli Department of Chemical Engineering, Indian Institute of Technology Hyderabad, Hyderabad, Telangana, India

Utkarsh Deep Tiwari Department of Chemical Engineering, Indian Institute of Technology Hyderabad, Telangana, India

Rashi Verma Department of Chemical Engineering, Indian Institute of Technology Hyderabad, Hyderabad, Telangana, India

Abbreviations

ABL	Atmospheric Boundary Layer
ADF	Augmented-Dickey-Fuller
ADM	Actuator Disc Model
AEP	Annual Energy Production
AI	Artificial Intelligence
ALM	Actuator Line Model
ANN	Artificial Neural Networks
ARIMA	Auto Regressive Integrated Moving Average
ASM	Actuator Surface Model
BDS	Brock-Dechert-Scheinkman
BIC	Bayesian Information Criterion
BO	Bayesian Optimization
BPNN	Back Propagation Neural Network
CAEs	Convolutional Auto Encoders
CCP	Chance Constrained Programming
CDF	Cumulative Density Function
CFD	Computational Fluid Dynamics
CNN	Convolutional Neural Networks
DBMOPP	Distance-based Visualisable Test Problems
DCGANs	Deep Convolutional Generative Adversarial Networks
DDPG	Deep Deterministic Policy Gradient
DQN	Deep Q Networks
DRIFT	Data-driven Robust optimization through Intelligent Fuzzy Transcription
DT	Decision Tree
EA	Evolutionary Algorithms
EDA	Exploratory Data Analysis
EEG	Electroencephalography
EMD	Empirical Mode Decomposition
EP	Expected Power
ES	Evolutionary Strategy

ETR	Extra Trees Regressor
FCCP	Fuzzy Chance Constrained Programming
FEVM	Fuzzy Expected Value Model
FLORIS	Flow Redirection and Induction in Steady State
FNN	Feed forward Neural Networks
FRO	Fuzzy Robust Optimization
GA	Genetic Algorithm
GANs	Generative Adversarial Networks
GP	Gaussian Process
GRU	Gated Recurrent Unit
GW	Giga Watts
HV	Hypervolume
iLQR	iterative Linear Quadratic Regulator
I-NFCM	Image-Neuro Fuzzy C Means clustering
INLP	Integer Nonlinear Programming
ITD	Inter Turbine Distance
kW	Kilo Watts
LES	Large Eddy Simulations
LHS	Latin Hypercube Sampling
LON	Local Optima Network
LQR	Linear Quadratic Regulator
LSSVR	Least Square Support Vector Regression
LSTM	Long Short-Term Memory
MAE	Mean Absolute Error
MINLP	Mixed Integer Nonl-Linear Programming Problem
MIP	Mixed-Integer Programming
ML	Machine Learning
MLP	Multi- Layer Perceptron
MOEA	Multi-Objective Evolutionary Algorithms
MOEA/D	Multi-Objective Evolutionary Algorithm based on Decomposition
MOO	Multi-Objective Optimization
MOP	Multi-Objective Optimisation Problem
MPC	Model Predictive Control
MSE	Mean Squared Error
MVNS	Multivariate Normal Sampling
MW	Mega Watts
NFCM	Neuro Fuzzy C Means clustering
NP	Non-deterministic Polynomial-time
NREL	National Renewable Energy Laboratory
NSGA II	Non-Dominated Sorting Genetic Algorithm II
NWP	Numerical Weather Prediction
OUU	Optimization Under Uncertainty
PDF	Probability Density Function
PFI	Permutation Feature Importance
P-LON	Pareto Local Optima Network

PLS	Pareto Local Search
PMF	Probability Mass Function
PSO	Particle Swarm Optimization
R^2	Coefficient of Regression
RANS	Reynolds-Averaged NavierNavier–Stokes
RFR	Random Forest Regressor
RL	Reinforcement Learning
RMSE	Root Mean Square Error
RNN	Recurrent Neural Network
RO	Robust Optimization
RSD	Relative Standard Deviation
RVEA	Reference Vector-based Evolutionary Algorithm
SARIMA	Seasonal Auto Regressive Integrated Moving Average
SD	Standard Deviation
SFS	Sub-Filter Scale
SHAP	SHapley Additive exPlanations
SOWFA	Simulator fOr Wind Farm Applications
SP	Stochastic Programming
SRM	Structural Risk Minimization
STL	Seasonal and Trend decomposition using LOESS
SVM	Support Vector Machines
WFLO	Wind Farm Layout Optimization
WFM	Wind Frequency Map
XAI	eXplainable AI

Chapter 1
Introduction to Wind Farm Micro-siting

Prateek Mittal and Kishalay Mitra

Abstract Global interest in green energy harvesting has surged due to its immense potential in generating electricity and reducing reliance on depleting fossil fuels to satisfy our energy needs. Amidst various forms of renewables, wind energy is one of the most extensively tapped, emerging, and economically profitable energy source. Despite its ongoing massive implementation, there is a huge scope for improving the current practice of wind farm layout design, as harvesting wind energy through wind farms involves a lot of contributing factors. This chapter presents a brief overview of the wind farm layout design task and the challenges involved in attaining the task goals. After motivating the reader about micro-siting, a detailed survey of different methods adopted in the literature, including the heuristic, meta-heuristic, single/multi-objective optimization, and machine learning, towards achieving efficient micro-siting, considering several practical aspects, is presented. Additionally, discussions on research gap areas and avenues for advancements in current state of the art are presented. Finally, the approach adopted in several chapters of this book covering basic building blocks of understanding micro-siting concepts is also outlined.

Keywords Wind energy · Wind farm layout optimization · Annual energy production · Machine learning · Optimization · Uncertainty analysis

---

P. Mittal
Department of Computer Science, University College London, London WC1E6BT, UK
e-mail: prateek.mittal@ucl.ac.uk

K. Mitra (✉)
Department of Chemical Engineering, Indian Institute of Technology Hyderabad, Hyderabad, Telangana 502284, India
e-mail: kishalay@che.iith.ac.in

Department of Climate Change, Indian Institute of Technology Hyderabad, Hyderabad, Telangana 502284, India

Department of Artificial Intelligence, Indian Institute of Technology Hyderabad, Hyderabad, Telangana 502284, India

K. Mitra et al. (eds.), *Optimization, Uncertainty and Machine Learning in Wind Energy Conversion Systems*, Engineering Optimization: Methods and Applications,
https://doi.org/10.1007/978-981-97-7909-3_1

1.1 Motivation

The growing population, ambitious plans for economic development, and the use of advanced technologies for various novel applications for human convenience have given rise to additional energy demand, which in turn led to the exorbitant consumption of fossil fuels and elevated emissions of greenhouse gases (Li et al. 2023; Qureshi and Warudkar 2023; Kirchner-Bossi and Porté-Agel 2024). According to the World Energy Consumption Report, a surge of 5% per year in energy demand is observed in the year 2022 and is expected to increase by 47% by 2050, following the ongoing trend (S&P Global Commodity Insights 2021; Global Energy Perspective 2023). This huge energy demand is mainly (82%) fulfilled by burning fossil fuels, which are not only present in finite quantity but also potential causes of increase (60% /year since 1990) in CO_2 emissions, leading to a significant toll on human health (Energy Economics 2023; Statista 2024b). However, the growing environmental concerns and dwindling fossil fuel reserves have encouraged the world to seek alternative low-carbon energy sources (renewable energy) for the survival of the human race. Amidst various forms of renewable sources of energy such as wind, solar, geothermal, and tidal, wind energy emerged as the most influential source of power and attracted considerable attention from researchers, industrialists, and federal policymakers due to its positive attributes of easy availability, cleaner production, and low operational cost. According to the Global Wind Energy Council, the global wind capacity has reached 906 GW with Year-on-Year growth of 15% observed in 2023, which is 1.8% higher than in 2022. It is further estimated that a new capacity of 680 GW will be added by 2027 and will increase fourfold by 2050, following the continuing trend (Global Wind Report 2023; Renewable Energy 2024; Statista 2024a).

In practice, wind energy is harnessed using wind turbines placed inside a wind farm, which convert the kinetic energy of wind into mechanical energy and finally into the electrical form. The standard approach to extract maximum energy is to place an optimal number of wind turbines in optimal configuration inside a wind farm, which is known as wind farm micro-siting or wind farm layout optimization (WFLO). However, the development of wind farms, with a target of maximum energy throughput, e.g. is not trivial and involves a variety of challenges. Following the common myth of more the number of turbines, more the energy generated, wind farm developers often instal turbines close to one another, causing a significant reduction in wind speed due to the inter-turbine wake interactions. Eventually, this reduces the overall power production by the wind farm and increases the installation cost (Mittal and Mitra 2017a; Borch et al. 2023). Also, the presence of different types of constraints, such as noise control, investment cost, wake interactions, the longevity of the turbine, visual impacts, etc., adds further complexity to the problem (Mittal and Mitra 2018b; Kitzing et al. 2020; Anagnostopoulos et al. 2023; Ellenbogen et al. 2023). Apart from these, practical issues such as the irregular shape of the wind farm, uncertain wind speed and direction, presence of human dwellings, natural habitats (rivers, lakes, forests, etc.), or forbidden zones (where, for different reasons, turbines cannot be placed), etc., enhance additional challenges for WFLO (Mittal and Mitra 2020b; Azlan et al. 2021; Reddy 2021a; Sickler et al. 2023). Some of these issues will be discussed next in further detail.

One of the efficient ways of performing micro-siting tasks is to make use of mathematical programming tools. From the mathematical point of view, it is equivalent to solving an optimization problem. In general, any optimization problem has three components: (i) Objective function(s), which a decision maker wishes to optimize (i.e. minimize/maximize) while attaining a goal; (ii) Decision variables, which are the handles to a decision maker to obtain several alternative designs before converging to an optimal one, and (iii) Constraints, i.e. certain restrictions, which the decision maker has to obey before accepting a design. From the micro-siting parlance, a simple example involving optimization can be finding the optimal coordinate and number of turbines (decision variables) inside a wind farm, which can lead to maximum energy output and minimum cost (objective functions) without violating the restrictions of a certain distance among the turbines (constraints). Optimization tools aim to find such values of decision variables, which help the decision maker to attain the objectives without violating the constraints. With this background, the task of determining the number of turbines and location coordinates together during micro-siting involves both binary and continuous decision variables (as shown in Fig. 1.1), which is known as mixed integer nonlinear programming problems (MINLP). Here, continuous decision variables are used to represent the locations of the turbines inside the wind farm, and the binary variables designate the presence or absence (represented by 1 or 0, respectively) of a turbine at those locations. The nonlinear association of decision variables with the objective functions (i.e. power calculation using location coordinates) imparts nonlinearity into the system. Due to the presence of binary variables, not only the search space is discrete, but also the solution procedure demands an intelligent way of finding an optimal solution (a particular combination of binary variables out of a very large set of all possible discrete combinations) without checking all possible combinations exhaustively. This aspect can make these problems non-deterministic polynomial-time (NP) hard. The nature of combinatorial explosion can be appreciated by the reported fact in the literature that even for placing 30 wind turbines inside a wind farm, 10^{44} potential solutions are possible (Feng and Shen 2015; Mittal and Mitra 2020a, b, c). In large wind farms, the allocation of thousands of wind turbines is quite common. Most of the existing techniques and software packages (windPRO 2023; Openwind 2023) simplify the WFLO problem by assuming the total number of turbines fixed, i.e. the rated capacity of the wind farm is known a priori, and the goal then becomes to determine only the optimal turbine locations (nonlinear optimization problem, NLP). However, this approach jeopardizes the purpose of tapping the maximum potential of the wind farm. This is because solving the problem with fixed rated capacity can either lead to un-attainable (if the rated capacity is assumed higher than the optimal value) or compromised energy yield (if the rated capacity is made lower than the optimal value). Therefore, it is necessary to come up with better algorithms and techniques to solve this complex optimization problem more efficiently and find the optimal

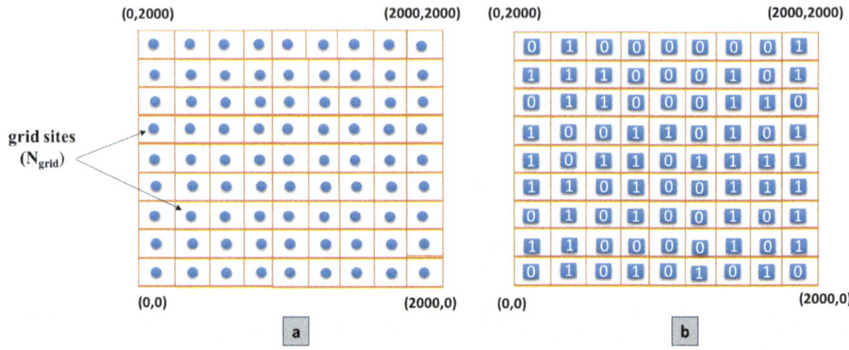

Fig. 1.1 An example representation of **a** grid locations and **b** the corresponding binary form of variables signifying the presence (binary value = 1) or absence (binary value = 0) of turbines at the grid locations for the $2000 \cdot 2000\ \text{m}^2$ wind farm

numbers as well as the location coordinates of turbines simultaneously to harness maximum energy out of a wind farm.

Extensive work has been carried out in the literature to determine the optimal layouts (numbers and locations) of the turbines inside a wind farm. These approaches are broadly classified into grid-based and grid-free methods. In the grid-based approaches, the wind farm area is partitioned into uniform grids (as shown in Fig. 1.1), considering these grid junctions or grid centres as the potential turbine locations (Mittal et al. 2016a; Shakoor et al. 2016). Depending on the grid resolution, the problem complexity is defined (Mittal et al. 2016a). For example, the possible number of distinct solutions $\left(N_{\text{pos}}\right)$ that have to be considered during WFLO in accordance with the grid points $\left(N_{\text{grid}}\right)$ can be given as:

$$N_{\text{pos}} = \frac{N_{\text{grid}}!}{N_{\text{fix}}!\left(N_{\text{grid}} - N_{\text{fix}}\right)!} \tag{1.1}$$

Here N_{fix} can be assumed as the fixed number of turbines that can be accommodated inside a wind farm. Therefore, according to Eq. (1.1), with the increase in the number of grid points in the search space (moving towards finer grid resolution), the size of the problem increases exponentially and could be unmanageable after a certain extent of fineness in the grid size. On the other hand, the grid-free approach (Gu and Wang 2013; Chen et al. 2014b; Mittal et al. 2017; Azlan et al. 2021) considers coordinate locations as continuous decision variables that can assume any value in the given wind farm coordinate bounds without grid restrictions, thereby creating an infinitely large number of possibilities for an optimizer and thus increasing the problem complexity enormously. Hence, it is important to maintain balance among the aspects of solution accuracy, execution time, and combinatorial complexity while developing solution methodologies for WFLO problems to observe the relative merits of these approaches.

Despite efficient exhaustive enumerative global solvers (Bayesian optimization (Forster et al. 2023), branch and bound (Tricoire et al. 2024), branch and model (Ma et al. 2023), etc.) being available for solving MINLPs, most of the reported works have approached the problem of WFLO using meta-heuristics class of optimization algorithms that provide acceptable solutions to complex problems in reasonable computation time (Tang 2022). The reasons behind the non-popularity of such global solvers for solving WFLO could be: (i) Systematic handling of solution branches emanating out of the tree generated to accommodate several possible combinations of binary variable realizations leading to very large combinations of solutions to be tried before declaring a global solution, hence could be time-consuming, (ii) Expressing conditional/logical nature of wake models while considering full/partial wake cases in the desired continuous form of models as demanded by the global solver environment. Popular meta-heuristic algorithms that are used to solve the WFLO problem include Genetic Algorithms (GA) (Mittal et al. 2015), Simulated annealing (SA) (Yang and Cho 2019), Particle Swarm Optimization (PSO) (Kunakote et al. 2022), etc., to name a few. Amidst these, GA and PSO are the most used evolutionary algorithms because of their ability to escape local optima while handling complex functions and handling both continuous and binary or integer types of variables in one go. However, their efficiencies often depend on the design of their variation and selection operators (exploitation and exploration in the search process). It is observed that the performance of these algorithms might deteriorate with the increase in problem size (expressed in terms of a number of decision variables, constraints and objective functions). However, it has been shown that by designing efficient variation operators, the efficiency of these algorithms can be significantly improved even for optimization problems of larger size (million variables) (Deb 2011; Mittal and Mitra 2018c). Several such encouraging facts lay the foundation for designing efficient evolutionary algorithms and exploring other novel optimization methods for solving WFLO problems.

Realistically, wind farm micro-siting demands many practicalities to be considered while formulating the optimization problem. One of the most important issues is the wake interference or wind speed deficit, known as the wake effect, caused by the upwind turbine(s) to their downwind counterparts. Due to this, a significant reduction in the power output of a wind farm can be observed as compared to the summation of the power generated by each turbine when operating as a standalone entity (Parada et al. 2017; Pujari et al. 2023a). Various wake models, e.g. analytical models based on empirical relations (such as Jensen, Frandsen, etc.) (Herbert-Acero et al. 2014), computational fluid dynamics (CFD)-based high-fidelity models (such as Reynolds Averaged Navier Stokes (RANS)) (Göçmen et al. 2016) are, hence, reported in the literature to simulate the wind speed deficit inside a wind farm. In fact, the precise power calculation of a wind turbine, which is connected with wind speed through a cubic relation, depends on how accurately the wind speed is evaluated at the turbine location. However, most of the studies have used the analytical Jensen wake model to evaluate the wind speed deficit for solving the WFLO problem, as this enables a facile integration into the problem and requires a low computational effort. Given the fact that optimization is an iterative exercise, which needs evaluation of such wake models several thousands of times before converging to the optimal layout configuration, the requirement of a wake model being computationally light is extremely favourable for such optimization exercises. Nevertheless, such computationally light wake models, victims of the accuracy versus their execution time trade-off, do not comply with the real field measurements and CFD-based dynamic simulations, as they are developed on many simplified assumptions (Parada et al. 2017). On the other hand, unrealistically large computational resources are required to perform WFLO if accurate and realistic wake simulations are utilized using high-fidelity CFD-based models in the iterative optimization framework, which seems unrealistic in a practical time frame. Balancing the accuracy of the target solutions and timely execution of the WFLO, as well as developing efficient and robust wake models, which can provide fairly accurate predictions with low computational cost, is highly necessary, parallel to the development of CFD models incorporating several realistic flow-related complexities. Here, the time-cheap models are going to be extremely useful for iterative optimization studies, whereas the high-fidelity models can always provide a check on the extent of inaccuracy involved in the time-cheap models.

With the low availability of land and strict norms and statutory regulations in place, the destructive environmental and human effects of wind farms are also gaining importance while determining the optimal layouts. Other than power generation, wind energy generators are home to various adverse negative impacts, such as noise propagation, shadow flickering, socio-economic impact, aesthetics, and land availability (presence of forbidden zones where turbines cannot be placed) that affect the nearby human health, the local environment, and the landscape value (Mittal and Mitra 2019a; Kitzing et al. 2020; Radun et al. 2022). Most of these impacts are deemed responsible for the long-time endurance of the wind project and its acceptance among local communities, thus becoming major objects of concern for the wind farm owners for the efficient operation of the wind farm. Among these, visual impacts like shadow flickering and aesthetics are site-specific effects affecting nearby human health and reducing the landscape value (Kitzing et al. 2020). These impacts are mostly addressed by performing proper wind farm layout planning and maintaining an appropriate setback between the observer locations and wind turbine(s) while designing a wind farm (Ahmed and Cameron 2014). On the other hand, noise is another major roadblock for wind farm owners as it adversely affects the nearby human health and environment (Mittal and Mitra 2017b; Mittal et al. 2024a). Most of the noise emitted in a wind farm is a result of either rotating machinery components or aerodynamic effects such as rotating speed, blade shape, and wake effects induced by the wind turbines in a wind farm (Herbert-Acero et al. 2014). Various single-objective and multi-objective optimization and control mitigation strategies, modifying or designing the turbine blade's structure, operating turbines at low or compromised performance, or optimally siting the wind turbines away from the habitats, have been adopted in the literature to reduce the adverse effects of noise on nearby settlements and to comply with the environmental impact assessment guidelines (Wu et al. 2020a; Cao et al. 2020; Mittal and Mitra 2020b). However, all of these impact mitigation approaches reduce the operating performance of the wind turbines, leading to the compromised energy yield of the wind farm (Hansen and Hansen 2020;

Azlan et al. 2021). Motivated by these efforts, it is evident that research advancement towards mitigating these adverse impacts and yielding more accurate and significant power output in the presence of such constraints will be encouraged.

Another practical issue that is mainly ignored while performing the WFLO is the stochastic nature of the wind as a source of energy (Mittal and Mitra 2020a; Pujari et al. 2020a; Pantula et al. 2022). The majority of studies in this regard assume a fixed local wind state (speed and direction) distribution without considering the uncertainty associated with the wind flow. The sources of uncertainties can be categorized as long-term (e.g. wind characteristics such as speed, direction, and time, environmental, and topographical which vary from month to month and year to year) and short-term (e.g. measurement errors, sudden small-scale turbulence, short-lived weather conditions, etc., which occurs due to the inevitable circumstances) conditions. Realistically, commercial wind farms designed for a longer lifetime (15–20 years) deserve serious consideration of uncertainties for an equivalent time frame (say, year-to-year variations) to generate realistic results (Mittal and Mitra 2019b; Wen et al. 2022). Owing to various realistic restraints such as stringent timelines, unavailability of data, etc., smaller duration wind measurements (1–2 years) are used to predict the approximate long-term wind distribution based on the short-term data (Messac et al. 2012; Jiang et al. 2019). This approach suffers from additional uncertainties due to regression and data collection methods and their adaptability for unknown newer conditions, resulting in the wrong prediction of the wind farm energy potential, return on investment, farm efficiency, and other factors related to the financial viability of wind projects (Burton et al. 2011). Studies towards design methodologies for a robust wind farm layout that is maximally insensitive to the uncertainties of the wind distribution over the long term are needed to generate consistent outputs.

In addition to the optimal positioning of turbines at locations that are less exposed to the wake and designing the new and improved wake models, in the recent days, cooperative wind farm control approaches have been adopted in WFLO to alleviate the effect of aerodynamic wake interactions among the turbines to achieve the maximized energy throughput of the wind farm (Kheirabadi and Nagamune 2019; Pujari et al. 2021). Based on the available degree of freedom of the individual wind turbines, the wind farm operator manipulates the wind field, such as yaw angle misalignment, blade pitch control, and axial induction factor of the wind turbines, to redirect the wake interactions and meet the desired wind farm performance level (Lopez et al. 2019; Pujari et al. 2021; Nash et al. 2021). Such works implementing control strategies can be broadly classified into model-based and model-free methods (Kheirabadi and Nagamune 2019; Yessef et al. 2022). The former methods are based on analytical wind farm models that are relatively easy to implement and can be employed to handle multiple control tasks. However, these approaches mostly rely on the accuracy of the models themselves and are much more sensitive to the stochastic environmental states (Kheirabadi and Nagamune 2019). On the other hand, model-free methods were developed to mitigate the limitations of model-based counterparts and are mostly driven by the measurement data to perform iterative learning and eventually acquire the optimal solutions (Dong et al. 2022). Despite showing the efficacies of control methods in multiple tasks, the inherent wind farm complexities, such as strong aerodynamic interactions, increased turbulent intensity, rising computational load, and fluctuating environmental states, restrict the wide-scale and long-term adaptability of these control objectives (Kheirabadi and Nagamune 2019; Dong et al. 2022). Thus, there is a need to improve the performance of these coordinated control methods for the accurate characterization of the wake flow interactions and to effectively handle the model complexities and fluctuating environmental states to increase the efficiency of the wind farm.

With the challenges mentioned above that are faced while performing an efficient WFLO, a laconic literature review around several aspects of it is presented next.

1.2 Literature Review

Over the years, diverse academicians, industrialists, and government stakeholders have devised various state-of-the-art techniques to conduct research in wind energy (Ahmed and Cameron 2014; Willis et al. 2018). Different avenues, such as (1) manufacturing and design of wind turbine blades, nacelle, gearbox, generators, and hub towers/foundations (Fuskele et al. 2022; Windisch et al. 2022; Essassi et al. 2023), (2) development of onshore/offshore wind farm and wind/wave flow characterization (Manwell et al. 2007; Donelan et al. 2012; Sirnivas et al. 2014), and (3) wind farm operations such as cooperative control (Yessef et al. 2022), maximum power production (Mittal and Mitra 2018a), reliability (Duer et al. 2022), structural health monitoring (SHM) (Khazaee et al. 2022), timely maintenance (Bakir et al. 2021), turbine blades anti-icing and de-icing strategies (Okulov et al. 2021), energy storage and smart grids transmission (Wu et al. 2020b; Khezri et al. 2022), etc., were targeted by the wind community to improve the overall performance of the wind farm. This section briefly overviews the prominent research works on different formulations and strategies of WFLO, including different siting optimization strategies, wake modelling and cooperative control methods for an accurate wake calculation and redirection, the impact of noise and the presence of practical constraints on wind farm output, and uncertain wind flow characterization, with the aim of harnessing the maximum potential out of the wind farm.

Wind farm design is an extremely involved activity. It includes a series of steps such as site selection (Latinopoulos and Kechagia 2015; Reddy 2021b), identifying wind flow behaviour on the selected site (Rad et al. 2022), adopting an accurate wake interaction model (Lundquist et al. 2019), assessments of environmental and economic impacts (Mittal et al. 2017; Guo et al. 2021), characteristics and specifications of wind turbines (Tao et al. 2020a), and uncertain or risk management (Mittal and Mitra 2020a). All these tasks are aimed at deploying the optimal number of wind turbines at optimal locations to maximize the energy output of the wind farm. This process is called wind farm layout optimization (WFLO) or wind farm micro-siting. Several mathematical formulations (Herbert-Acero et al. 2014; Mittal et al. 2016b; Azlan et al. 2021) combining single and multi-objective optimization techniques (such as Genetic algorithm (GA) and particle swarm optimization (PSO)) have been applied to analyse the effect of consideration of different objectives (such as maximizing annual energy production (AEP) (Mittal et al. 2015), minimizing the cost of energy (CoE) (Chen et al. 2014b), maximizing net present value (NPV) (Song et al. 2017), etc.) for efficient wind farm design.

Conventionally, the wind farms are designed by placing the wind turbines in an organized manner (in the same lines) with a sufficient distance between them (order of 6–10 rotor diameter apart). This is done with the expectation of resulting in low wake interference and high-power output. However, the higher cost and low availability of the land area and the expenses involved in inter-turbine cable layout restrict the wind farm owners from adopting this design. Moreover, optimizing wind farm layout enables the wind farm owners to construct wind farms with less likelihood of wake interferences or power losses by leveraging the merits of optimization techniques with the assigned objectives. Broadly, two ways of representing the wind turbine locations, i.e. grid-based and grid-free, dominate the current literature to target the WFLO problem. In the *grid-based approach*, the given wind farm area is discretized into a uniform set of grids and turbines are allowed to be placed either on the grid cross-sites or at the centre of these square grid boxes (see Fig. 1.1). (Mosetti et al. 1994) introduced the first WFLO study using a binary-coded genetic algorithm. In this work, the wind farm was divided into a uniform set of grids, and turbines were optimally placed at the centre of these grid boxes while minimizing the weighted function of cost and energy. By tuning the parameters of binary GA and modifying the objective function, (Grady et al. 2005) improved the quality of the solution obtained by (Mosetti et al. 1994). By discretizing the wind farm into various orientations of grids, such as square and triangular, the greedy algorithm was applied to analyse the effect of grids on the turbine positioning inside a wind farm (Chen et al. 2014a). Other prominent works have also been reported in the literature to solve the WFLO problem by adopting the grid-based approach (Kulkarni and Mittal 2014; Kunakote et al. 2022). (Kunakote et al. 2022) compared the performance of twelve different meta-heuristic approaches in solving the grid-based WFLO problem and reported the efficient performance of the moth-flame optimization (MFO) algorithm in an attempt to maximize the power output of the wind farm. In another study, (Kulkarni and Mittal 2014) developed a novel heuristic approach to simultaneously determine the optimal number of turbines and their optimal locations in a grid-based wind farm to maximize the net annual energy production (AEP) and minimize the wake losses in a wind farm. In the afore-mentioned grid-based works, due to the discrete representation of coordinates, many locations are not considered for placing turbines. Hence, the accuracy of this approach lies in the choice of 'fineness' of the grids (higher grid resolution). However, the number of candidate turbine locations increases exponentially with an increase in grid resolution, which increases the computational cost substantially. (Mittal and Mitra 2018c) addressed this issue by proposing a novel SAVE suite of algorithms that represents the grids in the form of indices. This work reduces the overall computational cost by attaining a better solution pool at the early set of generations using novel repair strategies and by rejuvenating the solution pool with fresh seeding. In another work, starting with the coarse grid representation, the balance between the solution quality and execution time is maintained using the proposed novel need-based grid resolution approach (Mittal and Mitra 2019c). Here, the issue of a huge computational burden using a fine-resolution grid is addressed by constructing a need-based grid refinement inside the starting coarse grid layout to improve the solution quality. The proposed approach has demonstrated an improvement of 9% in power production compared to the fixed-resolution fine grid technique. Although the above-described approaches tackled the problem of huge computational cost by improving the solution quality by attaining a better solution at an early stage, these results can motivate the researchers to show that there is a huge scope of improvement in solution quality possible while using such kind of flexible evolutionary algorithms for solving WFLO problems.

As an alternative to this discrete formulation, researchers used a grid-*free approach* to increase the freedom of movement of turbines beyond the grid points. In this approach, turbine coordinates (considered as continuous decision variables in optimization) can be placed freely inside the specified boundary, avoiding the need to mesh the wind farm. (Kusiak and Song 2010) proposed a bi-objective evolutionary algorithm to determine the turbine locations in a farm aiming highest energy harnessed and honouring the given constraints. (Wan et al. 2012) utilized a variant of swarm optimization, which allowed the placement of each turbine freely inside a given boundary to maximize the generated power. Different versions of ant colony optimization (ACO) (Eroĝlu and Seçkiner 2012) and particle filtering method (PFM) (Eroĝlu and Seçkiner 2013) are also utilized for optimally placing turbines inside wind farms. Not only annual energy production cost but also losses due to wake effect and the wind farm area are also attempted to be optimized by (Moreno et al. 2021) to design an efficient wind farm layout based on a grid-free approach. (Nagpal et al. 2021) introduced a way to improve energy output and cost of energy (levelized) from initial layouts generated using heuristics. Interestingly, utilizing both of these approaches (grid and grid-free) in tandem, WFLO problem has been shown to be solved more effectively. To demonstrate, a layout optimization problem has been designed as a two-echelon assignment and a combination of GA (grid) and PSO (grid-free) was utilized to solve the WFLO problem by (Long and Zhang 2015). Using the advantages of evolutionary and classical methods, (Mittal et al. 2016a) proposed a combined approach to handle the grid-based and grid-free formulations in a cyclic fashion and showed superior performance on the benchmark case studies (Mosetti et al. 1994; Grady et al. 2005). Following the trend, different optimization algorithms are also exploited towards improving wind farm efficiency considering with and without grid formulations (Azlan et al. 2021). Also, efforts were invested in considering additional features of wind farm design, e.g. different shapes (regular and irregular), terrains (flat and complex), and roughness of the wind farm(s), to analyse their effects on the wind farm power production (Gu and Wang 2013; Reddy 2021b). In addition, recent breakthroughs in machine learning (ML) approaches have also gained significant attention in the wind research community to determine the optimal turbine layouts by solving complex and highly nonlinear wind farms to maximize the energy yields (Marvuglia and Messineo 2012; Reddy 2021a; Shahzad 2022).

Works towards improving the wake models to mitigate the aerodynamic interference between the turbines and to enhance the yield of wind farms have also appeared in the literature. (De-Prada-Gil et al. 2015) have proposed a strategy to reduce the wake effects between turbines with an estimated ~ 5% increase in annual energy. (Parada et al. 2017) utilized a Gaussian wake model in contrast to the widely used Jensen model to minimize the cost of energy of a wind farm and achieve higher annual energy production at a low computational cost. (Tao et al. 2020b) presented a newly developed three-dimensional (3D) Gaussian wake model and showed its superior performance against both one and two-dimensional wake models while placing identical and multiple types of turbines inside a wind farm, respectively. Other works were also conducted by introducing new wake models, considering different velocity deficit profiles, and redirecting wake effects between turbines (Göçmen et al. 2016; Lopez et al. 2019; Nash et al. 2021). Most of these wake models do not involve high-fidelity CFD simulations as these models are to be integrated with an iterative optimization framework, where several thousands of times such model evaluations are involved (Witha 2021). Though the wind energy community has evolved with different kinds of wake models and redirection strategies, there is enormous scope for developing better wake models, as these models are reasonably poor in prediction compared to high-fidelity CFD models. (Kuo et al. 2016), for the first time, coupled the CFD model with the mixed integer programming (MIP) method to improve the prediction accuracy of the wake deficit calculations with the reduced number of required CFD simulations. Here, the hybrid model of CFD and MIP methods is utilized in an iterative manner and adaptively refined during the optimization process to determine the wake deficit only at the promising turbine locations to produce excellent layouts in complex terrains. In another work, a novel wake model as a solution to a simplified variation of Naiver–Stokes equations is proposed to reduce the computational cost while maintaining the accuracy (Kuo et al. 2018). It has been observed that the proposed model reduces the computational cost by two orders of magnitude compared to the full-scale CFD simulations, highlighting its potential for optimizing wind turbine layouts on complex terrains. Moreover, the recent trends of machine learning algorithms to compete with the iterative requirements of high-fidelity computationally expensive CFD wake models have encouraged researchers to utilize ML algorithms as an inexpensive surrogate wake model to predict the wake deficit accurately and efficiently with reduced computational cost (Ti et al. 2020, 2021; Nai-Zhi et al. 2022; Pujari et al. 2023a).

Though the wind characteristics are highly uncertain by nature, the above-mentioned works considered a fixed probability distribution function valid for the entire period of life while designing a wind farm. Only a few reported works have considered the variation in wind flow while performing the wind farm micro-siting. (Kirchner-Bossi et al. 2015) investigated the extent of deterioration in the wind turbine power output considering the long-term variation in the wind flow. (Messac et al. 2012) proposed parametric and nonparametric wind uncertainty models and used the Monte Carlo simulations method to evaluate the wind farm performance under uncertainty. (Mittal and Mitra 2020a) obtained the optimal wind turbine layouts (numbers and locations) that are immune to varying wind state conditions using the novel robust optimization methodology. The proposed approach provides a solution to choose a realistic layout that can be expected to generate power throughout the lifespan of wind farms, depending on the wind farm owner's preference. On a different note, (Wen et al. 2022) introduced a risk management method based on the conditional value-at-risk (CVaR) to increase the robustness of annual energy production under uncertain wind flow conditions. (Feng and Shen 2017) proposed a set of robustness metrics to quantify the wind farm's ability to produce power with high mean value and low variability under changing wind conditions. In the last few years, advances in machine learning methods have also gained significant attention in predicting new and future wind characteristics, behaviour, or flow patterns based on historical wind data while designing a wind farm (Demolli et al. 2019; Miriyala et al. 2020; Pujari et al. 2020b, 2023c, 2023b; Huang et al. 2021; Wang et al. 2022). Broadly, these forecasting techniques are classified into long-term (forecast with respect to a longer time, i.e. several days to 1 year) and short-term (forecast for a short period, i.e. one hour to several days) wind flow forecasting methods. (Pujari et al. 2023b) proposed a novel multi-objective optimization-driven deep learning tool using Long Short-Term Memory (LSTM) networks for long-term forecasting (2 years) of wind flow data based on the historical four years of the collected data with an accuracy of 97%. The proposed model is implemented on real wind farm data containing four turbines and shows an increment of 8% in power production with the forecasted wind conditions. (Demolli et al. 2019) performed long-term wind flow forecasting based on daily wind speed data and demonstrated the efficiency of different machine learning algorithms against four different wind farm locations, showing the benefits of forecasting wind speed and power in establishing a new wind plant at an unknown location. (Wang et al. 2022) predicted a short-term wind power using the proposed Gaussian mixed clustering-deep neural network probabilistic forecasting (GMC-DeepNN-PF) algorithm. The proposed algorithm utilizes the merits of each method to classify, extend, and forecast the wind data in a step-wise fashion and claims to provide better results than conventional machine learning algorithms with reduced cost. (Huang et al. 2021) presented two novel hybrid forecasting systems to capture the linear and nonlinear factors hidden in the wind speed data. At first, a decomposition algorithm was used to denoise the data, and then a linear autoregressive moving average model and nonlinear backpropagation neural network optimized by the cuckoo search algorithm models were employed to capture the linear and nonlinear hidden patterns in the data, respectively. The proposed model recorded a high prediction accuracy and greater stability than the comparison models, thus providing a new feasible solution for wind speed prediction and rational power-grid allocation.

On the other hand, efforts have also been invested in redirecting the aerodynamic wake interactions using cooperative control methods to enhance the energy output of the wind farm. These methods are broadly classified into model-based and model-free control methods (Dong et al. 2022). The former directly utilizes the analytical wind farm models to determine the optimal control solutions. (Nash et al. 2021) presented a novel passive concept wake steering control technique where an external 3D-printed wake deflector vane is installed between any two in-line turbines to steer the wake impact of the upstream turbines on downwind turbines. With the external wane installed, the proposed model does not sacrifice the performance of the upstream turbines and structural loads of turbines and provides a significant improvement of 15% in power production compared to the existing wake control techniques. (Park et al. 2013) employed the combination of the steepest descent algorithm and cooperative static GA to determine the optimal yaw angle and blade pitch angle control actions for improving the power production efficiency of the wind farm. (Goit et al. 2016) applied a receding horizon control method to optimize the thrust coefficient of each individual wind turbine and showed an increase of 7% in power generation compared to the greedy strategy. Additionally, a closed loop data-driven wind farm control (WFC) model, which exercises the different learning mechanisms and system identification tools to approximate the wind farm models, was implemented to determine the control inputs for extracting power gain out of the wind farm (Doekemeijer et al. 2019). Unlike model-based methods, which are very sensitive to wind farm model uncertainties and modelling errors, the model-free approaches were adopted in the literature to obtain more robust and adaptive solutions. (Marden et al. 2013) implemented a model-free game-theoretic distributed learning algorithm and depicted an increase of 25% in power production compared to the greedy control technique without explicitly modelling the wake between the turbines. (Ahmad et al. 2017) investigated a random search algorithm to determine the optimal control parameters of each individual turbine of the wind farm for maximizing the power production of the wind farm. In another work, (Park 2020) proposed a contextual Bayesian optimization method with a trust-region (CBOTR) approach to determine the optimum control inputs through the iterative learning and sampling procedure, which outperforms the conventional greedy control approach. Moreover, a dramatic increase in interest can be seen in reinforcement learning techniques in designing more accurate, robust, and inexpensive control approaches to intensify the power output of the wind farm (Pujari et al. 2021; Dong et al. 2022).

Based on the above studies, power maximization is observed as the most popularly adopted objective while performing layout optimization. Features like maintaining sufficient inter-turbine distance to avoid wake effects, factor for overall capacity of the farms are treated as popular constraints. Some of the relevant and realistic features that missed requisite attention is addressing the effect of noise on human habitats nearby, zones inside the wind farm where turbines cannot be installed due to practical difficulties, etc. One of the praiseworthy implementations of noise effects on wind farm goes back to the proposed solution presented by (Kwong et al. 2013), where noise–energy trade-off has been shown for different numbers of turbines in a farm obtained through grid-based formulation. (Mittal et al. 2017) improved these results by optimally setting the numbers and locations of turbines in a bi-objective setup. More realistic constraints, like land use, proximity, were considered by (Yamani Douzi Sorkhabi et al. 2018), in their energy-noise trade-off work. (Mittal and Mitra 2020b) showed the effect of noise propagation on wind turbine layouts and energy throughput of the wind farm when habitats and prohibited zones are present inside a farm and generated multiple avenues of solution set for wind farm owners based on different noise level restrictions. Recently, (Mittal et al. 2024a) have demonstrated the use of sound modulating acoustic metamaterials (Mittal et al. 2024b; Nampoothiri et al. 2024) in tackling the noise issue in wind farms without compromising their operating performance. For the first time, the study introduced the holographic manipulation of the phase and amplitude information of the wind farm sound towards the target location and recorded ~ 90% and ~ 50% noise level reduction for single and multiple-turbine scenarios, respectively. Studies handling the constraints of other types, such as socio-economic criteria and environmental consequences (Lundquist et al. 2019), local and land-owner participation (Kitzing et al. 2020), etc., for repowering or constructing wind farms were also reported in the literature.

1.3 Avenues for Future Research

Despite diversified issues of WFLO being addressed in the past or through ongoing research, there is a need to address the following niche areas for having realistic and practical solutions to the problem of wind farm micro-siting:

1. Improving the optimization formulations and approaches to ascertain more efficient outcomes with fewer model computations and without compromising the solution quality.
2. Solving the WFLO problem in a multi-objective framework by considering the multiple conflicting factors and practicalities to analyse their impacts on wind farm power generation and optimal turbine layouts is a more pragmatic approach. Most of the reported works were based on maximizing power as a single objective while considering the inter-turbine distance, characteristics of turbines, wind farm topology, etc. as constraints. Few works have incorporated other important factors such as noise, land availability, and other practical constraints as objectives. Moreover, most of the reported multi-objective problems in WFLO are targeted conventionally by decomposing them into multiple single-objective problems without exploiting the multi-objective search domain, thus adopting an ineffective approach.
3. Developing more accurate wake models to design the aerodynamic interactions among turbines under manageable resources such as computational cost, execution time, and data storage. High-fidelity CFD simulations are fairly accurate but unrealistic to be solved using manageable resources. To reduce the computational burden, most studies have utilized a simple and analytical Jensen wake model

to evaluate the wind speed deficit for solving the WFLO problem. However, to comply with the real field data, more robust and effective wake models are needed to balance the accuracy and computational burden under manageable resources.

4. Designing wind farm layouts which are robust and insensitive to the changes in wind state conditions. Though by nature, wind flow is extremely uncertain and possesses long-term and short-term variations, most of the reported works were based on a fixed probability distribution built using wind speeds measured over a short period of time (1–2 years). The assumption is that this distribution will be valid for the entire life span of the wind farm, which is unrealistic. Though some of the works have considered this aspect while designing a wind farm, the development of robust design methodologies that are maximally insensitive to the uncertain wind distribution over the long-term operation of the wind farms is yet to achieve maturity.
5. Recent breakthroughs in machine learning (ML) approaches have gained significant attention in the wind farm community to determine more accurate solutions to complex and highly nonlinear wind farm models with less computational effort, which was not possible earlier in solving the WFLO problem. Various works have shown the efficacy of machine learning-based approaches in designing more robust and accurate wind farm designs. With ML-based methods showing promising results, there lies a huge scope for designing more reliable and accurate designs for wind farms using ML-based techniques to harness the maximum benefit out of them.
6. Using cooperative wind farm control approaches to reduce the impact of wake interactions between turbines inside a wind farm. Several techniques have been developed to determine the optimal control actions with a target of achieving the maximum power gain. However, the model complexity, high nonlinearity, and the stochastic fluctuations of wind are significant barriers towards popularizing the wide applicability of these control systems. Therefore, the adaptability and robustness of these approaches under stochastic wind conditions provide a huge research scope that needs to be addressed.

1.4 Book Layout

The overall focus of this book, thus, is to take the readers through such essential components that are necessary for enabling maximum energy harnessing through wind farms under uncertain environment. As mentioned earlier in this chapter, the WFLO problem has been approached by several researchers using single and multi-objective optimization techniques. As micro-siting is broadly seen as an optimization problem, fundamentals of deterministic and stochastic multi-objective optimization and overview of several methodologies for solving such problems are outlined in Chapt. 2. Population-based evolutionary techniques are popularly used for solving multi-objective micro-siting problems. Though shown to solve such problems more efficiently than their classical counterparts, evolutionary techniques become computationally highly expensive if the objective functions and constraints evaluations are individually costly. Data-driven multi-objective optimization techniques that use expensive function evaluations judiciously by quantifying the uncertainty associated while using surrogates and further utilizing them in optimization have been covered in details in Chapt. 3. As accurate wind speed prediction is necessary for calculating wind farm power and other constraints, several predominant wake modelling approaches covering the CFD, Large Eddy Simulations (LES) and analytical routes are discussed in Chapt. 4. The role of key factors such as inter-turbine distance, yaw misalignment, atmospheric thermal stability, and wind direction in influencing wake dynamics within wind farms is discussed in Chapt. 5. Further, it covers a laconic introduction to the simulator for Wind Farm Applications (SOWFA) package, which can be used to efficiently model wake interactions inside a wind farm. Such high-fidelity CFD simulations are accurate in one hand but computationally expensive on the other. Building hybrid models, where a part of the complicated phenomena can be captured using supervised machine learning (ML) techniques keeping physics-based understanding to predict rest of the phenomena, is covered in Chapt. 6. Such hybrid models can be extremely useful for micro-siting as they are computationally light without compromising prediction accuracy to a great extent. Next, the focus is concentrated on forecasting of wind speed and direction, which plays an important role in building the wind frequency maps (WFM) used for wind farm power calculations. Long short-term memory network (LSTM) and support vector machine (SVM), ML techniques for modelling time series data, are utilized to generate accurate forecasting results for a realistic wind farm as demonstrated in Chapters 7 and 8, respectively. Utilizing the existing WFMs, based on past as well as forecasted data, approaches can be adopted to generate more unseen scenarios for future, which can in turn be used for robust wind farm designs immune to all such scenarios. One approach can be to cluster these WFMs, observe the patterns in them and generate new WFMs following the trend. Another approach can be to capture the embedded parent distribution from which different sampling can lead to the existing WFMs and generate more WFMs from the obtained distribution. Neuro Fuzzy C-means (NFCM) and Generative Adversarial Networks (GANs) are used for such demonstrations as shown in Chapters 9 and 10, respectively. Once the design of wind farm is optimized, operational optimization of wind turbines can be performed using several turbine parameters, such as yaw angle and blade pitch angle control actions, to enhance their performance towards generation of more power. Utilizing emerging techniques, control study on an optimally designed wind farm has been showcased in Chapt. 11 using reinforcement learning (RL). Model predictive control (MPC), one of the most successful techniques in the control domain, has been used next to see its applicability for the wind farm control problem and the results are discussed in Chapt. 12. Most of the literature reported wind speed and direction as the two most contributing factors in the wind farm design and control. As an effort to measure the sensitivity of different parameters on a relative scale, an exploratory data analysis has been performed on realistic wind farm data and reported in Chapt. 13. As artificial intelligence (AI) models are often criticized for not able to perform reasoning behind the decisions made, eXplainable

Fig. 1.2 Overall overview of the recent and novel techniques included in the subsequent chapters of the book addressing the wind farm micro-siting problem

AI (XAI) is becoming increasingly popular which has been utilized for a realistic wind farm data analysis and the results are reported in Chapt. 14 (Fig. 1.2).

Acknowledgements We would like to acknowledge the support provided by the SPARC project (SPARC/2018-2019/P1084/SL) sponsored by the Ministry of Education (MoE), Government of India.

References

Ahmad MA, Hao MR, Ismail RMTR, Nasir ANK (2017) Model-free wind farm control based on random search. Proceedings—2016 IEEE international conference on automatic control and intelligent systems, I2CACIS 2016, pp 131–134. https://doi.org/10.1109/I2CACIS.2016.7885302

Ahmed NA, Cameron M (2014) The challenges and possible solutions of horizontal axis wind turbines as a clean energy solution for the future. Renew Sustain Energy Rev 38:439–460. https://doi.org/10.1016/J.RSER.2014.06.004

Anagnostopoulos SJ, Bauer J, Clare MCA, Piggott MD (2023) Accelerated wind farm yaw and layout optimisation with multi-fidelity deep transfer learning wake models. Renew Energy 218:119293. https://doi.org/10.1016/j.renene.2023.119293

Azlan F, Kurnia JC, Tan BT, Ismadi MZ (2021) Review on optimisation methods of wind farm array under three classical wind condition problems. Renew Sustain Energy Rev. 135

Bakir I, Yildirim M, Ursavas E (2021) An integrated optimization framework for multi-component predictive analytics in wind farm operations & maintenance. Renew Sustain Energy Rev 138:110639. https://doi.org/10.1016/J.RSER.2020.110639

Borch K, Kirkegaard JK, Nyborg S (2023) Three wind farm developments, three different planning difficulties: cases from Denmark. Energies 16(12):4662. https://doi.org/10.3390/EN16124662

Burton T, Jenkins N, Sharpe D, Bossanyi E (2011) Wind energy handbook, Second edition. Wind energy handbook, Second edition. https://doi.org/10.1002/9781119992714

Cao JF, Zhu WJ, Shen WZ, Sørensen JN, Sun ZY (2020) Optimizing wind energy conversion efficiency with respect to noise: a study on multi-criteria wind farm layout design. Renew Energy 159:468–485. https://doi.org/10.1016/j.renene.2020.05.084

Chen K, Song M, Zhang X (2014a) The investigation on computational grids in wind turbine positioning optimization using greedy algorithm. Chin Sci Bull 59(26):3304–3313. https://doi.org/10.1007/S11434-014-0408-1/TABLES/10

Chen K, Song MX, Zhang X (2014b) Binary-real coding genetic algorithm for wind turbine positioning in wind farm. J Renew Sustain Energy 6(5):053115. https://doi.org/10.1063/1.4896599

Deb K (2011) Multi-objective optimisation using evolutionary algorithms: an introduction. Multi-Objective Evol Optim Product Des Manufact. https://doi.org/10.1007/978-0-85729-652-8_1

Demolli H, Dokuz AS, Ecemis A, Gokcek M (2019) Wind power forecasting based on daily wind speed data using machine learning algorithms. Energy Convers Manag 198:111823. https://doi.org/10.1016/J.ENCONMAN.2019.111823

De-Prada-Gil M, Alías CG, Gomis-Bellmunt O, Sumper A (2015) Maximum wind power plant generation by reducing the wake effect. Energy Convers Manag 101:73–84. https://doi.org/10.1016/J.ENCONMAN.2015.05.035

Doekemeijer BM, Hoek DCV der, Wingerden JW van (2019) Model-based closed-loop wind farm control for power maximization using Bayesian optimization: a large eddy simulation study. CCTA 2019—3rd IEEE Conference on control technology and applications. pp 284–289. https://doi.org/10.1109/CCTA.2019.8920587

Donelan MA, Curcic M, Chen SS, Magnusson AK (2012) Modeling waves and wind stress. J Geophys Res Oceans 117(C11):0–23. https://doi.org/10.1029/2011JC007787

Dong H, Xie J, Zhao X (2022) Wind farm control technologies: from classical control to reinforcement learning. Prog Energy 4(3):032006. https://doi.org/10.1088/2516-1083/AC6CC1

Duer S, Rokosz K, Bernatowicz D, Ostrowski A, Wozniak MW, Zajkowski K, Iqbal A (2022) Organization and Reliability Testing of a Wind Farm Device in Its Operational Process. Energies 15(17):6255. https://doi.org/10.3390/EN15176255

Ellenbogen JM, Kellam CB, Hankard M (2023) Noise-induced sleep disruption from wind turbines: scientific updates and acoustical standards. Sleep. https://doi.org/10.1093/SLEEP/ZSAD286

Energy economics | Home Statistical Review of World Energy. https://www.bp.com/en/global/corporate/energy-economics/statistical-review-of-world-energy.html. Accessed 20 Jan 2024

Eroĝlu Y, Seçkiner SU (2012) Design of wind farm layout using ant colony algorithm. Renew Energy 44:53–62. https://doi.org/10.1016/J.RENENE.2011.12.013

Eroĝlu Y, Seçkiner SU (2013) Wind farm layout optimization using particle filtering approach. Renew Energy 58:95–107. https://doi.org/10.1016/J.RENENE.2013.02.019

Essassi K, Ayachi A, Feki N, Bouguecha A, Chaari F, Haddar M (2023) Mechanical Characterization of glass fibers–reinforced composites for wind turbine blades applications. pp 901–908. https://doi.org/10.1007/978-3-031-14615-2_101/COVER
Feng J, Shen WZ (2015) Modelling wind for wind farm layout optimization using joint distribution of wind speed and wind direction. Energies 8:3075–3092. https://doi.org/10.3390/en8043075
Feng J, Shen WZ (2017) Wind farm power production in the changing wind: robustness quantification and layout optimization. Energy Convers Manag 148:905–914. https://doi.org/10.1016/j.enconman.2017.06.005
Forster T, Vázquez D, Guillén-Gosálbez G (2023) Algebraic surrogate-based process optimization using Bayesian symbolic learning. AIChE J 69(8):e18110. https://doi.org/10.1002/AIC.18110
Fuskele V, Baredar P, Sarviya RM, Lal S, Awasthi S (2022) Wind turbine nacelle cooling systems: a review. Wiley Interdiscip Rev Energy Environ. https://doi.org/10.1002/WENE.456
Global Energy Perspective (2023) Power outlook | McKinsey. https://www.mckinsey.com/industries/oil-and-gas/our-insights/global-energy-perspective-2023-power-outlook. Accessed 10 Oct 2023
Global Wind Report (2023) Global wind energy council. https://gwec.net/globalwindreport2023/. Accessed 19 Nov 2023
Göçmen T, van der Laan P, Réthoré PE, Diaz AP, Larsen GC, Ott S (2016) Wind turbine wake models developed at the technical university of Denmark: a review. Renew Sustain Energy Rev 60:752–769. https://doi.org/10.1016/J.RSER.2016.01.113
Goit JP, Munters W, Meyers J (2016) Optimal coordinated control of power extraction in LES of a wind farm with entrance effects. Energies 9(1):29. https://doi.org/10.3390/EN9010029
Grady SA, Hussaini MY, Abdullah MM (2005) Placement of wind turbines using genetic algorithms. Renew Energy 30(2):259–270. https://doi.org/10.1016/J.RENENE.2004.05.007
Gu H, Wang J (2013) Irregular-shape wind farm micro-siting optimization. Energy 57:535–544. https://doi.org/10.1016/j.energy.2013.05.066
Guo N, Zhang M, Li B, Cheng Y (2021) Influence of atmospheric stability on wind farm layout optimization based on an improved Gaussian wake model. J Wind Eng Ind Aerodyn. https://doi.org/10.1016/j.jweia.2021.104548
Hansen C, Hansen K (2020) Recent advances in wind turbine noise research. Acoustics 2(1):171–206. https://doi.org/10.3390/ACOUSTICS2010013
Herbert-Acero JF, Probst O, Réthoré PE, Larsen GC, Castillo-Villar KK (2014) A review of methodological approaches for the design and optimization of wind farms. Energies 7(11):6930–7016. https://doi.org/10.3390/EN7116930
Huang X, Wang J, Huang B (2021) Two novel hybrid linear and nonlinear models for wind speed forecasting. Energy Convers Manag. https://doi.org/10.1016/j.enconman.2021.114162
Jiang P, Li R, Li H (2019) Multi-objective algorithm for the design of prediction intervals for wind power forecasting model. Appl Math Model 67:101–122. https://doi.org/10.1016/J.APM.2018.10.019
Khazaee M, Derian P, Mouraud A (2022) A comprehensive study on Structural Health Monitoring (SHM) of wind turbine blades by instrumenting tower using machine learning methods. Renew Energy. https://doi.org/10.1016/J.RENENE.2022.09.032
Kheirabadi AC, Nagamune R (2019) A quantitative review of wind farm control with the objective of wind farm power maximization. J Wind Eng Ind Aerodyn 192:45–73. https://doi.org/10.1016/J.JWEIA.2019.06.015
Khezri R, Mahmoudi A, Haque MH (2022) Impact of optimal sizing of wind turbine and battery energy storage for a grid-connected household with/without an electric vehicle. IEEE Trans Industr Inform 18(9):5838–5848. https://doi.org/10.1109/TII.2022.3140333
Kirchner-Bossi N, Porté-Agel F (2024) Wind farm power density optimization according to the area size using a novel self-adaptive genetic algorithm. Renew Energy 220:119524. https://doi.org/10.1016/j.renene.2023.119524
Kirchner-Bossi N, García-Herrera R, Prieto L, Trigo RM (2015) A long-term perspective of wind power output variability. Int J Climatol 35(9):2635–2646. https://doi.org/10.1002/JOC.4161

Kitzing L, Jensen MK, Telsnig T, Lantz E (2020) Multifaceted drivers for onshore wind energy repowering and their implications for energy transition. Nat Energy 5(12):1012–1021. https://doi.org/10.1038/s41560-020-00717-1

Kulkarni K, Mittal P (2014) A fast and effective algorithm to optimize the total number and placement of wind turbines. In: 2014 IEEE Global humanitarian technology conference—South Asia Satellite, GHTC-SAS

Kunakote T, Sabangban N, Kumar S, Tejani GG, Panagant N, Pholdee N, Bureerat S, Yildiz AR (2022) Comparative performance of twelve metaheuristics for wind farm layout optimisation. Arch Comput Methods Eng 29(1):717–730. https://doi.org/10.1007/s11831-021-09586-7

Kuo JYJ, Romero DA, Beck JC, Amon CH (2016) Wind farm layout optimization on complex terrains—Integrating a CFD wake model with mixed-integer programming. Appl Energy 178:404–414. https://doi.org/10.1016/J.APENERGY.2016.06.085

Kuo J, Rehman D, Romero DA, Amon CH (2018) A novel wake model for wind farm design on complex terrains. J Wind Eng Ind Aerodyn 174:94–102. https://doi.org/10.1016/J.JWEIA.2017.12.016

Kusiak A, Song Z (2010) Design of wind farm layout for maximum wind energy capture. Renew Energy 35(3):685–694. https://doi.org/10.1016/J.RENENE.2009.08.019

Kwong WY, Zhang PY, Romero D, Moran J, Morgenroth M, Amon C (2013) Wind farm layout optimization considering energy generation and noise propagation. Proceedings of the ASME design engineering technical conference 3(PARTS A AND B), pp 323–332. https://doi.org/10.1115/DETC2012-71478

Latinopoulos D, Kechagia K (2015) A GIS-based multi-criteria evaluation for wind farm site selection. A regional scale application in Greece. Renew Energy 78:550–560. https://doi.org/10.1016/J.RENENE.2015.01.041

Li H, Jiang K, Lu Z, Hou J, Xu S, Qi W, Cai H (2023) Optimization of the offshore wind power grid-connected structure based on an improved genetic algorithm. Front Energy Res. https://doi.org/10.3389/fenrg.2023.1140711

Long H, Zhang Z (2015) A two-echelon wind farm layout planning model. IEEE Trans Sustain Energy 6(3):863–871. https://doi.org/10.1109/TSTE.2015.2415037

Lopez D, Kuo J, Li N (2019) A novel wake model for yawed wind turbines. Energy 178:158–167. https://doi.org/10.1016/j.energy.2019.04.120

Lundquist JK, DuVivier KK, Kaffine D, Tomaszewski JM (2019) Costs and consequences of wind turbine wake effects arising from uncoordinated wind energy development. Nat Energy 4(1):26–34. https://doi.org/10.1038/s41560-018-0281-2

Ma K, Rios LM, Bhosekar A, Sahinidis NV, Rajagopalan S (2023) Branch-and-model: a derivative-free global optimization algorithm. Comput Optim Appl 85(2):337–367. https://doi.org/10.1007/S10589-023-00466-3/FIGURES/8

Manwell JF, Elkinton CN, Rogers AL, McGowan JG (2007) Review of design conditions applicable to offshore wind energy systems in the United States. Renew Sustain Energy Rev 11(2):210–234. https://doi.org/10.1016/J.RSER.2005.01.002

Marden JR, Ruben SD, Pao LY (2013) A model-free approach to wind farm control using game theoretic methods. IEEE Trans Control Syst Technol 21(4):1207–1214. https://doi.org/10.1109/TCST.2013.2257780

Marvuglia A, Messineo A (2012) Monitoring of wind farms' power curves using machine learning techniques. Appl Energy 98:574–583. https://doi.org/10.1016/j.apenergy.2012.04.037

Messac A, Chowdhury S, Zhang J (2012) Characterizing and mitigating the wind resource-based uncertainty in farm performance. J Turbul. https://doi.org/10.1080/14685248.2012.661863

Miriyala SS, Chowdhury S, Pujari NSK, Mitra K (2020) Optimally designed variational autoencoders for efficient wind characteristics modelling. 2020 IEEE symposium series on computational intelligence. SSCI 2020:2869–2876. https://doi.org/10.1109/SSCI47803.2020.9308245

Mittal P, Mitra K (2018c) Determining layout of a wind farm with optimal number of turbines: a decomposition based approach. J Clean Prod 202:342–359. https://doi.org/10.1016/J.JCLEPRO.2018.08.093

Mittal P, Mitra K (2020a) In search of flexible and robust wind farm layouts considering wind state uncertainty. J Clean Prod. https://doi.org/10.1016/j.jclepro.2019.119195
Mittal P, Mitra K (2020b) Micrositing under practical constraints addressing the energy-noise-cost trade-off. Wind Energy 23(10):1905–1918. https://doi.org/10.1002/WE.2525
Mittal P, Kulkarni K, Mitra K (2015) A novel and efficient hybrid optimization approach for wind farm micro-siting. IFAC-PapersOnLine 48(8):397–402. https://doi.org/10.1016/J.IFACOL.2015.08.214
Mittal P, Kulkarni K, Mitra K (2016a) A novel hybrid optimization methodology to optimize the total number and placement of wind turbines. Renew Energy. https://doi.org/10.1016/j.renene.2015.07.100
Mittal P, Mitra K, Kulkarni K (2017) Optimizing the number and locations of turbines in a wind farm addressing energy-noise trade-off: a hybrid approach. Energy Convers Manag 132:147–160. https://doi.org/10.1016/j.enconman.2016.11.014
Mittal P, Christopoulos G, Subramanian S (2024a) Energy enhancement through noise minimization using acoustic metamaterials in a wind farm. Renew Energy 224:120188. https://doi.org/10.1016/J.RENENE.2024.120188
Mittal P, Nampoothiri KN, Jha A, Bansal S (2024b) Convergence of machine learning with microfluidics and metamaterials to build smart materials. Int J Interact Des Manufact. https://doi.org/10.1007/S12008-023-01707-9/TABLES/1
Mittal P, Mitra K (2017a) Decomposition based multi-objective optimization to simultaneously determine the number and the optimum locations of wind turbines in a wind farm. In: IFAC-PapersOnLine
Mittal P, Mitra K (2017b) Energy-noise trade-off to optimize the total number and the placement of wind turbines on wind farms: a hybrid approach. In: 2017 Indian control conference, ICC 2017—proceedings
Mittal P, Mitra K (2018a) On determining optimal number and layout of wind turbines using space decomposed cost-energy trade-off algorithm. In: 2018 Indian control conference, ICC 2018—proceedings
Mittal P, Mitra K (2018b) Decomposition-based multi-objective optimization of energy noise trade-off in a wind farm. In: https://services.igi-global.com/resolvedoi/resolve.aspx?doi=https://doi.org/10.4018/978-1-5225-2990-3.ch008. IGI Global, pp 177–205
Mittal P, Mitra K (2019a) Determination of optimal layout of wind turbines inside a wind farm in presence of practical constraints. In: 2019 5th Indian control conference, ICC 2019—proceedings
Mittal P, Mitra K (2019b) Robust wind farm layout optimization under uncertainty. In: 2019 6th Indian control conference, ICC 2019—Proceedings
Mittal P, Mitra K (2019c) Variable grid resolution based evolutionary multi-objective optimization towards micro-siting. In: 2019 IEEE Congress on evolutionary computation, CEC 2019—proceedings
Mittal P, Mitra K (2020) Efficient wind farm micro-siting using novel optimization approaches—Raiith. Doctoral dissertation, Indian Institute of Technology Hyderabad
Mittal P, Kulkarni K, Mitra K (2016b) Multi-objective optimization of energy generation and noise propagation: A hybrid approach. In: 2016 Indian control conference, ICC 2016—proceedings
Moreno SR, Pierezan J, dos Santos CL, Mariani VC (2021) Multi-objective lightning search algorithm applied to wind farm layout optimization. Energy. https://doi.org/10.1016/j.energy.2020.119214
Mosetti G, Poloni C, Diviacco B (1994) Optimization of wind turbine positioning in large windfarms by means of a genetic algorithm. J Wind Eng Ind Aerodyn 51(1):105–116. https://doi.org/10.1016/0167-6105(94)90080-9
Nagpal SV, Liu MV, Anderson CL (2021) A comparison of deterministic refinement techniques for wind farm layout optimization. Renew Energy 1(168):581–592. https://doi.org/10.1016/j.renene.2020.12.043

Nai-Zhi G, Ming-Ming Z, Bo L (2022) A data-driven analytical model for wind turbine wakes using machine learning method. Energy Convers Manag. https://doi.org/10.1016/j.enconman.2021.115130

Nampoothiri KN, Bansal S, Jha A, Mittal P (2024) Integration of microfluidics in smart acoustic metamaterials. Eur Phys J Special Topics 2024:1–9. https://doi.org/10.1140/EPJS/S11734-024-01112-7

Nash R, Nouri R, Vasel-Be-Hagh A (2021) Wind turbine wake control strategies: a review and concept proposal. Energy Convers Manag. https://doi.org/10.1016/j.enconman.2021.114581

Okulov V, Kabardin I, Mukhin D, Stepanov K, Okulova N (2021) Physical De-Icing Techniques for Wind Turbine Blades. Energies 14(20):6750. https://doi.org/10.3390/EN14206750

Openwind - Wind Farm Modeling and Layout Design Software | UL Solutions. https://www.ul.com/services/openwind-wind-farm-modeling-and-layout-design-software. Accessed 6 Aug 2023

Pantula PD, Miriyala SS, Mitra K (2022) A deep unsupervised learning algorithm for clustering of wind frequency maps. 2022 8th Indian control conference, ICC 2022—proceedings, pp 361–366. https://doi.org/10.1109/ICC56513.2022.10093581

Parada L, Herrera C, Flores P, Parada V (2017) Wind farm layout optimization using a Gaussian-based wake model. Renew Energy 107:531–541. https://doi.org/10.1016/J.RENENE.2017.02.017

Park J (2020) Contextual Bayesian optimization with trust region (CBOTR) and its application to cooperative wind farm control in region 2. Sustain Energy Technol Assess 38:100679. https://doi.org/10.1016/J.SETA.2020.100679

Park J, Kwon S, Law KH (2013) Wind farm power maximization based on a cooperative static game approach. pp 8688:204–218. https://doi.org/10.1117/12.2009618

Pujari NSK, Miriyala SS, Mitra K (2020b) Auto-tuned deep recurrent neural networks for application in wind energy conversion systems. 2020 IEEE symposium series on computational intelligence. SSCI 2020:3065–3072. https://doi.org/10.1109/SSCI47803.2020.9308478

Pujari KN, Miriyala SS, Mitra K (2023a) Jensen-ANN: a machine learning adaptation of Jensen wake model. IFAC-PapersOnLine 56(2):4651–4656. https://doi.org/10.1016/J.IFACOL.2023.10.979

Pujari KN, Miriyala SS, Mittal P, Mitra K (2023b) Better wind forecasting using evolutionary neural architecture search driven green deep learning. Expert Syst Appl 214:119063. https://doi.org/10.1016/J.ESWA.2022.119063

Pujari NSK, Miriyala SS, Mitra K (2023c) Comparative study of automated deep learning techniques for wind time-series forecasting. Stat Model Mach Learn Concepts Appl. https://doi.org/10.1016/B978-0-323-91776-6.00003-8

Pujari KN, Miriyala SS, Mittal P, Mitra K (2020a) Optimal long short term memory networks for long-term forecasting of real wind characteristics. In: IFAC-PapersOnLine

Pujari KN, Srivastava V, Miriyala SS, Mitra K (2021) Comparison of deep reinforcement learning techniques with gradient based approach in cooperative control of wind farm. 2021 7th Indian control conference, ICC 2021—proceedings, pp 400–405. https://doi.org/10.1109/ICC54714.2021.9703186

Qureshi TA, Warudkar V (2023) Wind farm layout optimization through optimal wind turbine placement using a hybrid particle swarm optimization and genetic algorithm. Environ Sci Pollut Res 30(31):77436–77452. https://doi.org/10.1007/S11356-023-27849-7/TABLES/4

Rad NN, Bekker A, Arashi M (2022) Enhancing wind direction prediction of South Africa wind energy hotspots with Bayesian mixture modeling. Sci Rep 12(1):11442. https://doi.org/10.1038/s41598-022-14383-8

Radun J, Maula H, Saarinen P, Keränen J, Alakoivu R, Hongisto V (2022) Health effects of wind turbine noise and road traffic noise on people living near wind turbines. Renew Sustain Energy Rev 157:112040. https://doi.org/10.1016/J.RSER.2021.112040

Reddy SR (2021a) A machine learning approach for modeling irregular regions with multiple owners in wind farm layout design. Energy 220:119691. https://doi.org/10.1016/J.ENERGY.2020.119691

Reddy SR (2021b) An efficient method for modeling terrain and complex terrain boundaries in constrained wind farm layout optimization. Renew Energy 165:162–173. https://doi.org/10.1016/j.renene.2020.10.076

Renewable energy (2024) World's renewable energy capacity grew at record pace in 2023 | Renewable energy | The Guardian. https://www.theguardian.com/environment/2024/jan/11/worlds-renewable-energy-capacity-grew-at-record-pace-in-2023. Accessed 10 Feb 2024

Shahzad U (2022) Application of machine learning for optimal wind farm location

Shakoor R, Hassan MY, Raheem A, Wu YK (2016) Wake effect modeling: a review of wind farm layout optimization using Jensen's model. Renew Sustain Energy Rev 58:1048–1059. https://doi.org/10.1016/J.RSER.2015.12.229

Sickler M, Ummels B, Zaaijer M, Schmehl R, Dykes K (2023) Offshore wind farm optimisation: a comparison of performance between regular and irregular wind turbine layouts. Wind Energy Science 8(7):1225–1233. https://doi.org/10.5194/WES-8-1225-2023

Sirnivas S, Musial W, Bailey B, Filippelli M (2014) Assessment of offshore wind system design. Safety Oper Stand. https://doi.org/10.2172/1122306

Song M, Wen Y, Duan B, Wang J, Gong Q (2017) Micro-siting optimization of a wind farm built in multiple phases. Energy 137:95–103. https://doi.org/10.1016/J.ENERGY.2017.06.127

S&P Global Commodity Insights (2021) Global energy demand to grow 47% by 2050, with oil still top source: US EIA | https://www.spglobal.com/commodityinsights/en/market-insights/latest-news/oil/100621-global-energy-demand-to-grow-47-by-2050-with-oil-still-top-source-us-eia. Accessed 15 Sep 2023

Statista (2024a) Fernández Lucía global wind power market—statistics & facts | Statista. https://www.statista.com/topics/4564/global-wind-energy/#topicOverview. Accessed 15 Jan 2024

Statista (2024b) Greenhouse gas emissions worldwide—Statistics & Facts. https://www.statista.com/topics/5770/global-greenhouse-gas-emissions/#topicOverview. Accessed 22 Jan 2024

Tang XY (2022) Optimizing wind farm layouts. Handbook of wind energy aerodynamics, pp 1223–1249. https://doi.org/10.1007/978-3-030-31307-4_61

Tao S, Xu Q, Feijóo A, Zheng G, Zhou J (2020b) Wind farm layout optimization with a three-dimensional Gaussian wake model. Renew Energy 159:553–569. https://doi.org/10.1016/j.renene.2020.06.003

Tao S, Xu Q, Feijóo A, Zheng G, Zhou J (2020a) Nonuniform wind farm layout optimization: a state-of-the-art review. Energy 209

Ti Z, Deng XW, Yang H (2020) Wake modeling of wind turbines using machine learning. Appl Energy. https://doi.org/10.1016/j.apenergy.2019.114025

Ti Z, Deng XW, Zhang M (2021) Artificial Neural Networks based wake model for power prediction of wind farm. Renew Energy 172:618–631. https://doi.org/10.1016/j.renene.2021.03.030

Tricoire F, Parragh SN, Gansterer M (2024) The hiking tourist problem. CEJOR 2024:1–25. https://doi.org/10.1007/S10100-023-00895-9

Wan C, Wang J, Yang G, Gu H, Zhang X (2012) Wind farm micro-siting by Gaussian particle swarm optimization with local search strategy. Renew Energy 48:276–286. https://doi.org/10.1016/J.RENENE.2012.04.052

Wang Q, Pan L, Wang H, Wang X, Zhu Y (2022) Short-term wind power probabilistic forecasting using a new neural computing approach: GMC-DeepNN-PF. Appl Soft Comput 126:109247. https://doi.org/10.1016/j.asoc.2022.109247

Wen Y, Song M, Wang J (2022) Wind farm layout optimization with uncertain wind condition. Energy Convers Manag. https://doi.org/10.1016/j.enconman.2022.115347

Willis DJ, Niezrecki C, Kuchma D, Hines E, Arwade SR, Barthelmie RJ, DiPaola M, Drane PJ, Hansen CJ, Inalpolat M, Mack JH, Myers AT, Rotea M (2018) Wind energy research: state-of-the-art and future research directions. Renew Energy 125:133–154. https://doi.org/10.1016/J.RENENE.2018.02.049

Windisch T, Hensel E, Drossel WG (2022) Reducing noise and vibrations in wind power plants with DFIG and planetary gearbox through improved rotor current control. Vibroeng Proced 42:82–88. https://doi.org/10.21595/VP.2022.22529

windPRO—The premier software package for design of wind farms and PV projectsEMD International. https://www.emd-international.com/windpro/. Accessed 13 Jun 2023

Witha B (2021) CFD-type wake models. Handbook of wind energy aerodynamics, https://doi.org/10.1007/978-3-030-05455-7_51-1

Wu X, Hu W, Huang Q, Chen C, Jacobson MZ, Chen Z (2020a) Optimizing the layout of onshore wind farms to minimize noise. Appl Energy. https://doi.org/10.1016/j.apenergy.2020.114896

Wu Y, Zhang S, Wang R, Wang Y, Feng X (2020b) A design methodology for wind farm layout considering cable routing and economic benefit based on genetic algorithm and GeoSteiner. Renew Energy 146:687–698. https://doi.org/10.1016/J.RENENE.2019.07.002

Yamani Douzi Sorkhabi S, Romero DA, Beck JC, Amon CH (2018) Constrained multi-objective wind farm layout optimization: novel constraint handling approach based on constraint programming. Renew Energy 126:341–353. https://doi.org/10.1016/J.RENENE.2018.03.053

Yang K, Cho K (2019) Simulated annealing algorithm for wind farm layout optimization: a benchmark study. Energies 12(23):4403. https://doi.org/10.3390/EN12234403

Yessef M, Bossoufi B, Taoussi M, Lagrioui A, Chojaa H (2022) Overview of control strategies for wind turbines: ANNC, FLC, SMC, BSC, and PI controllers. Wind Eng. https://doi.org/10.1177/0309524X221109512

Chapter 2
Multi-objective Optimisation with Uncertainty: Some Considerations for Wind Farm Optimisation

Jonathan Fieldsend

Abstract In this chapter we introduce the multi-objective optimisation task and highlight its distinctions from the single objective optimisation task. We employ a running example of a wind farm layout optimisation problem, and illustrate how the search landscape changes depending on the objective function used, and when shifting from a single to multiple objectives. We also outline some of the different forms of uncertainty which an optimisation task may commonly experience—particularly in the context of windfarm optimisation. The chapter then continues by describing how such uncertainty can affect the performance of standard multi-objective optimisation approaches. We describe mitigation approaches which may be employed, casting the task as a robust optimisation, with different wind profile scenarios, and illustrate the use of a simple multi-objective optimiser developed for this formulation. This chapter sets the scene for later applied chapters in this volume.

Keywords Fitness landscape · Multiple objectives · Problem representation

2.1 Introduction

There are a number of different objectives that might be considered when designing a wind farm (Chen et al. 2013; Rodrigues et al. 2016; Li et al. 2017; Biswas et al. 2018), which can include:

1. maximising power output;
2. minimising build cost;
3. minimising running/maintenance cost; and
4. reducing environmental impact (among others).

These criteria may be broken down into further separate objectives, for instance the environmental cost could include minimising the footprint of the farm, the use of particular materials, but also aspects such as noise pollution (Amon et al. 2012).

J. Fieldsend (✉)
Department of Computer Science, University of Exeter, Rennes Dr, Exeter, EX4 4RN, UK
e-mail: j.e.fieldsend@exeter.ac.uk

K. Mitra et al. (eds.), *Optimization, Uncertainty and Machine Learning in Wind Energy Conversion Systems*, Engineering Optimization: Methods and Applications,
https://doi.org/10.1007/978-981-97-7909-3_2

Multi-objective optimisation of wind farms involves finding the best trade-off between two or more such objectives. We are concerned with trade-offs here, as typically these objectives are in conflict—so we have to consider and compare designs which may be better on one criterion, but achieve this by sacrificing performance on another criterion. Consider for instance power: if we increase the density of turbines the resultant power of the farm is likely to increase, however this will likely also make the wind farm more costly to manufacture and install, and the environmental impact will also usually be increased. Overall efficiency may also be impacted by increased wake effects (González-Longatt et al. 2012). Note however that in some circumstances modifying a particular design may cause objectives to *improve together*—changing the material of a component may improve cost *and* performance *and* reduce the environmental impact (if sourcing such a material is relatively less environmentally harmful than the one it is replacing). Such circumstances notwithstanding, typically when tackling such multi-objective problems there are a set of trade-off solutions identified where improvements in one objective cannot be achieved without compromising one or more other performance measures. Therefore a choice needs to be made as to which trade-off design to implement, given the decision maker's preferences. This selection process falls into the decision science area of multi criteria decision making (Miettinen 1999). For the remainder of this chapter we will concern ourselves principally with the task of finding a set of trade-off solutions which leads to an *a posteriori* approach to final design identification (a trade-off set is found and then a decision is made on to which to use, rather than trying to decide on the trade-off location before or interactively during optimisation). However other contributions in this volume will explore further on multi-criterion decision making aspects.

An additional and important property of the wind farm optimisation task is that it it often exhibits *uncertainty*. Wind speed and direction can vary greatly over time, making it difficult to predict how much power a wind farm will produce. There are also cyclical seasonal effects—dominant wind directions and strengths often vary at locations as the seasons change, however over longer time periods changes in climate can modify prevailing wind directions and intensities also. Beyond this source of uncertainty, there can also be regulatory and financial uncertainties. This can be in terms of, e.g. incentives, and fluctuations of realised energy prices—whose predictions steer investment decisions. These (and other) sources of uncertainty can make it challenging to optimise the wind farm for multiple objectives, and they affect different objectives to different extents.

Beyond these issues, the task of optimising wind farms themselves is inherently complex. Wind farms are composed of many interconnected components, such as turbines, power converters, and control systems, alongside their arrangement spatially in some typically contiguous geographical region. Aspects such as land availability and access to the power grid induce constraints on designs, and can limit the accessibility of particular configurations.

All the aforementioned aspects combine to make optimisation tasks in this domain non-trivial and complex.

The chapter will progress as follows. We formally set out the standard multi-objective optimisation task in Sect. 2.2 and some of the notation which will be common across the book, and relate this to wind farm design. In Sect. 2.3 we set out a small example, using the wind farm optimisation benchmark of Wilson et al. (2018), to illustrate the complexities of the multi-objective search landscapes of such problems. In Sect. 2.4 we discuss further the sources and forms of uncertainty that may be experienced in this application domain, and recast our running example as a robust optimisation task, to address the uncertainty in the wind profiles. In Sect. 2.5 we conclude the chapter with a discussion.

2.2 The Multi-objective Optimisation Task

Without loss of generality we cast the multi-objective wind farm optimisation task in terms of minimisation.[1] Given a feasible search space $\mathcal{X}$ (e.g. all possible placements of a set of wind turbines, which adhere to any geographical and legal constraints in the farm configuration), a design (often called a decision vector) from this space (for instance a set of placement locations) is denoted by $\mathbf{x}$. We denote by the term *solution* the tuple of a decision vector, and its evaluation under the vector of objective functions, i.e. $s = (\mathbf{x}, \mathbf{f}(\mathbf{x}))$. A decision vector $\mathbf{x}$ said to *dominate* another decision vector $\mathbf{x}'$—denoted as $\mathbf{x} \prec \mathbf{x}'$—if it is no worse on all M objectives, $f_i(\mathbf{x})$ (i.e. $f_i(\mathbf{x}) \leq f_i(\mathbf{x}')\ \forall i$), and better on at least one. A design $\mathbf{x}$ is said to *weakly dominate* another design $\mathbf{x}'$, denoted $\mathbf{x} \preceq \mathbf{x}'$, if it is no worse on all assessment criteria (it is possible therefore for $\mathbf{x} \preceq \mathbf{x}'$ and $\mathbf{x}' \preceq \mathbf{x}$ to hold, if the objective vectors are identical, however it is not possible for both $\mathbf{x} \prec \mathbf{x}'$ and $\mathbf{x}' \prec \mathbf{x}$ to hold in any circumstance).

The dominance relationship allows us to put a partial order on designs, as we can identify those which are better or worse than each other, and those which are functionally equivalent (all objective values are the same), even if their configurations differ. However, without any prior preference on objectives, we will also have the situation where decision variables are *incomparable*, i.e. where $\mathbf{x} \not\preceq \mathbf{x}'$ and $\mathbf{x}' \not\preceq \mathbf{x}$. Such pairs of solutions are called mutually non-dominating, and a set of solutions where all members are mutually non-dominating with one another is known as a *non-dominated set*.

If it is not possible to improve a decision vector's performance on one objective by modifying it, without also sacrificing performance on one or more other objectives, then the solution is known as *Pareto optimal*.

The set of Pareto optimal decision vectors (the *Pareto set*) is defined as $\mathcal{P} = \{\mathbf{x} \in \mathcal{X} \mid \nexists \mathbf{x}' \prec \mathbf{x},\ \mathbf{x}' \in \mathcal{X}\}$. The Pareto set is therefore also a non-dominated set, but it is worth emphasising here that not all non-dominated sets (sets of mutually non-dominating solutions) are the Pareto set.

[1] Note of course a maximisation function can trivially be converted to a minimisation version by multiplying the function by -1.

The image of $\mathcal{P}$ under $\mathbf{f}$ is known as the *Pareto Front*, $\mathcal{F}$. Note that, due to many-to-one mappings, the cardinality of the Pareto Set may be (much) larger than that of the Pareto Front.

As a multi-objective optimiser searches across $\mathcal{X}$ it typically maintains an approximation to $\mathcal{P}$, called its approximation set (or Pareto archive, A)—such an approximation is typically a non-dominated set. The task of maintaining this set is commonly referred to as the dynamic non-dominance problem (Schütze 2003), and is summarised in Algorithm 1.

Algorithm 1 Updating a non-dominated archive A with $\mathbf{x}$ (no duplicates)

```
Require: A                                     ▷ The current non-dominated set of solutions
Require: x                                     ▷ A new solution to check against A
1: if ∄x′ ∈ A | x′ ⪯ x then                    ▷ If x is not dominated or operationally equal
2:     A := A \ {x′ ∈ A | x ≺ x′}              ▷ Remove any members of A dominated by x
3:     A := A ∪ {x}                            ▷ Add x to non-dominated set
4: return A
```

Typically for such optimisation problems metaheuristics such as evolutionary algorithms (EAs) are used (Coello Coello et al. 2007), this is because there is often not a differentiable form of the cost function(s) available. EAs maintain and modify a population of solutions iteratively, which makes them particularly well-suited to multi-objective optimisation problems, where a set of trade-off solutions is sought.

We will now consider an example of a multi-objective wind farm optimisation problem, and the complexity of the search landscape such problems embody, both when considered with a single objective, and with multiple conflicting objectives.

2.3 Example Wind Farm Optimisation Search Landscape

We draw on the simplified benchmark functions from the 2014 and 2015 Wind Farm Layout Optimization Competition at ACM GECCO, which are open source and freely available.[2]

First, let us consider the search landscape of a single objective wind farm layout optimisation task for these problems. In the 2015 competition configuration the cost of energy in the layout (production price of a kilowatt) was the target to minimise, which aimed to balance turbine costs and energy capture. Specifically (from Wilson et al. 2018), given the number of turbines n, turbines per substation m and farm energy output P, the energy cost of a farm is calculated as:

[2] https://github.com/d9w/WindFLO.

$$\texttt{EnergyCost} = \frac{(\texttt{Cost}_{\texttt{const}} \times \texttt{EoS}) + \texttt{Cost}_{\texttt{op}}}{\texttt{Interest}} \times \frac{1}{\texttt{Output}} \times \texttt{SizeCoefficient} \tag{2.1}$$

where the construction cost $\texttt{Cost}_{\texttt{const}}$ is calculated as

$$\texttt{Cost}_{\texttt{const}} = c_t \times n + c_s \times \lfloor n/m \rfloor \tag{2.2}$$

with turbine cost c_t of $750,000 and substation cost c_s of $8 million used in the benchmark. The economies of scale $\texttt{EoS}$ are calculated as

$$\texttt{EoS} = \frac{2}{3} + \frac{1}{3} \times e^{-0.00174n^2} \tag{2.3}$$

and the yearly operating costs $\texttt{Cost}_{\texttt{op}}$ are

$$\texttt{Cost}_{\texttt{op}} = \$20{,}000 \times n \tag{2.4}$$

the interest (assuming a farm lifetime of 20 years and interest rate of 3%) is

$$\texttt{Interest} = (1 - 1.03^{20})/0.03 \tag{2.5}$$

a yearly power output of

$$\texttt{Output} = 8760 \times P \tag{2.6}$$

and a farm size coefficient of

$$\texttt{SizeCoefficient} = \frac{0.1}{n}. \tag{2.7}$$

To gain an understanding of the behaviour of such a problem, we now illustrate the landscapes using a compressed network representation, the local optima network.

2.3.1 *Local Optima Networks*

A local optima network (LON) compresses the information contained in the search space of an optimisation problem into a weighted directed graph. The model is adapted from the inherent networks of energy landscapes in physical-chemistry and can be used to as a way visualise and apply complex network analysis tools to combinatorial optimisation problems (Ochoa et al. 2008; Tomassini et al. 2008), and more recently continuous problems (Adair et al. 2019).

The vertices (also known as nodes) of the graph represent the local optima under the particular neighbourhood operator used, and the size of the vertex is proportional to the (hyper)volume of the search space which *supports* this optima. In LONs these *basins* are composed of all solutions where a hill-climb from the solutions (sequentially moving to an immediate neighbour which is better) terminates at that particular local optima. He we use greedy hill-climbing (see Algorithm 2), which considers all neighbours and selects the best quality one when making each hill-climb move.

Algorithm 2 Greedy (best) improvement hill-climbing

```
Require: x^start                                       ▷ Initial sample
x^end := x^start
converged := false
while converged = false do
    better_found := false
    for x' ∈ N(x^end) do
        if f(x) < f(x^end) then                        ▷ If neighbour is better
            x^end = x                                  ▷ Replace with better neighbour
            better_found := true
    if better_found = true then
        goto 4                                         ▷ Better neighbour found, so climb again
    else
        converged := true                              ▷ No better neighbour found
return x^end                                           ▷ Return path start and end locations
```

An edge between two vertices can be defined in different ways, but usually it represents whether a transition between the two basins of attraction is possible (the existence of an edge), and what the probability of this is (the width of the edge). One popular approach for edge construction is the basin transition (see e.g. Herrmann et al. 2018), which uses the union of the neighbourhoods of all basin members. It then constructs edges and calculates their weights based on the destinations of all hill-climbs from this set. Depending on the size and shape of the basin, this can result in a large *self-loop* (as many basin members' neighbours will be other members of the same basin, and therefore the hill-climb will lead to the same optima), but some walks will lead to *other* adjacent basins (if the starting point is a neighbour of a member of the basin periphery). The other regularly used formulation is that of the edge escape (Vérel et al. 2012). This is generally quicker to compute than the basin transition, and relies on using a wider initial neighbourhood function, $\tilde{N}()$. This is applied to the local optima location to form a wider set of initial neighbours, and then the hill-climb process is applied as usual with the standard neighbourhood applied. For instance, for a binary string represented problem, $N()$ may be the hamming distance one neighbourhood (i.e. single bit-flip), whereas $\tilde{N}()$ may be the hamming distance two neighbourhood.

2.3.2 Problem Instance Search Landscape Visualisation

We use the first scenario from the test suite with obstacles, i.e. the problem defined in `obs_00.xml`. The search space of the problem is too large to directly enumerate as it is continuous. Its direct use is also subject to a variable length encoding (1 to n turbines may each be placed anywhere in the farm, meaning 2 to $2 \times n$ real values define the turbine placements). There are also some constraints built into the problems—turbines cannot be placed so close as to risk physical contact between their rotors, and some regions of the farm are identified as illegal for turbine placement (simulating situations such as roads or existing structures which impose location constraints).

Following the convention used in the example genetic algorithm optimiser the competition organisers provide (Wilson et al. 2018), we convert the continuous and variable length problem into a fixed size Boolean encoding. This is accomplished by changing the representation to a mesh (grid) imposed on the windfarm geographic location, such that the mesh completely covers the legal placement area. Depending on the fidelity employed in the mesh, this leads to a different search space sizes. Using this binary matrix representation a '1' indicates a single turbine at that cell, and a '0' indicates that cell is not used. For a rectangular region bounding the farm location, the size of the solution space is $2^{nx \times ny}$, where nx is the number of cells in the x-dimension and ny is the number of cells in the y-dimension.

For computational tractability when representing the entire landscape we use a range of coarse meshes in our illustrations below, from 2×2 through to 5×5. Even such a coarse bit-string representations of the problem have between $2^4 = 16$ to $2^{25} = 33{,}554{,}432$ unique solutions. (Note, however a few of these will violate obstacle constraints, and one will be the trivial solution of a matrix of zeros—no turbines.) We use the Hamming distance 1 neighbourhood (i.e. single bit-flip) in the hill-climbs. As such the set of neighbours of a configuration are all designs which have turbines in the same position and one additional turbine, and also those which are identical in placement apart from one turbine which is absent. A decision vector therefore has $nx \times ny$ neighbours.

Note that a network visualisation of the mesh employed in their genetic algorithm which has a 23×46 mesh is therefore likely to lead to a substantially more complex network than the largest shown here, but is infeasible to enumerate, as it has $2^{23 \times 46} \approx 3 \times 10^{318}$ unique configurations to evaluate.

Figure 2.1 shows the corresponding LONs for these coarse representations of the problem using the single objective described in (2.1). Interestingly there is a single basin of attraction for 2×2–5×5 meshes—all hill-climbs lead in each mesh resolution lead to the same solution.

Figure 2.2 shows the LONs corresponding to the same domains and neighbourhoods, but employing the 2014 competition objective function, which was to maximise the wake free ratio:

$$\texttt{WakeFreeRatio} = \texttt{Output}/(\texttt{WakeFreeEnergy} \times n) \tag{2.8}$$

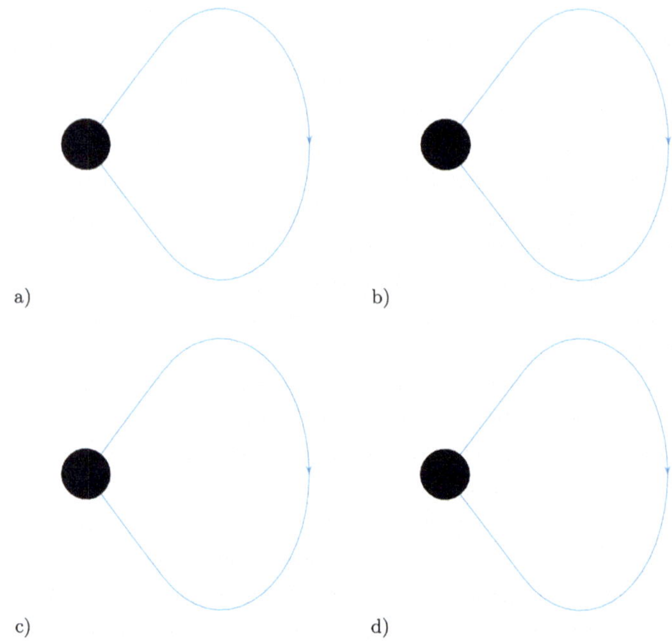

Fig. 2.1 LONs for scenario `objs_00.xml` using a 2 × 2 (**a**) through to an 5 × 5 mesh (**d**) and a single objective, `EnergyCost`, defined in (2.1)

As can be seen the landscapes illustrated by the networks are much more rugged, with multiple local optima, even on the coarsest encoding. The number of local optima increases with the mesh fidelity—with the 5 × 5 mesh resulting in 745 distinct local optima. It is interesting to note that the landscape induced by the 2014 competition objective function is much more complex than that used in 2015 (at least for these coarse representations) and would appear substantially more difficult to optimise.

Figure 2.3 shows scatter plots of the pairs of objective values (energy cost and wake free ratios) obtained by enumerating each of the mesh resolutions shown in Figs. 2.1 and 2.2. The Pareto front for each of these domains is therefore the top-left edge of these point clouds, as energy cost is minimised and wake free ratio is maximised.

We recently developed analogues for LONs for multi-objective problems, *Pareto-LONs* (P-LONs) (Fieldsend and Alyahya 2019), and these are shown for the different mesh resolutions considered in Fig. 2.4 when using both objectives together, i.e. energy cost (2.1) to be minimised, and the wake free ratio (2.8) to be maximised.

P-LONs employ Pareto Local Search (Paquete et al. 2007), which is a hill-climbing analogue in the multi-objective setting (see Algorithm 3). Full details of the P-LON construction may be found in Fieldsend and Alyahya (2019). Succinctly: each vertex in the P-LON is a unique set of mutually non-dominated solutions, as returned by Pareto Local Search, whose set of immediate neighbours are dominated. The vertex size denotes how many PLS hill-climbs lead to that set (i.e. reflecting the basin size,

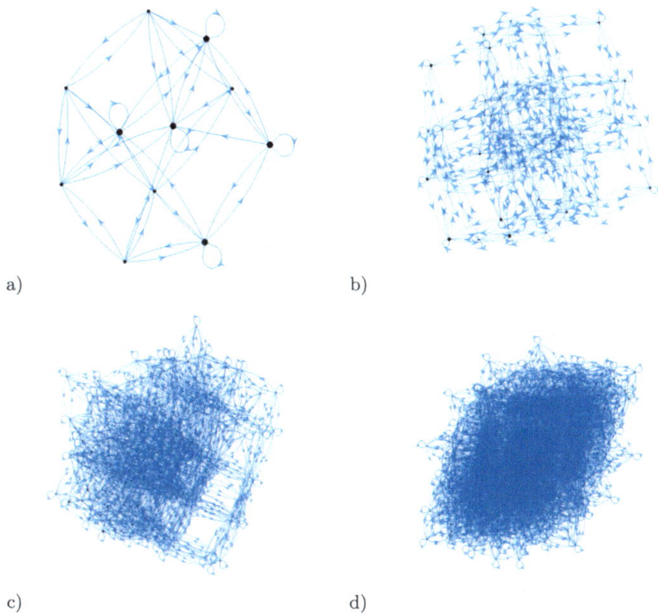

Fig. 2.2 LONs for scenario `objs_00.xml` using a 2×2 (**a**) through to an 5×5 mesh (**d**) and a single objective, `WakeFreeRatio`, defined in (2.8)

as in the single objective LON case), and as with LONs the edges between vertices denote the probability of transition between the connected vertices. Here the shading of the vertex denotes the proportion of the non-dominated set residing at that optima which are Pareto optimal. For the 2×2 and 3×3 meshes all non-dominated set vertices are composed entirely of Pareto optimal solutions (and are coloured black to reflect this) however in the 4×4 network all the large basin vertices are grey, the black vertices having much smaller diameters (indicating that their basin sizes are significantly smaller).

As can be seen, on the same coarse search domains, the two single objective landscapes differ considerably, and the the bi-objective P-LON in Fig. 2.4 shows a different structure again. Table 2.1 list the number of optima observed in the LONs/PLONs for the different formulations and search space sizes. Such network visualisations of the landscapes show that even without considering uncertainty the multi-objective optimisation of such wind farm layout problems is complex and non-trivial.

2.3.3 Multi-objective Optimisation

We now illustrate the multi-objective optimisation of this problem with a simple optimiser. We use the fine mesh which is employed in Wilson et al. (2018). This

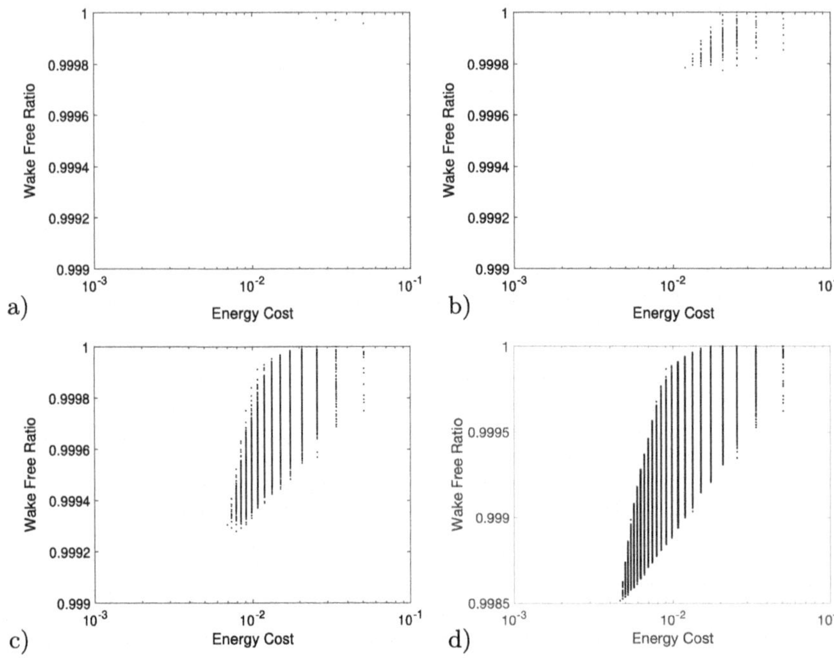

Fig. 2.3 Scatter plots of `EnergyCost` versus `WakeFreeRatio` objectives obtained by enumerating all design combinations possible on a 2×2 mesh (**a**) through to an 5×5 mesh (**d**) for scenario `objs_00.xml`. Log scale on x-axis

Algorithm 3 Pareto Local Search (PLS) of Paquete et al. (2007)

Require: $\mathbf{x}^{start}$ ▷ Initial sample
$F := \{\mathbf{x}^{start}\}$ ▷ Initial set to process
$X_{PLO} := \emptyset$ ▷ Set holding estimated Pareto local optima (PLO)
while $F \neq \emptyset$ **do** ▷ While still locations to process
  $\mathbf{x} := random\_draw(F)$
  $W := \emptyset$
  for all $\mathbf{x}' \in \mathcal{N}(\mathbf{x})$ **do**
    if $\nexists \mathbf{x}'' \in W \cup F \cup X_{PLO} \,|\, \mathbf{x}'' \preceq \mathbf{x}'$ **then**
      $W := W \cup \mathbf{x}'$ ▷ Not dominated by current estimate
  $X_{PLO} := \{\mathbf{x}' \,|\, \mathbf{x}' \in X_{PLO} : \nexists \mathbf{x}'' \in W \,|\, \mathbf{x}'' \prec \mathbf{x}'\}$
  $F := F \setminus \{\mathbf{x}\} \cup W$
  $F := \{\mathbf{x}' \,|\, \mathbf{x}' \in F : \nexists \mathbf{x}'' \in W \,|\, \mathbf{x}'' \prec \mathbf{x}'\}$
return X_{PLO} ▷ Return path start location and end set

leads to the possible placement locations shown in Fig. 2.5 (the obstacle locations in middle of the farm and top right-hand corner for this problem are clear to see, as no turbines are permitted to be placed in these regions). This configuration results in a binary string decision vector of 1029 bits.

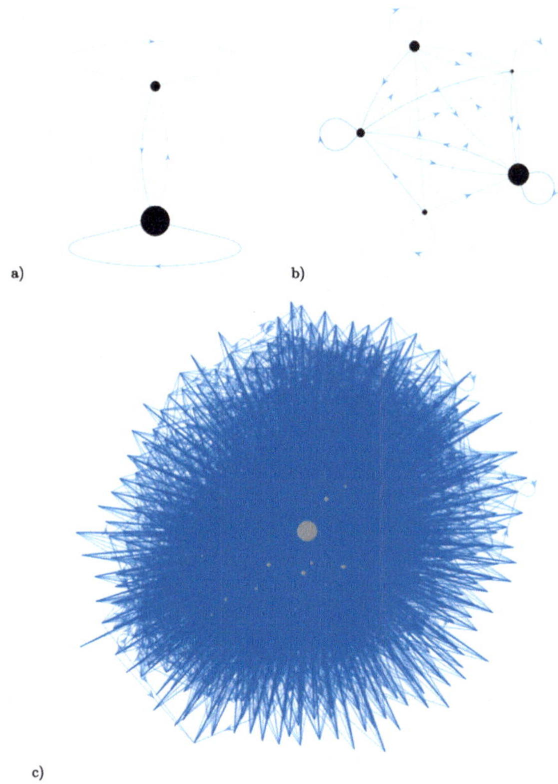

Fig. 2.4 Pareto-LONs for scenario `objs_00.xml` using a 2×2 (**a**) through to a 4×4 mesh (**c**) and the `EnergyCost` and `WakeFreeRatio` objectives

Table 2.1 Number of local optima under the different single and bi-objective landscapes

	Grid resolution			
	2×2	3×3	4×4	5×5
	Number of local optima			
LON, energy cost	1	1	1	1
LON, wake free ratio	11	58	259	745
PLON, energy cost and wake free ratio	2	5	531	–

Note values for the exhaustive P-LON for 5×5 is not provided as a matrix to store the domination relations ($2^{25} \times 2^{25}$) was required which exceeded the machine memory limits

We employ a simple evolution strategy (ES), which draws solutions to recombine and mutate from a maintained archive. Specifically it is a $(|A| + 1)$–ES. Parents are drawn from the current best approximation to the Pareto set (the archive A), and a single child is generated at each iteration and compared to the A for updating. Algorithm parameters are set out in Table 2.2, and the optimiser at a high-level is listed in Algorithm 4.

We run the algorithm 21 times, with a budget of 10,000 function evaluations. The empirical attainment function (Fonseca et al. 2011) from the runs is shown

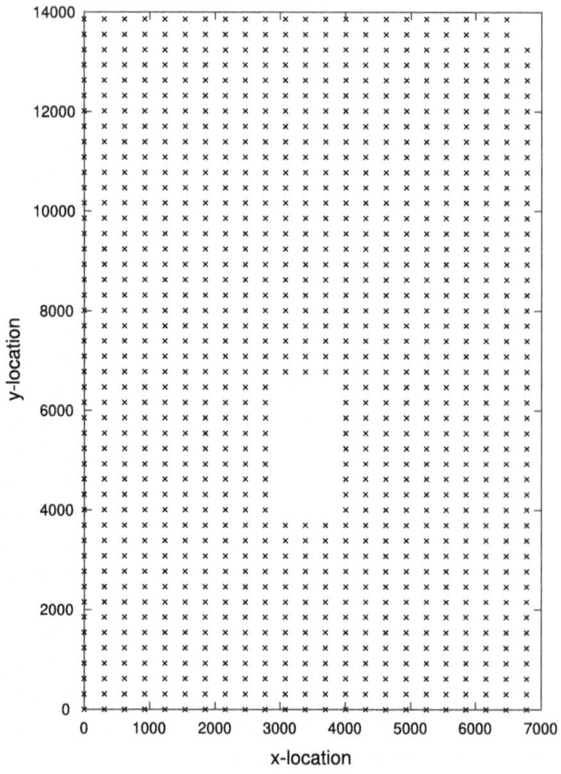

Fig. 2.5 Illustration of the legal placement locations for turbines in the wind farm, using the finer mesh in our optimisation runs

Table 2.2 Parameters used in the $(|A| + 1)$–ES optimiser

Variable	Value
Crossover type	Uniform
Probability of crossover	0.8
Rate of crossover	0.4
Rate of mutation (probability of flipping each bit)	0.05
Number of initial random solutions	100

in Fig. 2.6—this shows the probability of attainment of combinations of the two objectives by the collection of different runs. Locations coloured white are attained (that is the objective vector defined at that location is equalled or dominated) by one or more members of each archive returned by every run. Those coloured black are not attained by any run, and those shaded between these extremes have been attained by some runs but not others (see the colourmap on the right-hand side of the figure for scaling information on this shading). As can be seen, the trade-off front for Wake Free Ratio values at or below 0.95 has been consistently found by

Algorithm 4 $(|A| + 1)$–ES for the multi-objective optimisation task

Require: p^{mut} ▷ Probability of mutation
Require: p^{cross} ▷ Probability of crossover
Require: n ▷ Number of initial random starting solutions
Require: max_evals ▷ Evaluation budget
$X :=$ `initialise`(n) ▷ Generate n initial decision vectors
$Y :=$ `evaluate`(X) ▷ Evaluate initial decision vectors
$S := \{X^i, Y^i\}_{i=1}^n$ ▷ Construct the set of initial solutions
$A :=$ `nondom(S)` ▷ Identify non-dominated subset of S and put in A
$i := n$
while $i < max\_evals$ **do**
  $child :=$ `evolve`(A, p^{mut}, p^{cross}) ▷ Create child decision vector
  $s := (child,$ `evaluate`$(child))$ ▷ Construct evaluated solution
  $A :=$ `nondom`$(A \cup \{s\})$ ▷ Update archive
  $i := i + 1$
return A ▷ Return non-dominated set approximating the Pareto set

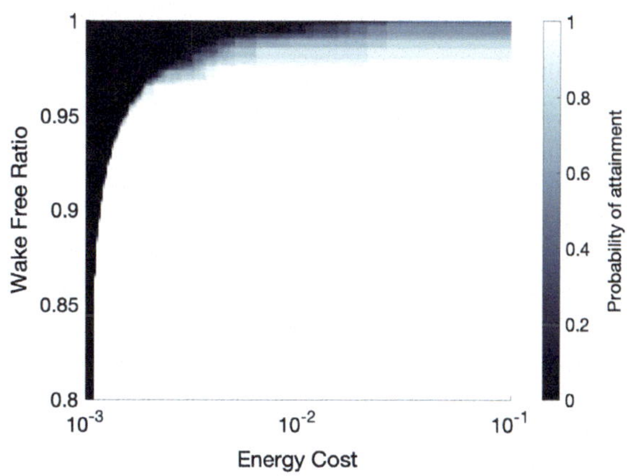

Fig. 2.6 Empirical attainment front from the 21 runs. By 10,000 function evaluations all runs are performing similarly in the left hand portion of the front, though not all have converged on energy costs for wake free ratio values above 0.95

all 21 runs by 10,000 function evaluations (irrespective of the random starting set and stochastic operations) however not all runs have converged to the same trade-off front for higher values, indicating a longer run is required for consistent performance of this particular optimiser on the problem.

In this section we have explored how even when using a simplified model (and a coarse representation) the multi-objective optimisation task of selecting a wind farm layout induces complex optimisation landscapes—with potentially very many local optima. We have also shown the multi-objective optimisation of the layout task in practice. However, as mentioned in the introduction, often in optimisation relating to wind farms we also have to deal with uncertainty in various forms. This complicates the process further, and we discuss such issues more in following section.

2.4 Effects of Uncertainty, and Optimising Under Uncertainty

Uncertainty can significantly impact the multi-objective optimisation of wind farms. This is because wind speed and direction, which are the primary drivers of power output in wind farms, can vary greatly over time, making it difficult to accurately predict how much power the wind farm will produce. There may also be other uncertainties which would affect results, for instance future interest rates which may affect profitability and investment decisions, and realised purchase costs. Such uncertainties can affect the optimisation process in a number of ways.

Uncertainty in wind speed and direction can make it challenging to find the best trade-off between different objectives, such as maximising power output and minimising environmental impact. For example, increasing turbine density may increase power output, but it may also increase the environmental impact. However, the optimal density may change depending on the specific wind conditions, which are uncertain. Similarly, uncertainty in the future cost of equipment, maintenance, etc., may affect the relative cost per kilowatt of different designs, and change rankings between them.

Such considerations relate to design *robustness*. Robustness refers to the ability of an optimised design to remain effective under different scenarios or operating conditions. If the wind conditions are uncertain, the optimisation results may be less robust and particular designs may be more likely to fail under certain conditions. As such the different scenarios/conditions need to be taken into consideration when optimising and ranking solutions.

Depending on the wind farm optimisation task, there may be uncertainty in the *evaluation* of the design itself, $\mathbf{f}(\cdot)$, such that repeated evaluations lead to different values being returned. An example might be for turbine blade performance evaluation using computational fluid dynamics (CFD)—different initial conditions of the solver leading to slightly different performance values.

Using the running example in this chapter, we now formulate an optimise a multi-objective robust problem, where the worse case performance is optimised under a limited set of wind profile scenarios.

2.4.1 *Robust Optimisation Example*

The ACM GECCO wind farm layout test problems are configured with the same farm location, but different wind scenarios. Therefore we can consider it in a robust optimisation context, with 10 different scenarios, as set out in `objs_00.xml`-`objs_09.xml`.

In such a formulation, assuming minimisation of the objectives functions, when optimising for the worst case performance over scenarios (the uncertainty set), $U = \{\mathbf{u}_1, \ldots, \mathbf{u}_{10}\}$, the task becomes

$$\mathbf{x}^\star = \underset{\mathbf{x}\in\mathcal{X}}{\operatorname{argmin}} \max_{\mathbf{u}_i\in U} \mathbf{f}(\mathbf{x}, \mathbf{u}_i). \tag{2.9}$$

As with the previous section, where the objective are competing it is unlikely that a single objective vector will satisfy this, and a set of solutions will be sought.

A naive approach would be to evaluate every scenario $\mathbf{u}_i$ for each decision vector considered, however, this scales poorly with the cardinality of U. Indeed, for our running example, this will make the search process an order of magnitude larger as $|U| = 10$. Instead, taking inspiration from the accumulative sampling approaches to noisy optimisation (Fieldsend and Everson 2015), we can select elements of U initially at random to approximate the rhs of (2.9). As the search progresses, we can evaluate *additional* scenarios for members of our current non-dominated set, in order to refine our objective estimates. At the limit, once all 10 scenarios are considered for a particular $\mathbf{x}$, there is no longer any uncertainty regarding its worst performance over U.

When comparing two partially evaluated solutions (i.e. whose maximum performance on objectives is estimated from proper subsets of U) there is some uncertainty on the ordering derived from Pareto dominance, as the estimated worst objective values may be in error. However as we are dealing with the worst case performance, if $\mathbf{x}$ is been evaluated on all members of U, and it is seen to dominate $\mathbf{x}'$ we can say with certainty that $\mathbf{x} \prec \mathbf{x}'$, *irrespective of how many members of* U *that* $\mathbf{x}'$ *has been evaluated on*. This is because with worst case performance, the calculation of the maximum objective values associated with $\mathbf{x}'$ can only ever stay the same or get worse (larger) when evaluated under additional scenarios. This is a useful property we can exploit in our optimisation framework.

The approach we take is set out in Algorithm 5. Again, for this simple illustration we employ a $(|A| + 1)$–ES. An initial set of designs are evaluated, each drawing a single scenario from U at random. The non-dominated solutions are identified and placed in A, and the members of U they have been evaluated on recorded. However, rather than discarding those solutions which are marked as dominated, they are stored in a reserve pool D, as reevaluation of designs in A with other members of U may lead to a re-ranking of designs, and therefore a reappraisal of the relative performance of some members of D. Note however that once members of A have been evaluated on all of U, then any members of D they dominate (and new solutions considered that they dominate) may be immediately discarded. The algorithm therefore balances evaluating new decision vectors, and reevaluating members of A on scenarios they have not yet been assessed with. Over time the average number of scenarios that the archive members are assessed on grows, and at a much faster rate than the average of all solutions considered (see Fig. 2.7). This targeted budget allocation of scenario evaluations means that considerably more decision vectors can be considered than if all solutions where fully evaluated (see Fig. 2.8)—our total per run budget for these experiments was 40,000, so we would only have managed to consider 4000 configurations if all were evaluated fully.

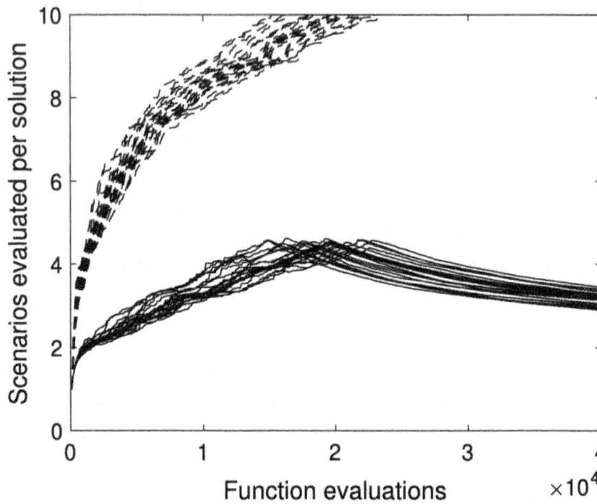

Fig. 2.7 Number of scenarios evaluated per solution for the 21 runs of the robust multi-objective optimiser. The dashed lines show this value for the approximation set A, which can be seen to saturate at 10 (corresponding to U) at 18,000–22,000 function evaluations. The solid line shows this value for all solutions considered in the optimisation process. This rises up to where the A values saturate, and then falls down again, as there is much less demand for additional scenarios needing evaluating to refine previous designs

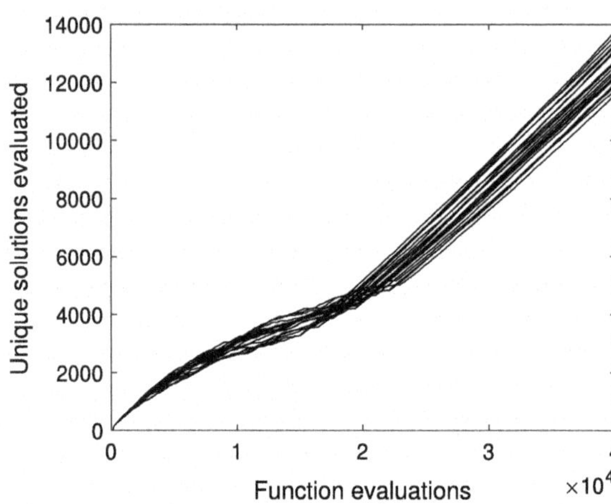

Fig. 2.8 Number of unique solutions considered over each of the 21 runs of the robust multi-objective optimiser. Accumulating reevaluations on additional uncertainty set members over the duration of the run preferentially for those deemed higher quality allows a much larger number of designs to be considered than if all designs are assessed on all uncertainty set members (typically over $3\times$ more in the example here)

Algorithm 5 $(|A| + 1)$–ES for the robust multi-objective optimisation task

Require: p^{mut} ▷ Probability of mutation
Require: p^{cross} ▷ Probability of crossover
Require: n ▷ Number of initial random starting solutions
Require: max_evals ▷ Evaluation budget
Require: U uncertainty set of operating scenarios
$i := 1$
$A^0 := \emptyset$ ▷ initial empty archive
$D := \emptyset$ ▷ Initial set of estimated dominated solutions
while $i \leq n$ **do**
  $\mathbf{x} :=$ `initialise(1)` ▷ Generate initial decision vector
  $\mathbf{u} :=$ `select`(U) ▷ Draw uncertainty set member
  $s := \{\mathbf{x}, \mathbf{f}(\mathbf{x}, \mathbf{u}), \{\mathbf{u}\}\}$ ▷ Construct solution, inc. scenarios considered
  $A^i :=$ `nondom`$(A^{i-1} \cup s)$ ▷ Identify non-dominated
  $D := (A^{i-1} \cup D \cup \{s\}) \setminus A^i$ ▷ Updated estimated dominated
$j := n$
while $i < max\_evals$ **do**
  $child :=$ `evolve`$(A^j, p^{mut}, p^{cross})$ ▷ Create child decision vector
  $\mathbf{u} :=$ `select`(U) ▷ Draw uncertainty set member
  $s := (child,$ `evaluate`$(child, \mathbf{u}), \{\mathbf{u}\})$ ▷ Construct evaluated solution
  $A^{j+1} :=$ `nondom`$(A^j \cup \{s\})$ ▷ Update archive (estimated non-dominated)
  $D := (A^j \cup D \cup \{s\}) \setminus A^{j+1}$ ▷ Update estimated dominated
  $i := i + 1$
  if `mean_revals`$(A^{j+1}) < |U|$ **then**
    repeat
      $s :=$ `draw_evaled_on_fewest_scenarios`(A^{j+1}) ▷ Get archive member whose minmax performance is most uncertain
      $\mathbf{y} :=$ `get_objective_vector`(s)
      $\tilde{U} :=$ `get_evaluated_scenarios`(s) ▷ Uncertainty set members evaluated with s so far
      $\mathbf{u} :=$ `select_random_member`$(U \setminus \tilde{U})$ ▷ Draw uncertainty set member
      $s := \{\mathbf{x}, \max(\mathbf{f}(\mathbf{x}, \mathbf{u}), \mathbf{y}), \tilde{U} \cup \{\mathbf{u}\}\}$ ▷ Update minmax estimate for s with an additional uncertainty set member draw
      $A^{j+1} :=$ `nondom`$(A^j \cup \{s\} \cup D)$ ▷ Update archive—need to include D due to potential move in objective vector location of s
      $D := (A^j \cup D \cup \{s\}) \setminus A^{j+1}$
      $i := i + 1$ ▷ Track number of evalutations
    until `mean_revals`$(A^{j+1}) >$ `mean_revals`(A^j)
  $j := j + 1$ ▷ Track number of loops, which corresponds to the total number of unique designs considered
return A ▷ Return non-dominated set approximating the Pareto set

We also exploit the distribution of worst performing scenarios in our algorithm. Once we reach a point where solutions have been fully evaluated, then we can track which particular scenarios a solution is performing worst on for each objective, and keep track of these statistics (see Fig. 2.9). We use this to sort the order in which scenarios are processed for designs (the ones considered most likely to have the worst performance selected first, given a random objective).

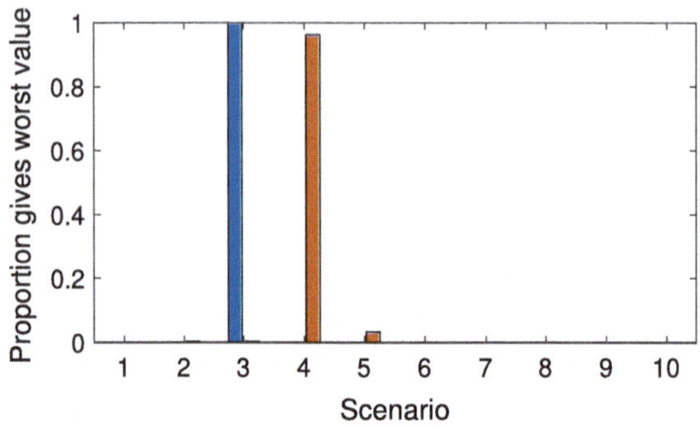

Fig. 2.9 Probability of a scenario causing the worst score for each objective from an indicative run (derived from fully evaluated solutions). By the end of this run approximately 400 solutions had been fully evaluated on all members of the uncertainty set. Scenario 3 always led to the worst performance on the first objective. Scenario 4 nearly always led to the worst performance on the second objective, though sometimes it was scenario 5, and once apiece it was scenarios 2 and 3

Fig. 2.10 Empirical attainment front from the 21 runs of the robust optimisation. By 10,000 function evaluations all runs are performing similarly in the lower left hand portion of the front, though as with the previous formulation not all have converged on energy costs for wake free ratio values above 0.95

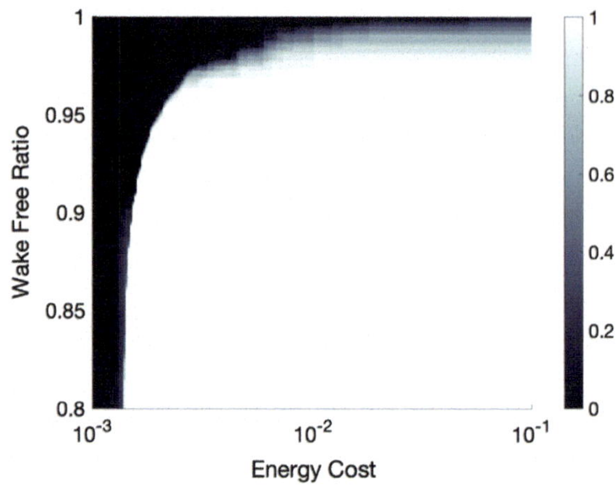

Figure 2.10 show the attainment front of the archive at the end of the robust optimisation runs. Unsurprisingly this is a bit to the right of the previous experiments where we were concerned with a single scenario, not the worst performance across 20, however the shape is similar, as is the variation for high wake free ratio values, indicating the runs haven't fully converged yet.

2.5 Conclusion

In this chapter we have introduced the multi-objective optimisation task, and related it to various aspects of wind farm optimisation. We have used a running example of wind farm layout optimisation, and shown how the optimisation landscape varies depending on the choice of objective, and when multiple objectives are considered in parallel. We have shown the multi-objective optimisation of a farm layout using a simplified model, and also shown how this may be treated to mitigate for the uncertainty of multiple plausible wind profiles using a simple optimisation pipeline. In later chapters the reader will be exposed to more advanced optimisation approaches, and a range of wind farm optimisation related problems, which build on some of the core concepts laid out here.

Acknowledgements The support of the Ministry of Education (MoE), Government of India provided through the SPARC project # SPARC/2018–2019/P1084/SL has been acknowledged by the author, along with the UK-India Education and Research Initiative (UKIERI).

References

Adair J, Ochoa G, Malan KM (2019) Local optima networks for continuous fitness landscapes. In: Proceedings of the genetic and evolutionary computation conference companion, GECCO'19. Association for Computing Machinery, New York, NY, USA, pp 1407–1414

Amon C, Kwong WY, Moran J, Morgenroth M, Romeo D, Zhang PY (2012) Multi-objective optimization of wind farm layouts under energy generation and noise propagation. In: Proceedings of the ASME 2012 international design engineering technical conferences and computers and information in engineering conference. Volume 3: 38th design automation conference, parts A and B. ASME, pp 323–332

Biswas PP, Suganthan PN, Amaratunga GAJ (2018) Decomposition based multi-objective evolutionary algorithm for windfarm layout optimization. Renew Energy 115:326–337

Chen Y, Li H, Jin K, Song Q (2013) Wind farm layout optimization using genetic algorithm with different hub height wind turbines. Energy Convers Manag 70:56–65

Coello Coello CA, Lamont GB, Van Veldhuizen DA (2007) Evolutionary algorithms for solving multi-objective problems. Springer

Fieldsend JE, Alyahya K (2019) Visualising the landscape of multi-objective problems using local optima networks. In: Proceedings of the genetic and evolutionary computation conference companion, GECCO'19. Association for Computing Machinery, New York, NY, USA, pp 1421–1429

Fieldsend JE, Everson RM (2015) The rolling tide evolutionary algorithm: a multiobjective optimizer for noisy optimization problems. IEEE Trans Evol Comput 19(1):103–117

Fonseca CM, Guerreiro AP, López-Ibáñez M, Paquete L (2011) On the computation of the empirical attainment function. In: Takahashi RHC, Deb K, Wanner EF, Greco S (eds) Evolutionary multi-criterion optimization. Springer Berlin Heidelberg, Berlin, Heidelberg, pp 106–120

González-Longatt F, Wall P, Terzija V (2012) Wake effect in wind farm performance: steady-state and dynamic behavior. Renew Energy 39(1):329–338

Herrmann S, Ochoa G, Rothlauf F (2018) Pagerank centrality for performance prediction: the impact of the local optima network model. J Heurist 24:243–264

Li W, Özcan E, John R (2017) Multi-objective evolutionary algorithms and hyper-heuristics for wind farm layout optimisation. Renew Energy 105:473–482

Miettinen K (1999) Nonlinear multiobjective optimization. Volume 12 of international series in operations research & management science. Springer Science & Business Media

Ochoa G, Tomassini M, Vérel S, Darabos C (2008) A study of NK landscapes' basins and local optima networks. In: Proceedings of the 10th annual conference on genetic and evolutionary computation, GECCO'08. ACM, New York, NY, USA, pp 555–562

Paquete L, Schiavinotto T, Stützle T (2007) On local optima in multiobjective combinatorial optimization problems. Ann Oper Res 156(1):83

Rodrigues S, Bauer P, Bosman PAN (2016) Multi-objective optimization of wind farm layouts—complexity, constraint handling and scalability. Renew Sustain Energy Rev 65:587–609
Schütze O (2003) A new data structure for the nondominance problem in multi-objective optimization. In: Fonseca CM, Fleming PJ, Zitzler E, Thiele L, Deb K (eds) Evolutionary multi-criterion optimization. Springer Berlin Heidelberg, Berlin, Heidelberg, pp 509–518
Tomassini M, Vérel S, Ochoa G (2008) Complex-network analysis of combinatorial spaces: the NK landscape case. Phys Rev E 78:066114
Vérel S, Daolio F, Ochoa G, Tomassini M (2012) Local optima networks with escape edges. In: Hao J-K, Legrand P, Collet P, Monmarché N, Lutton E, Schoenauer M (eds) Artificial evolution. Springer Berlin Heidelberg, Berlin, Heidelberg, pp 49–60
Wilson D, Rodrigues S, Segura C, Loshchilov I, Hutter F, Buenfil GL, Kheiri A, Keedwell E, Ocampo-Pineda M, Özcan E, Peña SIV, Goldman B, Rionda SB, Hernández-Aguirre A, Veeramachaneni K, Cussat-Blanc S (2018) Evolutionary computation for wind farm layout optimization. Renew Energy 126:681–691

Chapter 3
Data-Driven Multi-objective Optimisation with Uncertainty Quantification

Atanu Mazumdar and Tinkle Chugh

Abstract Many real-world optimisation problems often feature multiple conflicting objectives that must be optimised simultaneously. For a subset of these problems, there are no analytical forms or numerical simulation models for the objective functions. Yet, there exist datasets relating the input parameters to the associated objective values. In such cases, the only realistic option is to deploy surrogates, also known as metamodels or emulators of the objective functions, alongside a multi-objective evolutionary algorithm (MOEA) to locate a good approximation of the Pareto front. Since there is no opportunity to gather new data, it is important that we quantify the uncertainties associated with surrogate predictions and utilise them during optimisation and communicating solutions to decisions-makers. In this chapter, we summarise the challenges of using surrogates with MOEAs to solve data-driven optimisation problems. We mainly focus on the probability of selection criterion in decomposition-based MOEAs and show its potential on several benchmark problems with different settings.

Keywords Pareto optimality · Bayesian optimisation · Decision-making · Surrogates

3.1 Introduction

Many real-world optimisation problems involve more than one objective to be optimised. Such problems are known as multi-objective optimisation problems (MOPs). A MOP can be defined as:

A. Mazumdar
Department of Electrical Engineering and Automation, Aalto University, Espoo, Finland
e-mail: atanu.mazumdar@aalto.fi

T. Chugh (✉)
Department of Computer Science, University of Exeter, Exeter, UK
e-mail: t.chugh@exeter.ac.uk

K. Mitra et al. (eds.), *Optimization, Uncertainty and Machine Learning in Wind Energy Conversion Systems*, Engineering Optimization: Methods and Applications,
https://doi.org/10.1007/978-981-97-7909-3_3

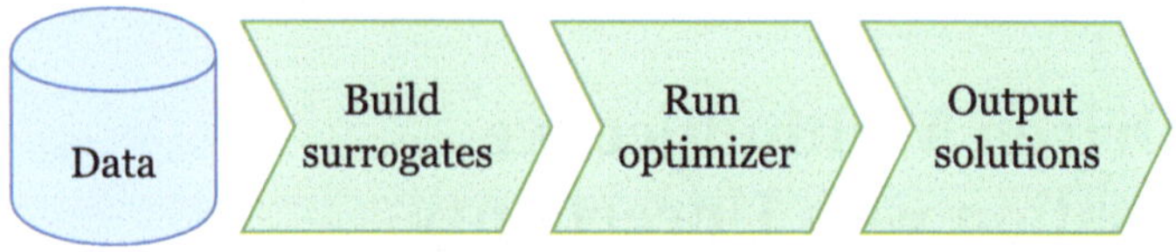

Fig. 3.1 Flowchart of a generic data-driven multi-objective optimisation approach

$$\begin{aligned} &\text{minimise } \{f_1(\mathbf{x}), \ldots, f_M(\mathbf{x})\} \\ &\text{subject to } \mathbf{x} \in \Omega, \end{aligned} \tag{3.1}$$

where $M \geq 2$ and Ω represent the total number of objectives and the feasible region of the decision space, respectively. For a feasible decision vector $\mathbf{x}$, the corresponding objective vector is $\mathbf{f}(\mathbf{x})$, that comprises of the underlying objective (function) values $\{f_1(\mathbf{x}), \ldots, f_M(\mathbf{x})\}$. A solution $\mathbf{x}_1 \in \Omega$ dominates another solution $\mathbf{x}_2 \in \Omega$ if $f_k(\mathbf{x}_1) \leq f_k(\mathbf{x}_2)$ for all $k = 1, \ldots, M$ and $f_k(\mathbf{x}_1) < f_k(\mathbf{x}_2)$ for at least one $k = 1, \ldots, M$. An MOP's solution is considered to be non-dominated if no other feasible solution dominates it. The Pareto optimal solutions are those nondominated solutions of (3.1) that exist within the domain of Ω. To find a set of approximately Pareto optimal solutions, optimisation algorithms such as multi-objective evolutionary algorithms (MOEAs) can be employed. The set of solutions in the decision space is referred to as the Pareto set, while their image in the objective space is referred to as the Pareto front.

In many real-world MOPs, it is not possible to perform further experiments or simulations during the optimisation process. The starting point of the optimisation process is data that have been acquired previously from the experiments or simulations. Approximation models known as surrogates (or metamodels or emulators) can be utilised using the provided data, and their predictions can assist an optimisation algorithm in finding solutions. These types of problems are referred to as data-driven multi-objective optimisation problems (Jin et al. 2019).

An illustration of a generic approach for tackling data-driven MOPs is presented in Fig. 3.1. According to Jin et al. (2019) and Wang et al. (2019), the optimisation process can be divided into three stages: (a) collecting the data, (b) modelling, and (c) optimisation. The process begins with a dataset consisting of N samples, each of which is a tuple containing a decision vector $\mathbf{x}$ and its corresponding objective vector $\mathbf{f}(\mathbf{x})$ represented as two matrices: (X, Y), where $X \in \mathbb{R}^{N \times D}$ and $Y \in \mathbb{R}^{N \times M}$. The rows in X represent decision vectors, while the rows in Y represent objective vectors. In the first stage, preprocessing the data is an option. Next, surrogate(s) are built using the provided data. Surrogates such as neural networks (Bishop 2006), Kriging (Forrester et al. 2008), and support vector machines (Bishop 2006) are popular choices for solving data-driven MOPs. Lastly, an optimisation algorithm, such as an MOEA (Deb 2001), is used to solve the MOP by considering the surrogates as objectives.

In practice, the surrogate models do not perfectly imitate the underlying objective functions; therefore, solutions produced by optimisation algorithms utilising surrogates may need to be more accurate. In other words, the solutions by utilising

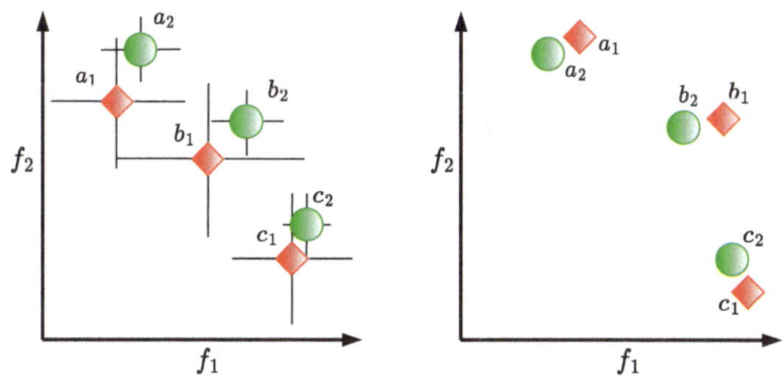

Fig. 3.2 Six solutions with uncertainties in the surrogate objective space (left) and underlying objective space (right) for a biobjective minimization problem. The error bars show the 95% confidence interval of the objective values approximated by the $\mathcal{GP}$ surrogates

surrogates as objectives can be worse when applied in the real world. Surrogates such as Kriging or Gaussian process ($\mathcal{GP}$) regression (Forrester et al. 2008; Rasmussen and Williams 2006) also provide the uncertainty in the predictions that can be used in guiding the optimisation process. The solutions with higher uncertainties in the surrogate space have a higher probability of having an error. This phenomenon can be better understood from Fig. 3.2. The figure shows six solutions evaluated with surrogates and real (or underlying) objectives. The error bars show the 95% confidence interval (or uncertainty in predictions) by $\mathcal{GP}$ surrogates. In the surrogate objective space, the red (diamond) solutions dominate the green (circle) ones and have higher uncertainty. When these solutions are evaluated with the underlying objectives (figure on the right), two out of three green/circle solutions dominate the red/diamond ones. Therefore, uncertainty quantification is essential when using surrogates in the optimisation process.

Previous works that utilise uncertainty in predictions (Hughes 2001; Mazumdar et al. 2019; Rahat et al. 2018) are not suitable for problems with more than three objectives. Those methods rely on the dominance between solutions, and utilising dominance between solutions usually is not meaningful with more than three objectives (also called many-objective optimisation problems). In this chapter, we provide probabilistic selection approaches (Mazumdar et al. 2022) that utilise uncertainty in the optimisation process and are suitable for multi and many-objective optimisation problems. We build probabilistic models on the objective functions which are used in the selection criteria of decomposition-based MOEAs. The proposed approach is generic to decomposition-based MOEAs and provides a plug-and-play feature. The rest of the chapter is structured as follows. Section 3.2 provides an overview of $\mathcal{GP}$ regression, probabilistic selection in single-objective optimisation and two well-known decomposition-based MOEAs. In Sect. 3.3, we describe the proposed selection criteria with uncertainty quantification for multi and many-objective optimisation problems, followed by numerical testing in Sect. 3.4. Finally, we provide conclusions and future research directions in Sect. 3.5.

3.2 Background

This section covers the background on Gaussian process regression, probabilistic selection in single-objective optimisation and two well-known decomposition-based MOEAs.

3.2.1 *Gaussian Processes Regression*

Gaussian process ($\mathcal{GP}$) regression models are renowned models because of their advantage in providing uncertainty information in predictions. A $\mathcal{GP}$ can be defined with a multivariate normal distribution (Rasmussen and Williams 2006):

$$\mathbf{f} \sim \mathcal{N}(\boldsymbol{\mu}, \Sigma),$$

where $\boldsymbol{\mu}$ represents the mean vector and Σ is the covariance matrix. Here, we assume a zero mean prior without the loss of generality. A covariance function or kernel (such as Gaussian or Matern kernel) is employed to determine the covariance matrix. A Gaussian (or RBF or squared exponential) kernel with automatic relevant determination is expressed as:

$$\kappa(\mathbf{x}, \mathbf{x}', \Theta) = \sigma_f^2 \exp\left(-\frac{1}{2}\sum_{j=1}^{n}\frac{|x_j - x_j'|^2}{l_j^2}\right) + \sigma_n^2 \delta_{\mathbf{x}\mathbf{x}'},$$

where $\Theta = (\sigma_f, l_1, \ldots, l_n, \sigma_n)$ is the vector of hyperparameters and $\delta_{\mathbf{x}\mathbf{x}'}$ is the Kronecker delta function. The Euclidean distance between x_j and x_j' is represented by $|x_j - x_j'|$. The following marginal likelihood function (Rasmussen and Williams 2006) can be maximised to estimate the hyperparameters:

$$p(\mathbf{f}|X, \Theta) = \frac{1}{|2\pi\Sigma|^{\frac{1}{2}}} \exp\left\{-\tfrac{1}{2}\mathbf{f}^{\mathrm{T}}\Sigma^{-1}\mathbf{f}\right\}.$$

The posterior predictive distribution of the trained $\mathcal{GP}$ regression model is also Gaussian. The predicted distribution at new point $\mathbf{x}^*$ is:

$$p\left(f^*|\mathbf{x}^*, X, \mathbf{f}, \Theta\right) = \mathcal{N}\Big(\kappa(\mathbf{x}^*, X)\Sigma^{-1}\mathbf{f}, \kappa(\mathbf{x}^*, \mathbf{x}^*) - \kappa(\mathbf{x}^*, X)^{\mathrm{T}}\Sigma^{-1}\kappa(X, \mathbf{x}^*)\Big).$$

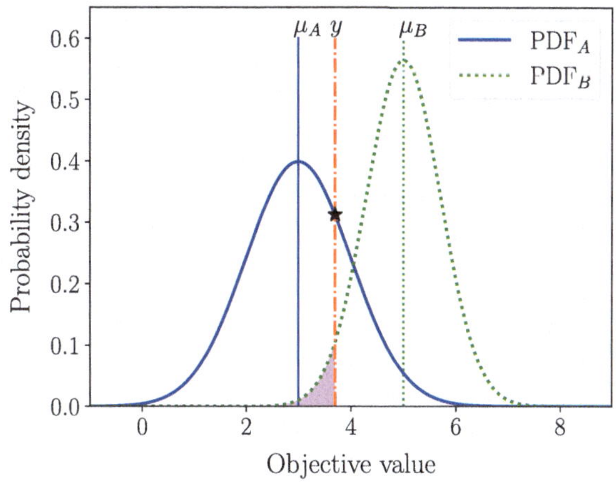

Fig. 3.3 The PDF of two individuals with uncertain objective values (approximated by the surrogate) for a single objective minimization problem

3.2.2 Probabilistic Selection in Single-Objective Optimisation

This section provides the importance of quantifying uncertainty in single-objective optimisation problems. We illustrate this with an example, where we aim to select a solution with the minimum objective function value. Let us consider two solutions, A and B, that are uncertain. The uncertainty can come from different sources, such as noise and measurement error. For simplicity, we assume that this uncertainty can be described with a Gaussian distribution. The probability density functions (PDFs) of two solutions are shown in Fig. 3.3, where μ_A and μ_B are means of two distributions.

When a random sample y is drawn from PDF_A (denoted by star), we can see that the value of A is higher than that of B. Therefore, it is possible that an algorithm can select B over A if the whole distribution or uncertainty is not considered during optimisation.

The area in the shaded region of PDF_B, or cumulative density function (CDF) of B (denoted by $\text{CDF}_B(y)$) is the total probability of selecting the wrong solution, i.e. B over A. The total probability of a random sample y drawn with probability $\text{PDF}_A(y)$ being greater than a random sample drawn from PDF_B is $\text{PDF}_A(y) \cdot \text{CDF}_B(y)$. The probability of wrongly selecting B over A is (Hughes 2001):

$$P_{\text{wrong}}(A > B) = \int_{-\infty}^{\infty} \text{PDF}_A(A - y) \cdot \text{CDF}_B((A - y) > (B - y)) \, \text{d}y. \quad (3.2)$$

CDF_B can be replaced as an integral of PDF_B as:

$$P_{\text{wrong}}(A > B) = \int_{-\infty}^{\infty} \left(\text{PDF}_A(y) \cdot \int_{-\infty}^{y} \text{PDF}_B(\mu) \text{d}\mu \right) \, \text{d}y. \quad (3.3)$$

Once the probability of wrong selection is known between two solutions, a rank can be given to all solutions, e.g. in a population of size $|P|$ of an MOEA. Ranking a population of with uncertain solutions based on probabilities of wrong selection is:

$$R_i = \sum_{l=1}^{|P|} P_{\text{wrong}}(I_l > I_i) - 0.5, \tag{3.4}$$

where R_i is the ranking score given to the ith solution. The solution with the lowest R value is the one with the best among all solutions and has the lowest probability of wrong selection.

3.2.3 Decomposition-Based MOEAs

Decomposition-based MOEAs are popular for their ability to solve problems with more than three objectives (also known as many-objectives) (Deb and Jain 2014). These algorithm divide the objective space into a number of sub-spaces using weight (or reference) vectors (Das and Dennis 1998). In this chapter, we focus on two well-known decomposition-based MOEAs, MOEA/D, and RVEA and embed our proposed probabilistic selection approach in them.

MOEA/D (Zhang and Li 2007) is one of the first decomposition-based MOEA. After generating the initial population and a set of weight vectors, it creates a neighbourhood for each weight vector and performs the search within the neighbourhood. The offspring are generated with crossover and mutation in each neighbourhood and a selection criterion, e.g. penalty boundary intersection (PBI) or Tchebycheff scalarizing function, is used to select the population for the next generation. In this chapter, we focus on PBI, which is defined as:

$$g_{\mathbf{x}}^{\text{PBI}} = d_1 + \rho d_2, \tag{3.5}$$

where parameter ρ is the penalty term that balances between convergence and diversity, $d_1 = \left\|(\mathbf{z}^{\min} - \mathbf{f}(\mathbf{x}))^\top \mathbf{v}_j\right\| / \left\|\mathbf{v}_j\right\|$ and $d_2 = \left\|\mathbf{f}(\mathbf{x}) - (\mathbf{z}^{\min} - d_1\mathbf{v}_j)\right\|$, respectively, $\mathbf{v}_j$ is the jth reference vector in the neighbourhood and $\mathbf{z}^{\min}$ is the vector of minimum objective function values at the current generation. MOEA/D uses PBI values in each neighbourhood to update the solution by checking $g_{\mathbf{x}'}^{\text{PBI}} \leq g_{\mathbf{x}}^{\text{PBI}}$, then $\mathbf{x}'$ replaces the $\mathbf{x}$ and is assigned to the weight vector.

RVEA uses a set of weight vectors and differentiates from MOEA/D in selection criterion and does not update the population in a generation sequentially. It first translates the objective vectors as $\mathbf{f}'(\mathbf{x}) = \mathbf{f}(\mathbf{x}) - \mathbf{z}^{\min}$ for each solution. It assigns the solutions to a vector based on the cosine angle between the objective vector of the solution and weight vectors. The cosine value between the jth reference vector $\mathbf{v}_j$ and the translated objective vector $\mathbf{f}'$ is given by:

$$\cos\,\theta_j = \frac{\mathbf{f}'(\mathbf{x}) \cdot \mathbf{v}_j}{\|\mathbf{f}'(\mathbf{x})\|}, \tag{3.6}$$

where $\|\mathbf{f}'\|$ is the Euclidean norm. A solution with minimum θ_j (or highest $\cos\,\theta_j$ value) is assigned to the jth weight vector. Once solutions are assigned to different weight vectors, in each vector, RVEA uses a selection criterion called angle penalised distance (APD) for selecting one solution. The APD is defined as:

$$g_{\mathbf{x}}^{\mathrm{APD}} = (1 + O(\theta_j)) \cdot \left\|\mathbf{f}'(\mathbf{x})\right\|, \tag{3.7}$$

where $O(\theta_j) = K \cdot (t/t_{\max})^{\alpha} \cdot \theta_j/\gamma_{\mathrm{v}}$ is the penalty function, t is the current generation index, and $t_{\max}$ is the maximum number of generations, $\gamma_{\mathrm{v}} = \min_{i \in 1,\ldots,N_{\mathrm{v}}, i \neq j} \langle \mathbf{v}_i, \mathbf{v}_j \rangle$, is the smallest angle between reference vector $\mathbf{v}_j$ and the other reference vectors and α controls the rate of change of $O(\theta_j)$.

3.3 Probabilistic Selection in Decomposition-Based MOEAs

The probabilistic selection approaches for decomposition-based MOEAs (Mazumdar et al. 2022) quantify the uncertainty and use it during the optimisation for solving data-driven MOPs. The approaches reformulate the probability of wrong selection and probabilistic ranking for single objective optimisation as shown in (3.2) and (3.3), respectively, for the selection criterion specific to the decomposition-based MOEA. As mentioned in the introduction, the approaches are generic to other decomposition-based MOEA and provide a plug-and-play feature. In this chapter, we show the effectiveness of these approaches when using RVEA and MOEA/D.

The steps of a decomposition-based MOEA utilising probabilistic selection approaches are set out in Algorithm 1. The algorithm inputs are the dataset containing the decision vector and objective vector with N samples. The other parameters used are the maximum number of surrogate function evaluations as the termination criterion and the number of samples used for kernel density estimation (KDE). Other relevant parameters specific to the MOEA are set according to their respective recommended values. In the algorithm, we build $\mathcal{GP}$ for each objective function (step 1). The given data are the initial population for the MOEA (step 2). A set of uniformly distributed unit weight vectors, V_0 of size N_{v}, is generated. In MOEA/D, a neighbourhood for each vector is also defined (steps 3 and 4). Next, the offspring population is generated using crossover and mutation (step 6), which are then evaluated with the surrogates (step 7) and the number of function evaluations is updated. Next, we explain the probabilistic ranking using two selection criteria. The solutions are then selected (steps 8–10) using the probabilistic ranking of the selection criterion of the MOEA. The selected solutions are used as parents for the next generation, and the MOEA is run until the termination criterion is met.

Algorithm 1: Data-driven probabilistic decomposition-based MOEA

Input: Data of size N; Number of reference vectors = N_v; Number of Monte-Carlo samples to be used for estimating the distributions = S; Maximum number of function evaluations with surrogates = FE_{max}

Output: Approximated nondominated solutions

1 Use the given data to build $\mathcal{GP}$ surrogates per objective
2 Initialize the population using the given data;
3 Initialize a uniformly distributed unit reference vectors' set V_0 of size N_v
4 For each unit reference vector, find its neighbourhood
5 **while** *Termination criterion is not met* **do**
6   Apply crossover and mutation on the current population and to produce offspring individuals
7   Evaluate the offspring individuals with the surrogates and append them to the population
8   Use Monte-Carlo sampling to draw S samples from the predicted posterior distribution of the $\mathcal{GP}$ surrogates for each individual
9   Assign each individual to a sub-population according to the MOEA
10   Use probabilistic selection criteria to select individuals
11 end

Probabilistic Ranking of Selection Criterion

In RVEA, a fixed number of Monte Carlo samples is generated from the posterior predictive distribution of a solution (multivariate Gaussian distribution for more than one objective). Each of these samples is assigned to reference vectors according to (3.6). A solution is assigned to the weight vector to which most samples are assigned (or by a voting scheme). In MOEA/D, there is no need for assignment as the algorithm creates a neighbourhood of weight vectors. This completes steps 8 and 9 in Algorithm 1.

After all the solutions (each with their multivariate Gaussian distribution and Monte Carlo samples) are probabilistically assigned to weight vectors, the selection criterion values (APD in RVEA and PBI in MOEA/D) of all the samples are calculated using (3.7) or (3.5). To assign a ranking to different solutions, we approximate the distribution of selection criteria using kernel density estimation (KDE). The kernel used for KDE is Gaussian, and the bandwidth parameter is chosen according to Silverman's rule of thumb (Silverman 1986). After estimating the distributions, the solutions are ranked using (3.4) in RVEA. In MOEA/D, as the selection process is performed in the neighbourhood, the solutions are ranked directly using (3.3). One solution with the best rank is then selected from each weight vector for the next generation. An illustration of distributions of APD and PBI selection criteria estimated with KDE is shown in Fig. 3.4. The figure shows probability density functions and empirical CDF values of both selection criteria.

In RVEA, there is a pairwise computation of ranks between distributions of APD solutions for a weight vector. Thus, the overall computation cost of selection using probabilistic APD is $O(|G|^2)$, where $|G|$ is the size of the subpopulation. An efficient way to reduce the computation cost is the perform the pairwise computation of ranks only once. In MOE/D, the computation of P_{wrong} is one (offspring) to many (parents in a neighbourhood). The computational cost is further reduced by calculating the

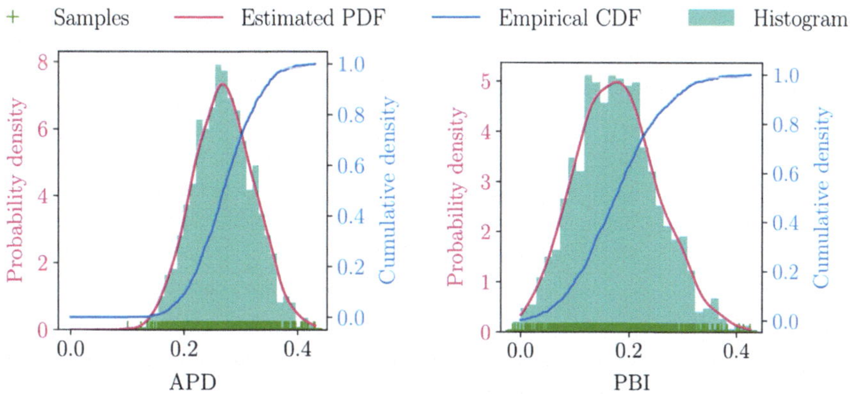

Fig. 3.4 Samples drawn represented with '+', histogram, estimated probability density function (PDF), and empirical cumulative distribution function (CDF) of the selection criterion for APD and PBI

empirical CDF instead of using integral to find CDF in (3.3). The lower limit of the integral is changed to zero instead of $-\infty$ since APD and PBI values as always ≥ 0. Also, all the P_{wrong} values (for different weight vectors in RVEA and different neighbourhoods in MOEA/D) are calculated in parallel to reduce the computation cost further.

In Fig. 3.5, we show an example of probabilistic selection in RVEA in a biobjective minimisation problem, where each solution has some uncertainty. The goal is to select one solution for each weight vector. Let us assume two solutions are assigned to each weight vector. The numbers on the solution represent the weight vector it has been assigned to. On the left, we use only the means of the posterior predictive distribution (defined as the generic approach in the introduction), and on the right, we use the distribution of APD and rank the solutions for each weight vector. The selected solutions (in green/solid error bars) are different in both approaches. This is because the probabilistic selection considers the uncertainties. For example, the generic approach selects the solution with better mean value (and ignore the uncertainty) for reference vector '2'. On the other hand, the probabilistic approach selects the solution with lower uncertainties (with high mean values). A flowchart of the optimisation process with probabilistic selection is shown in Fig. 3.6.

3.4 Numerical Experiments

In this section, we compare generic and probabilistic versions of RVEA and MOEA/D (four in total). We also use a recently proposed transfer learning based approach (Yang et al. 2020) for comparison. We test these algorithms on distance-based visualisable test problems (DBMOPP) (Fieldsend et al. 2019, 2022). One of the main advantages

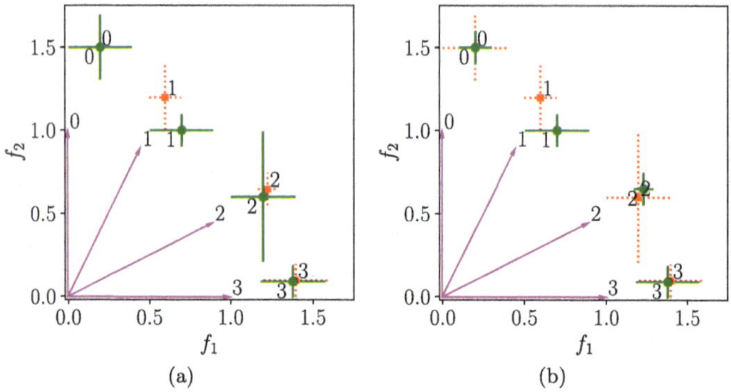

Fig. 3.5 Solutions selected in a biobjective minimization problem using **a** posterior mean and **b** probabilistic APD. Error bars show the 95% confidence interval of the distribution of objective values approximated by the surrogates. The solutions in green/circle with solid error bars are selected and the solutions in red/square with dotted error bars are not selected

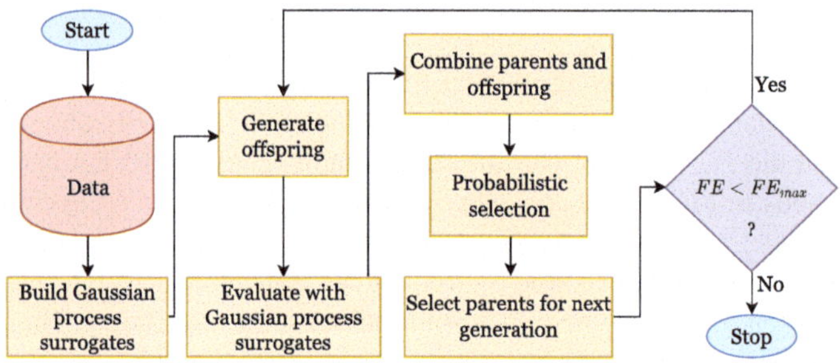

Fig. 3.6 Flowchart showing the algorithm with probabilistic selection

of using these problems is that one can visualise the search during the optimisation in both decision and objective spaces for any number of dimensions. Moreover, these problems provide a control over the features of the problem and one can change them to reflect the needs of the real-world optimisation problems. Data samples of size 109 were generated using Latin hypercube sampling (LHS) and multivariate normal sampling (MVNS) (Mazumdar et al. 2022). The two problems were P1 and P2 with a total of 28 instances (with different combinations of features, sampling strategy, and objectives) for testing different algorithms. The features of problems used are provided in Table 3.1. Each approach was independently executed 31 times and the the parameters for RVEA and MOEA/D were kept the same as in Cheng et al. (2016) and Zhang and Li (2007), respectively.

For measuring the quality of solutions, we used hypervolume as a measure of convergence and diversity and root mean squared error (RMSE) as a measure of the accuracy of solutions when using different selection approaches. Figure 3.7 shows the statistical comparison (Wilcoxon signed rank with Bonferroni correction) of probabilistic versions of RVEA and MOEA/D (represented by Prob-RVEA and Prob-

Table 3.1 Configuration of the DBMOPP problems used

Problem name	Configuration	Dimension (D)	Objectives (M)
DBMOPP (P1)	number_of_disconnected_set_regions = 0, number_of_local_fronts = 0, number_of_dominance_resistance_regions = 0, number_of_discontinuous_regions = 0	10	2, 3, 4, 5, 6, 8, and 10
DBMOPP (P2)	number_of_disconnected_set_regions = 1, number_of_local_fronts = 0, number_of_dominance_resistance_regions = 0, number_of_discontinuous_regions = 0	10	2, 3, 4, 5, 6, 8, and 10

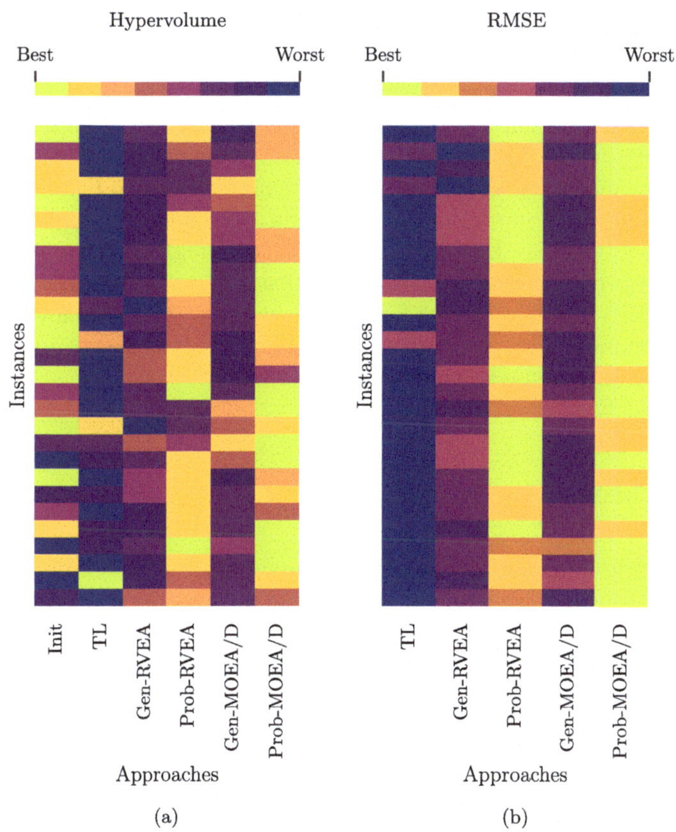

Fig. 3.7 Heatmaps of: **a** hypervolume and **b** RMSE of solutions obtained by initial sampling, transfer learning, and the generic and probabilistic approaches for RVEA and MOEA/D, respectively, for DBMOPP problem instances

MOEA/D, respectively) on 28 instances of DBMOPP problems using hypervolume and RMSE. The other approaches compared are generic RVEA (Gen-RVEA) and generic MOEA/D (Gen-MOEA/D) and the transfer learning approach (TL). The Init is the initial dataset generated with LHS and MVNS sampling. The experimental results showed the superior performance of the probabilistic selection approaches in both hypervolume and RMSE.

The progress of the solution process for two instances (with median hypervolume) is shown in Fig. 3.8. The circle in the plot with points denote the Pareto front. It can be observed that the non-dominated solutions (shown in '+') produced by the probabilistic approaches are closer to the Pareto front. When running the algorithm, we estimated the hypervolume of solutions during the optimisation. We denote this hypervolume as HV (Surrogate). To measure the differences in solutions on these benchmark problems, we evaluated solutions in the algorithm produced by probabilistic selection approaches with underlying or real objective functions and measured their hypervolume. We denote this hypervolume as HV (Underlying). Both HV (Surrogate) and HV (Underlying) and RMSE values with the number of function evaluations on one instance of DBMOPP problem are shown in Fig. 3.9. It can be observed that the HV (Surrogate) for the generic approaches consistently increases with function evaluations. However, their HV (Underlying) decreases over function evaluations, whereas, in the probabilistic selection approaches, both HV (Surrogate) and HV (Underlying) increase with function evaluations. This is because the generic approaches optimise without considering the uncertainties in the solutions, leading to solutions with worse underlying objective values.

3.5 Conclusions

Many real-world optimisation problems rely on existing datasets. Surrogates or metamodels can be used in an optimisation algorithm to find the solutions. However, using surrogates increases the risk of inaccurate solutions; therefore, the solutions obtained after running the optimisation algorithm may need more reliability. This chapter attempted to address this challenge by quantifying the uncertainty when using surrogates and further utilising it in optimisation. We utilised probabilistic selection approaches in decomposition-based MOEAs and tested them on benchmark problems. We showed that these approaches produced solutions closer to the Pareto optimal solutions and with higher accuracy compared to approaches that did not use any uncertainty information. We demonstrated the potential of these approaches in two well-known decomposition-based MOEAs, RVEA and MOEA/D. By building Gaussian process surrogates for each objective function, drawing Monte Carlo samples from the multivariate normal distribution of the posterior predictive distribution, and estimating the distributions of selection criteria (APD in RVEA and PBI in MOEA/D) with kernel density estimation, we performed the probabilistic ranking of solutions. The approaches are generic to any decomposition-based MOEA and provide a plug-and-play feature.

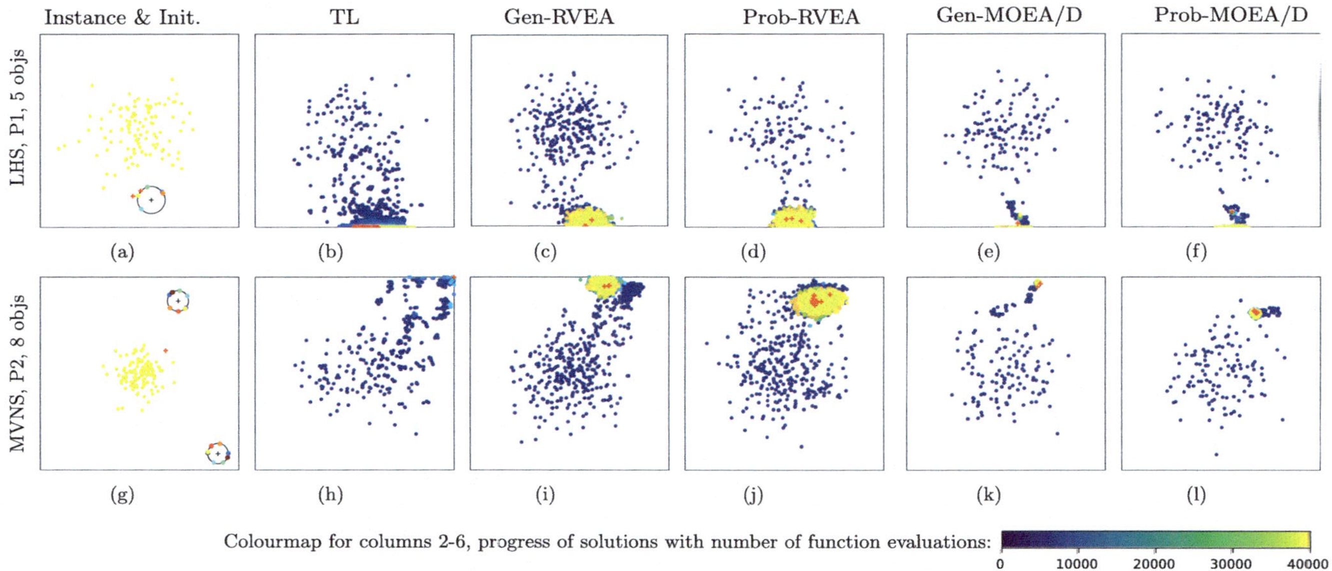

Fig. 3.8 Progress of the solution process with median hypervolume values. DBMOPP P1 with $M = 5$ with LHS and P2 with $M = 8$ with MVNS. The colour code represents the number of function evaluations during the solution process

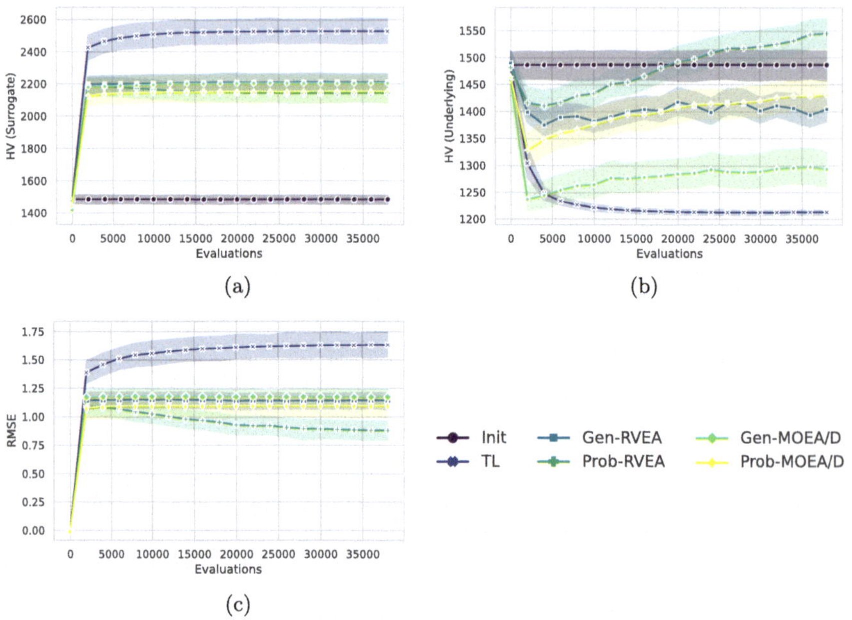

Fig. 3.9 Progress of the solution process with median hypervolume values. DBMOPP P1 with $M = 5$ and LHS showing the variation of hypervolume (HV) in surrogate and underlying objective spaces, and RMSE with function evaluations

It is essential to consider the correlation among objective functions during the optimisation. The approaches in this chapter assume that the objective functions when building the Gaussian process surrogates are independent. We will extend these approaches to consider correlation among objective functions in the future. Moreover, using Monte Carlo samples can increase computation cost, and we plan to alleviate this cost by using an efficient alternative sampling. In addition, testing these approaches on real-world optimisation problems is also in our future works.

Acknowledgements The support of the Ministry of Education (MoE), Government of India, provided through the SPARC project SPARC/2018-2019/P1084/SL has been acknowledged by the authors.

References

Bishop C (2006) Pattern recognition and machine learning. Springer-Verlag, New York

Cheng R, Jin Y, Olhofer M, Sendhoff B (2016) A reference vector guided evolutionary algorithm for many-objective optimization. IEEE Trans Evol Comput 20:773–791

Das I, Dennis JE (1998) Normal-boundary intersection: a new method for generating the pareto surface in nonlinear multicriteria optimization problems. SIAM J Optim 8(3):631–657. https://doi.org/10.1137/S1052623496307510

Deb K (2001) Multi-objective optimization using evolutionary algorithms. Wiley, Chichester

Deb K, Jain H (2014) An evolutionary many-objective optimization algorithm using reference-point-based nondominated sorting approach, part I: solving problems with box constraints. IEEE Trans Evol Comput 18:577–601

Fieldsend JE, Chugh T, Allmendinger R, Miettinen K (2019) A feature rich distance-based many-objective visualisable test problem generator. In: Proceedings of the genetic and evolutionary computation conference. ACM, pp 541–549

Fieldsend JE, Chugh T, Allmendinger R, Miettinen K (2022) A visualizable test problem generator for many-objective optimization. IEEE Trans Evol Comput 26(1):1–11. https://doi.org/10.1109/TEVC.2021.3084119

Forrester A, Sobester A, Keane A (2008) Engineering design via surrogate modelling. Wiley

Hughes EJ (2001) Evolutionary multi-objective ranking with uncertainty and noise. In: Zitzler E, Thiele L, Deb K, Coello Coello CA, Corne D (eds) Evolutionary multi-criterion optimization, proceedings. Springer, pp 329–343

Jin Y, Wang H, Chugh T, Guo D, Miettinen K (2019) Data-driven evolutionary optimization: an overview and case studies. IEEE Trans Evol Comput 23:442–458

Mazumdar A, Chugh T, Hakanen J, Miettinen K (2022) Probabilistic selection approaches in decomposition-based evolutionary algorithms for offline data-driven multiobjective optimization. IEEE Trans Evol Comput 26:1182–1191

Mazumdar A, Chugh T, Miettinen K, López-Ibáñez M (2019) On dealing with uncertainties from Kriging models in offline data-driven evolutionary multiobjective optimization. In: Deb K, Goodman E, Coello Coello CA, Klamroth K, Miettinen K, Mostaghim S, Reed P (eds) Evolutionary multi-criterion optimization, proceedings. Springer, pp 463–474

Rahat AAM, Wang C, Everson RM, Fieldsend JE (2018) Data-driven multi-objective optimisation of coal-fired boiler combustion systems. Appl Energy 229:446–458

Rasmussen CE, Williams CKI (2006) Gaussian processes for machine learning. The MIT Press

Silverman B (1986) Density estimation for statistics and data analysis. Chapman & Hall/CRC, London

Wang H, Jin Y, Sun C, Doherty J (2019) Offline data-driven evolutionary optimization using selective surrogate ensembles. IEEE Trans Evol Comput 23:203–216. https://doi.org/10.1109/TEVC.2018.2834881

Yang C, Ding J, Jin Y, Chai T (2020) Offline data-driven multiobjective optimization: knowledge transfer between surrogates and generation of final solutions. IEEE Trans Evol Comput 24(3):409–423. https://doi.org/10.1109/TEVC.2019.2925959

Zhang Q, Li H (2007) MOEA/D: a multiobjective evolutionary algorithm based on decomposition. IEEE Trans Evol Comput 11:712–731

Chapter 4
Wake Models

Niranjan S. Ghaisas, Naveen N. Kethavath, Jay A. Patel, and Kingshuk Mondal

Abstract The rapid expansion of wind energy globally underscores its pivotal role in the global energy transition. However, wind farm wake losses, resulting from complex interactions between turbines, hinder the overall efficiency. The wake of a wind turbine, characterized by reduced wind speed and increased velocity fluctuations downstream, significantly impacts the wind farm performance due to interactions with the upstream turbines. Accurate modelling of these interactions is essential for designing efficient wind farms. This chapter provides details of two design-oriented analytical wake modelling frameworks, namely the top-down and bottom-up approaches. Top-down models focus on large wind farms and their interactions with the atmospheric boundary layer. In contrast, bottom-up models focus on the wakes of individual turbines and construct models for the entire wind farm using different wake-merging methods. Different variants on these broad themes are discussed. The performance of these models is compared to predictions obtained from high-fidelity large-eddy simulations. The efficacy of analytical wake models in reproducing the most important features of the flow field at a fraction of the cost is demonstrated.

Keywords Wind energy · Wake model · Wind farm wake losses · Large eddy simulation · Atmospheric boundary layer · Surface heterogeneity · Wake merging

---

N. S. Ghaisas (✉) · N. N. Kethavath · J. A. Patel · K. Mondal
Department of Mechanical and Aerospace Engineering, Indian Institute of Technology Hyderabad, Hyderabad, Telangana 502284, India
e-mail: nghaisas@mae.iith.ac.in

N. N. Kethavath
e-mail: me20resch11010@iith.ac.in

J. A. Patel
e-mail: me21resch11002@iith.ac.in

K. Mondal
e-mail: me21m23p100001@iith.ac.in

N. S. Ghaisas
Department of Climate Change, Indian Institute of Technology Hyderabad, Hyderabad, Telangana 502284, India

K. Mitra et al. (eds.), *Optimization, Uncertainty and Machine Learning in Wind Energy Conversion Systems*, Engineering Optimization: Methods and Applications,
https://doi.org/10.1007/978-981-97-7909-3_4

4.1 Introduction

The wake of a wind turbine is the region downstream of it that has reduced wind speed and increased velocity fluctuations. A wind turbine typically operates in the wake of several upstream turbines in a wind farm, depending upon the layout and wind direction. Wind farm performance is adversely affected by such wake interactions. The wake interactions are governed by complex processes, simulating which requires resolving all the length and time scales of the turbulent flow, and is extremely expensive. Hence, modelling of the wake interactions, or development of reduced wake models, is critically important. The primary requirement that must be satisfied by a wake model is reasonable accuracy in predicting the quantity of interest at a very small computational cost.

Wake models can be broadly classified based on their purpose or utility as either design-oriented or control-oriented. Design-oriented models aim to predict the time-averaged velocity field in a wind farm, which can be used for optimizing the farm layout at the pre-construction or site-selection stage. Conversely, control-oriented wake models must predict the instantaneous velocity profile in a wind farm and help in developing and deploying active control strategies based on pitch/yaw/tilt modulation for the purposes of maximizing wind power, and for reducing turbine-to-turbine and/or temporal fluctuations in the generated power.

This chapter describes two design-oriented wake modelling frameworks, the so-called top-down (Sect. 4.2) and bottom-up (Sect. 4.3) models. These modelling frameworks are selected because they are significantly distinct from each other, and other design-oriented wake models may be viewed as being variants or combinations of these two broad categories.

4.2 Top-Down Models

Top-down models are similar to the so-called single-column models commonly referred to in the atmospheric sciences literature. These models provide expressions for the horizontally-averaged mean streamwise velocity in the interior portion of very large wind farms. Here, the phrase 'very large' refers to wind farms whose horizontal extent is at least one order of magnitude larger than the vertical height of the atmospheric boundary layer (ABL). Assuming the height of the ABL to be of the order of 1 km, the mean velocity profile inside a wind farm that spans 10 km or more in the horizontal directions is predicted reasonably by top-down models. Far from the wind farm edges, the entrance/exit effects are unimportant and the atmospheric boundary layer is fully adjusted to the forcing applied by the array of wind turbines. It is reasonable to expect that the only parameters that govern the wind distribution and power extraction are the inter-turbine spacings along streamwise and spanwise directions, the alignment or staggering of subsequent rows with respect to the direction of the mean incoming wind, surface aerodynamic roughness, and turbine parameters such as hub-height, rotor diameter and thrust coefficient.

To illustrate the principles behind the top-down models, we assume a very large wind farm sited on flat ground with surface aerodynamic roughness $z_{0,\mathrm{lo}}$. The aerodynamic roughness height is a measure of the frictional force exerted by the underlying surface through small undulations or roughness elements that are too small to resolve. Mathematically, it is taken as the height at which the velocity profile extrapolated from the logarithmic region sufficiently far away from the wall would go to zero. The mean wind direction is along the positive x axis. The spanwise direction is along y and vertical direction is along the z axis. The flow is driven by a streamwise pressure gradient, $\mathrm{d}P/\mathrm{d}x = -u_{*,\mathrm{hi}}^2/H$. Here, H is the height of the boundary layer and $u_{*,\mathrm{hi}}$ is the friction velocity at the bottom surface in the absence of wind turbines. All turbines are of diameter D, have a hub-height z_h and operate with a thrust coefficient C_T. The turbines are spaced $s_x D$ and $s_y D$ distances apart along the x and y directions, respectively. These parameters can be clubbed together into a wind farm loading coefficient, $c_\mathrm{ft} = \pi C_\mathrm{T}/(4 s_x s_y)$.

In the absence of wind turbines, the mean pressure gradient forcing would be balanced by friction forces at the bottom wall. By definition of the friction velocity, the wall shear stress, divided by the (constant) density is $-u_{*,\mathrm{hi}}^2$. Assuming that the eddy-viscosity scales with the distance from the wall gives rise to the classical logarithmic law-of-the-wall profile for the mean velocity.

$$-u_{*,\mathrm{hi}}^2 = -\nu_\mathrm{t}\frac{\mathrm{d}u}{\mathrm{d}z} = -\left(u_{*,\mathrm{hi}}\kappa z\right)\frac{\mathrm{d}u}{\mathrm{d}z} \Rightarrow \frac{\mathrm{d}u}{\mathrm{d}z} = \frac{u_{*,\mathrm{hi}}}{\kappa z} \Rightarrow u(z) = \frac{u_{*,\mathrm{hi}}}{\kappa z}\ln\frac{z}{z_0} \tag{4.1}$$

The derivation of the top-down model of Calaf et al. (2010) follows from three assumptions.

A. The domain can be divided vertically into four regions. Region 1 is from the ground ($z = z_{0,\mathrm{lo}}$) to the bottom tip of the turbines ($z = z_\mathrm{h} - D/2$). Regions 2 and 3 are from the bottom tip to the hub-height ($z = z_\mathrm{h} - D/2$ to z_h) and from the hub-height to the top tip of the rotor ($z = z_\mathrm{h}$ to $z_\mathrm{h} + D/2$), respectively. Region 4 is from the top tip of the turbines to the top of the ABL ($z = z_\mathrm{h} + D/2$ to H).
B. Regions 1 and 2 are marked by a constant stress, with friction velocity $u_{*,\mathrm{lo}}$ and Regions 3 and 4 are constant stress regions with a friction velocity $u_{*,\mathrm{hi}}$. Regions 1 and 2 are marked by the same aerodynamic roughness height as that of the ground surface, i.e. $z_{0,\mathrm{lo}}$. Regions 3 and 4 see a different aerodynamic roughness due to the presence of the wind turbines, i.e. $z_{0,\mathrm{hi}}$.
C. Regions 1 and 4 (i.e. below and above the wind turbine rotors) are equilibrium regions where the eddy-viscosity scales with the vertical height. Regions 2 and 3 (i.e. within the wind turbine rotor region) have a further augmented eddy-viscosity. Thus, the eddy- viscosities in the four regions are $u_{*,\mathrm{lo}}\kappa z$, $\left(u_{*,\mathrm{lo}}\kappa z + \nu_\mathrm{w}\right)$, $\left(u_{*,\mathrm{hi}}\kappa z + \nu_\mathrm{w}\right)$ and $u_{*,\mathrm{hi}}\kappa z$, respectively.

From these three assumptions, expressions can be derived for the velocity gradient, $\mathrm{d}u/\mathrm{d}z$, which can be integrated in each region separately. The final expressions for the horizontally-averaged streamwise velocity profiles in the four regions are

$$u(z) = \begin{cases} \frac{u_{*,\mathrm{lo}}}{\kappa} \ln\left(\frac{z}{z_{0,\mathrm{lo}}}\right), & z_{0,\mathrm{lo}} < z < z_\mathrm{h} - D/2 \\ \frac{u_{*,\mathrm{lo}}}{\kappa} \ln\left[\left(\frac{z}{z_\mathrm{h}}\right)^{1-\beta}\left(\frac{z_\mathrm{h}}{z_{0,\mathrm{lo}}}\right)\left(1-\frac{D}{2z_\mathrm{h}}\right)^{\beta}\right], & z_\mathrm{h} - D/2 < z < z_\mathrm{h} \\ \frac{u_{*,\mathrm{hi}}}{\kappa} \ln\left[\left(\frac{z}{z_\mathrm{h}}\right)^{1-\beta}\left(\frac{z_\mathrm{h}}{z_{0,\mathrm{hi}}}\right)\left(1+\frac{D}{2z_\mathrm{h}}\right)^{\beta}\right], & z_\mathrm{h} < z < z_\mathrm{h} + D/2 \\ \frac{u_{*,\mathrm{hi}}}{\kappa} \ln\left(\frac{z}{z_{0,\mathrm{hi}}}\right), & z_\mathrm{h} + D/2 < z < H \end{cases} \tag{4.2}$$

In the above equations, $\beta = \nu_\mathrm{w}^*/(1+\nu_\mathrm{w}^*)$ and $\nu_\mathrm{w}^* = \nu_\mathrm{w}/(u_*\kappa z)$ is the non-dimensionalized wake-eddy-viscosity. Three unknowns have been introduced into the modelling framework at this stage, namely $u_{*,\mathrm{lo}}$, $z_{0,\mathrm{hi}}$, ν_w. An empirical model is used to specify the eddy-viscosity augmentation, $\nu_\mathrm{w}^* \approx 28\sqrt{c_\mathrm{ft}/2}$. Two further constrains are needed to determine the remaining two unknowns. The first is a kinematical constraint, i.e. continuity of the velocity profile at the hub-height when computed from Regions 2 and 3, which yields

$$\frac{u_{*,\mathrm{hi}}}{u_{*,\mathrm{lo}}} = \frac{\ln\left[\left(\frac{z_\mathrm{h}}{z_{0,\mathrm{lo}}}\right)\left(1-\frac{D}{2z_\mathrm{h}}\right)^{\beta}\right]}{\ln\left[\left(\frac{z_\mathrm{h}}{z_{0,\mathrm{hi}}}\right)\left(1+\frac{D}{2z_\mathrm{h}}\right)^{\beta}\right]}$$

The second is a dynamic constraint, i.e. the net momentum transported by the shear stresses between Regions 1 and 4 must be balanced by the thrust force exerted by the turbines. This last constraint can be written as

$$u_{*,\mathrm{hi}}^2 - u_{*,\mathrm{lo}}^2 = \frac{c_\mathrm{ft}}{2} u_\mathrm{turb}^2 \tag{4.3}$$

Calaf et al. (2010) further assumed that the velocity at the turbine is the same as the horizontally-averaged velocity at the hub-height, i.e. $u_\mathrm{turb} = u(z_\mathrm{h})$. With these assumptions in place, explicit relations can be derived for the unknown lower friction velocity and effective aerodynamic roughness,

$$z_{0,\mathrm{hi}} = z_\mathrm{h}\left(1+\frac{D}{2z_\mathrm{h}}\right)^{\beta} \exp\left(-\left[\frac{c_\mathrm{ft}}{2\kappa^2} + \left(\ln\left[\frac{z_\mathrm{h}}{z_{0,\mathrm{lo}}}\left(1-\frac{D}{2z_\mathrm{h}}\right)^{\beta}\right]\right)^{-2}\right]^{-1/2}\right) \tag{4.4}$$

This modelling framework can be easily related to other top-down models. Neglecting the eddy-viscosity augmentation in Regions 2 and 3 by setting $\nu_\mathrm{w} = 0$ gives the Frandsen (1992) and Frandsen et al. (2006) model. With minor modifications, the framework described above was extended to the developing region of wind farms by Meyers and Meneveau (2012). Stevens et al. (2015) coupled the top-down model of Calaf et al. (2010) with a wake (or bottom-up) model to propose the coupled wake boundary layer (CWBL) model, which is applicable to very large wind farms. Xie et al. (2017) extended the framework to incorporate effects of a layer of vertical-axis wind turbines below the turbine rotor region.

Zhang et al. (2021) suggested an improvement to the Calaf et al. (2010) model by incorporating effects of heterogeneity in the velocity field at the hub-height. Specifically, the velocity used to calculate the force in Eq. (4.3) was considered to be $u_{\mathrm{turb}} = \alpha u(z_{\mathrm{h}})$, with the heterogeneity factor $\alpha \neq 1$. An iterative procedure, following from the same idea proposed for the CWBL model by Stevens et al. (2015), was devised to determine the heterogeneity factor α. A simpler, empirical procedure was proposed by Ghaisas et al. (2024) to prescribe α based on observed correlations with either the loading factor, c_{ft}, or with the effective roughness height, $z_{0,\mathrm{hi}}$. As an example, the empirical formula for the heterogeneity factor based on the loading factor is $\alpha = \mathrm{e}^{0.505}\left(c'_{\mathrm{ft}}\right)^{0.127}$. Here, c'_{ft} is the loading coefficient based on the local thrust coefficient (C'_{T}) instead of the nominal thrust coefficient (C_{T}) with the two being related by $C_{\mathrm{T}} = 16C'_{\mathrm{T}}/\left(4 + C'_{\mathrm{T}}\right)^2$.

The performance of the top-down models is illustrated next. Large-eddy simulations of very large wind farms are performed and the data is used to assess the top-down models. Each LES is performed in a domain of size $(\pi \times \pi \times 1)H$ where $H = 1000$ m is the height of the boundary layer. Periodic boundary conditions are assumed for all horizontal boundaries, thus ensuring that the fully-developed wind turbine array boundary layer caused by a very large wind farm is simulated. Wind farms with different wind turbine loading (c_{ft}) are examined. Different values for the loading are obtained by varying the inter-turbine spacing (or, equivalently, the number of wind turbines in the domain) or the thrust coefficient of the individual turbines. Figure 4.1 shows contours of the instantaneous streamwise velocity field at representative wind farm loadings. A wide range of scales and the wake behind each turbine is clearly seen in Fig. 4.1.

Averaging the streamwise velocity values over the horizontal directions and over time gives a profile along the vertical coordinate. These profiles are shown in Fig. 4.2 for several turbine loading values. It is clear that the averaged velocity reduces as the loading coefficient is increased. Furthermore, the regions above and below the wind turbine rotor region (shaded portion) are clearly seen to be following different logarithmic laws. The profile within the rotor region is better mixed than the regions above or below it. This suggests that the profile in the rotor region will be predicted well on using an augmented eddy-viscosity.

Figures 4.3 and 4.4 show quantitative comparisons between the LES results and the top-down model predictions. Here, we focus on the model of Calaf et al. (2010), and the improvements proposed in Zhang et al. (2021) and Ghaisas et al. (2024). These three versions of the top-down models are referred to as 'TD-2010', 'TD-2021' and 'TD-2024', respectively. Figure 4.3 shows the vertical profiles of the averaged

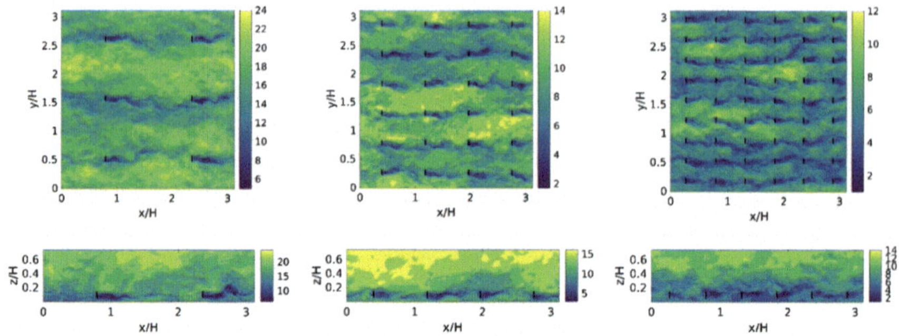

Fig. 4.1 Contours of the instantaneous streamwise component of velocity in the plan view (top panels) and in the side view (bottom panels) for wind farms with different turbine loading coefficients (c_{ft}) obtained by changing the inter-turbine spacing. Left panels: $c_{\mathrm{ft}} = 0.0036$; middle panels: $c_{\mathrm{ft}} = 0.0143$; right panels: $c_{\mathrm{ft}} = 0.0322$. In each wind farm, the thrust coefficient is $C_{\mathrm{T}} = 0.75$, and the surface roughness is $z_{0,\mathrm{lo}}/H = 10^{-6}$

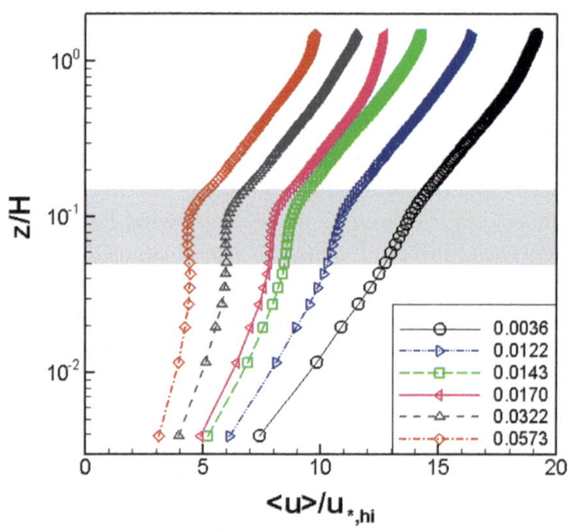

Fig. 4.2 Vertical profiles of the time- and horizontally-averaged streamwise velocity in wind farms with different loading coefficients (denoted in the legend) and with surface aerodynamic roughness $z_{0,\mathrm{lo}}/H = 10^{-4}$. The shaded region is the rotor region, $z_{\mathrm{h}} - D/2 \leq z \leq z_{\mathrm{h}} + D/2$

streamwise velocity for different combinations of the turbine loading coefficient and surface roughness values. All three top-down models are seen to capture the broad effect of c_{ft} and z_0 accurately. Among the three models, the TD-2024 predictions are seen to be most accurate for most of the cases shown in Fig. 4.3.

The differences between the three models for a larger number of cases are shown more quantitatively and succinctly in Fig. 4.4. Three quantities of interest, i.e. the lower log-law friction velocity ($u_{*,\mathrm{lo}}$) the effective roughness height ($z_{0,\mathrm{hi}}$) and the

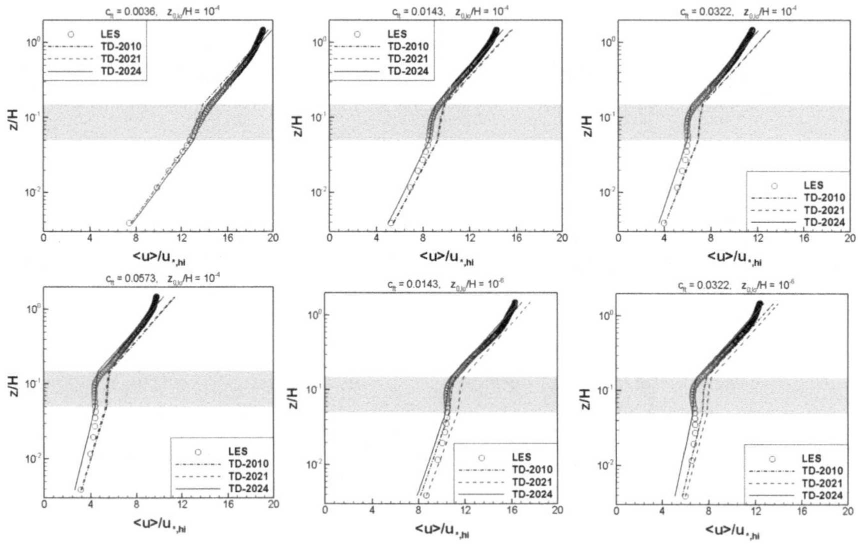

Fig. 4.3 Vertical profiles of the time- and horizontally-averaged streamwise velocity for wind farms with different loading coefficients and different aerodynamic surface roughness values. TD-2010, TD-2021 and TD-2024 refer to the top-down models of Calaf et al. (2010), Zhang et al. (2021) and Ghaisas et al. (2024), respectively

wind speed at the hub-height ($u(z_h)$), are compared in Fig. 4.4 across multiple cases. The predictions of the effective roughness height are of similar accuracy between the three models. The lower friction velocity and the hub-height wind speed are much better predicted by the model of Ghaisas et al. (2024). Thus, this model is more robust than the other two since it is more accurate over a larger range of turbine loading coefficient and surface roughness values.

To summarize, different versions of top-down models have been developed over the years, with differing fidelity to LES results. These models are useful to study the mean (time-and horizontally-averaged) interactions between large wind farms and the atmospheric boundary layer. In particular, the effects of turbine density per unit area, turbine operating thrust coefficients on surfaces with different aerodynamic roughness heights, can be investigated very efficiently. While not discussed above, with small modifications, these models can also be used to determine first-order effects on, e.g., the optimal spacing between turbine rows in the streamwise direction as a function of the aerodynamic surface roughness height and the turbine diameter to the ABL height ratio, to study kinetic energy entrainment in multi-layer wind farms.

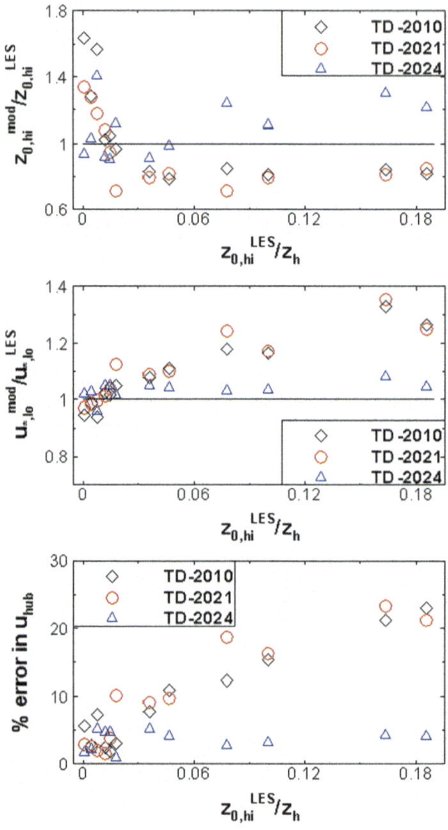

Fig. 4.4 Comparison of three versions of top-down models against LES results. Top panel: effective roughness height predictions. Middle panel: lower-level friction velocity predictions. Bottom panel: error in hub-height wind speed predictions. TD-2010, TD-2021 and TD-2024 refer to the top-down models of Calaf et al. (2010); Zhang et al. (2021) and Ghaisas et al. (2024), respectively

4.3 Bottom-Up Models

Bottom-up (wake) analytical models are useful for the prediction of the velocity field behind a single wind turbine. These models are coupled to different wake-merging strategies and can be used to predict the velocity field in wind farm clusters or farms. Unlike top-down models, these bottom-up models do not require horizontal homogeneity and do not predict the horizontally-averaged flow field. Instead, the three-dimensional, time-averaged flow field is predicted by bottom-up wake models. Thus, they are applicable to the region near the edges of large wind farms, or to finite-sized wind farms. The bottom-up wake models primarily account for interactions among adjacent turbines on each other, primarily the effect of upstream turbines on downstream ones. Apart from a few exceptions (e.g. Stevens and Meneveau 2014, Stevens et al. 2016), most of bottom-up models do not account for vertical interactions between the wind farm and the atmospheric boundary layer.

Bottom-up models generally attempt to model the velocity deficit behind a wind turbine. In most studies, the velocity deficit at any given point is defined as the difference between the velocity upstream of a turbine at the same radial location with respect to the turbine and the velocity at that point. For example, with the mean wind direction being along the x axis, the deficit is given by $\Delta u(x, y, z) = u(x, y, z) - u_\infty(y, z)$. Here, u_∞ is the velocity at some upstream location with respect to the turbine, $u_\infty(y, z) = u\big(x_{\mathrm{up}}, y, z\big)$, with say $x_{\mathrm{up}} = x_{\mathrm{t}} - 1D$. In situations where the flow field changes along the streamwise direction even in the absence of wind turbines, this definition of 'upstream' and, hence, this definition of velocity deficit is found to be inappropriate for analytical modelling purposes. Alternative definitions of the upstream velocity are needed in such situations. Examples of such situations include wind farms sited on heterogeneously rough surfaces or on uneven terrain (Kethavath et al. 2022; Mondal et al. 2023; Kethavath and Ghaisas 2024).

A. Isolated-Turbine Models

The simplest analytical wake model for an isolated turbine was introduced by Jensen (1984) by applying the conservation of mass to a control volume located downstream of the wind turbine. The model is not applicable to the region immediately downstream of a wind turbine, say up to 4 turbine diameters from the turbine location, where tip-vortices and root-vortices are generated, propagate downstream and quickly break down into small-scale turbulence. The Jensen (1984) model utilizes the theory of Betz (1926) to establish a relationship between the wind speed immediately behind the rotor and the turbine thrust coefficient C_{T} (Katic et al. 1986). Furthermore, the model assumes a top-hat (i.e. radially uniform) distribution for the velocity deficit in the wake. The normalized velocity deficit is given by,

$$\frac{\Delta u(x, y, z)}{U_\infty} = \frac{\left(1 - \sqrt{1 - C_{\mathrm{T}}}\right)}{(1 + 2k_{\mathrm{w}}x/D)^2} \tag{4.5}$$

Here, k_{w} is the wake expansion coefficient, D is the wind-turbine diameter, C_{T} is the thrust coefficient of the turbine, and x is the distance downstream of the turbine. The deficit is seen to be dependent only on the streamwise distance from the turbine, which reflects the top-hat shape in the $y - z$ plane. The wake expansion coefficient (k_{w}) is constant which implies that the wake width is assumed to grow linearly with the downstream distance from the turbine. Jensen (1984) originally used a constant value ($k_{\mathrm{w}} = 0.1$) for the wake growth rate. Later studies suggested different values for k_{w}, e.g., 0.04 or 0.05 for off-shore wind farms and 0.075 for an onshore wind farm (Barthelmie et al. 2006, 2007; Barthelmie and Jensen 2010; Cleve et al. 2009). A further improvement was to relate the value of k_{w} to the ratio of the friction velocity to the streamwise velocity component at the turbine hub-height for the incoming boundary layer (Frandsen 1992). Assuming a logarithmic incoming wind profile, this approximation yields $k_\mathrm{w} \approx 0.5/\ln(z_\mathrm{h}/z_0)$, where z_h is the turbine hub-height and z_0 is the aerodynamic surface roughness length.

Frandsen et al. (2006) considered the top-hat distribution of the velocity deficit in the wake and employed mass and momentum conservation over a control volume around a turbine to derive an expression for the wake velocity deficit,

$$\frac{\Delta u(x, y, z)}{U_\infty} = \frac{1}{2}\left(1 - \sqrt{1 - \frac{2C_\mathrm{T}}{\beta + \alpha x D}}\right), \tag{4.6}$$

where $\beta = \left(1 + \sqrt{1 - C_\mathrm{T}}\right)/\left(2\sqrt{1 - C_\mathrm{T}}\right)$ and α is of the order of $10k_\mathrm{w}$.

The top-hat models of Jensen (1984) and Frandsen et al. (2006) do not precisely represent the velocity field behind a wind turbine. Specifically, due to the assumption of a top-hat profile for downstream wake velocity deficit profiles, these models tend to underestimate the velocity deficit at the centre of the wake and overestimate it at the wake edges. An incorrect estimate of the velocity field in a wind farm might result in large errors in the energy output projection, as the energy extracted from the wind is proportional to the cube of its velocity. Thus, as demonstrated by Bastankhah and Porte-Agel (2014), the top-hat model power predictions exhibit unrealistically high sensitivity to the lateral positioning of turbines as compared to actual wind farms.

A significant improvement over models that assume a top-hat radial profile are models that assume a Gaussian radial profile for the velocity deficit. These models are consistent with wind-tunnel measurements (Chamorro and Porté-Agel 2009; Medici and Alfredsson 2006; Markfort et al. 2012; Zhang et al. 2013; Porté-Agel et al. 2011) and field observations (Gaumond et al. 2014; Nygaard et al. 2013) which concluded that, after some downstream distances, the velocity deficit in the wake assumes an essentially Gaussian distribution. Bastankhah and Porté-Agel (2014) proposed a new analytical model to estimate the velocity deficit distribution downwind of an isolated wind turbine by applying mass and momentum conservation over a control volume and assuming a self-similar Gaussian shape for the velocity deficit. The well-known expression derived by Bastankhah and Porté-Agel (2014) is

$$\frac{\Delta u(x, y, z)}{U_\infty} = \left(1 - \sqrt{1 - \frac{C_\mathrm{T}}{8(\sigma/d)^2}}\right) \exp\left(\frac{-r^2}{2\sigma^2}\right), \tag{4.7}$$

where $r = \sqrt{(y - y_\mathrm{t})^2 + (z - z_\mathrm{t})^2}$ is the radial distance from the centre of the wind turbine located at $(x_\mathrm{t}, y_\mathrm{t}, z_\mathrm{t})$ and σ is the wake width. The simplest model for the wake width is that it grows linearly with distance from the turbine,

$$\sigma = \sigma_0 + k^*(x - x_\mathrm{t}). \tag{4.8}$$

Here, σ_0 is the initial wake width (i.e. wake width immediately downstream of the turbine) and k^* is the wake growth rate. This wake growth rate parameter in the

Bastankhah and Porté-Agel model is slightly different from the wake expansion parameter k_w used in the Jensen (1984) model and has different numerical values. The original model of Bastankhah and Porté-Agel (2014) assumed that the wake growth rate (k^*) is identical in both lateral (y) and vertical (z) directions. Later studies (Abkar and Porté-Agel 2015; Xie and Archer 2015) demonstrated that assigning different values in the lateral and vertical directions (implying an elliptic rather than a circular wake cross-section) might be more accurate. The different values of k^* help account for effects of the mean velocity vertical shear due to the ground surface and thermal stratification. The elliptic cross-sectional version of the Gaussian wake model can be written as

$$\frac{\Delta u(x,y,z)}{U_\infty} = \left(1 - \sqrt{1 - \frac{C_T}{8(\sigma_y \sigma_z / D^2)}}\right) \exp\left(\frac{-1}{2}\left[\left(\frac{z - z_t}{\sigma_z}\right)^2 + \left(\frac{y - y_t}{\sigma_y}\right)^2\right]\right), \tag{4.9}$$

where σ_z and σ_y are the wake width along vertical (z) and lateral (y) directions, respectively. These wake widths can be further written as linear fits, $\sigma_y = \sigma_{y,0} + k_y^*(x - x_t)$ and $\sigma_z = \sigma_{z,0} + k_z^*(x - x_t)$, with parameters σ_y, σ_z, k_y^*, k_z^*. Recent work has shown that the wake does not grow perfectly linearly and that the wake growth rate in fact reduces far away from the turbine. Studies such as Cheng and Porté-Agel (2018), Vahidi and Porté-Agel (2022), have been proposed for the evolution of the wake width $\sigma(x)$ without resorting to a linear fit.

The original Gaussian model of Bastankhah and Porté-Agel (2014) neglects several terms in the momentum deficit conservation equation and, as a result, is applicable starting a few diameters (say 3–4D) downstream of the wind turbine. The study by Kethavath et al. (2022) improved upon the original Gaussian model by considering all terms in the momentum deficit conservation equation. This study also allowed for the background velocity (u_∞) to be a function of the streamwise and vertical coordinate directions. With these two improvements, the model of Kethavath et al. (2022) is given by

$$\frac{\Delta U}{U_{0,\mathrm{h,up}}} = \left[1 - \sqrt{1 - \frac{1 - F(x)}{(\sigma/d)^2}\left[\frac{C_T}{8} + \frac{\gamma}{\pi}\right]}\right] \times \exp\left(-\frac{r^2}{2\sigma^2}\right) + \left(\frac{U_0}{U_{0,\mathrm{h,up}}} - 1\right) \tag{4.10}$$

Setting the factor γ, the term $F(x)$ and the last term to zero ensures that Eq. (4.10) reduces to the original Gaussian model of Bastankhah and Porté-Agel (2014).

We evaluate different isolated wind turbine wake models in Figs. 4.5 and 4.6. To outline the broad features of the wake models, we have focused on the Jensen (1984) top-hat model, the original Gaussian model of Bastankhah and Porté-Agel (2014) and the corrected-Gaussian model of Kethavath et al. (2022). The wake model predictions are compared against data from high-fidelity LES of an isolated turbine. Figure 4.5 shows the axial evolution of the centerline velocity deficit and Fig. 4.6 shows the radial profiles at different distances downstream of the turbines. The incoming flow velocity profile is controlled by the surface roughness, z_0, and the turbine operation is parameterized by the thrust coefficient, C_T. Model comparisons are demonstrated for representative values of these parameters, $(z_0/D, C_T) = \left(3.75 \times 10^{-3}, 0.6\right)$.

We observe that the top-hat model has an incorrect radial shape of the velocity deficit. However, the axial variation of the centerline velocity deficit is reasonably accurate, particularly beyond $x/d = 6$. The original Gaussian model offers a huge improvement in the prediction of the radial shape as well as the centerline velocity deficit over the Jensen (1984) top-hat model. The corrected-Gaussian model is slightly better than the original Gaussian model, particularly close to the turbine, and in predicting the axial decay of the centerline velocity deficit.

B. Wake-Merging Strategies

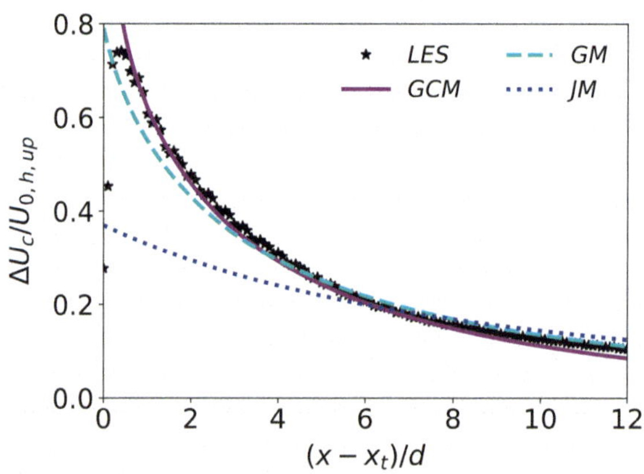

Fig. 4.5 Axial profiles of centerline velocity deficit behind an isolated turbine with $(z_0/D, C_T) = \left(3.75 \times 10^{-3}, 0.6\right)$ from LES and three bottom-up wake models. 'JM' denotes the Jensen (1984) model, 'GM' denotes the Gaussian model of Bastankhah and Porté-Agel (2014) and 'GCM' denotes the corrected-Gaussian model of Kethavath et al. (2022)

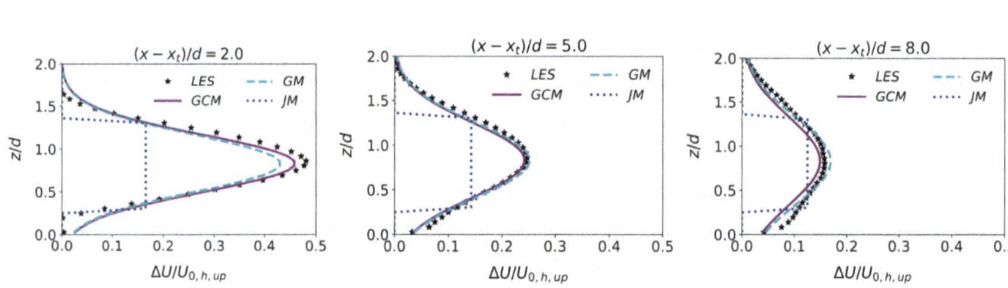

Fig. 4.6 Radial profiles of the velocity deficit at 2, 5 and 8 diameters downstream of an isolated turbine obtained from LES and three bottom-up wake models. 'JM' denotes the Jensen (1984) model, 'GM' denotes the Gaussian model of Bastankhah and Porté-Agel (2014) and 'GCM' denotes the corrected-Gaussian model of Kethavath et al. (2022)

The aforementioned wake models are employed to predict the velocity deficit in the wakes of individual or isolated wind turbines. When estimating the power production of large wind farms, we must consider the presence of multiple wind turbines and the interactions among multiple wakes. Therefore, superposition methods are utilized to model overlapping of multiple wakes. The two most commonly used or simplest superposition techniques are the so-called Linear and Quadratic wake-merging.

Linear merging involves simply adding up the velocity deficits at a point due to multiple turbines considered one at a time. In a wind farm comprised of N_t turbines, the combined velocity deficit as per the linear wake-merging technique is

$$\Delta u(x, y, z) = \sum_{k=1}^{N_t} \Delta u_k(x, y, z), \tag{4.11}$$

where $\Delta u_k(x, y, z)$ is the velocity deficit due to turbine k at the point (x, y, z). Quadratic wake-merging is the second simplest superposition technique. The combined velocity deficit due to multiple turbines is taken to be the square-root of the sum of squares of the deficits due to the individual turbines,

$$\Delta u(x, y, z) = \sqrt{\sum_{k=1}^{N_t} \left[\Delta u_k(x, y, z)\right]^2} \tag{4.12}$$

The above two wake-merging techniques assume that the background velocity field (u_∞) is homogeneous, or unchanging in the streamwise and vertical directions. A wake-merging technique that explicitly accounts for a varying background velocity field is the one proposed by Lanzilao and Meyers (2022). The combined wake due to multiple turbines is expressed as

$$\Delta u(x, y, z) = U_\infty(x, z) \left[1 - \prod_{k=1}^{N_t} (1 - \Delta U_k(x, y, z)/U_{k-1}(x, z)) \right] \tag{4.13}$$

To demonstrate the predictions of these three models, we conduct LES of extended aligned wind farms with 10 turbine rows and 3 turbine columns. The turbines are spaced 5D apart in the streamwise direction and 4D apart in the spanwise direction. Contours of the streamwise velocity at a representative time instant are shown in Fig. 4.7. The wakes of multiple upstream turbines can be seen to interact with each other and affect the performance of the downstream turbines in Fig. 4.7.

To quantitatively assess the wake-merging methodologies, predictions of the velocity deficit along the centerline of six model combinations are shown in Fig. 4.8. Two isolated-turbine models, namely the original Gaussian (Bastankhah and Porté-Agel 2014) and the Gaussian-corrected (Kethavath et al. 2022) models are combined with the three wake-merging methodologies outlined in Eqs. (4.11), (4.12) and (4.13) above. The LES results are averaged in time and along the three columns. The models

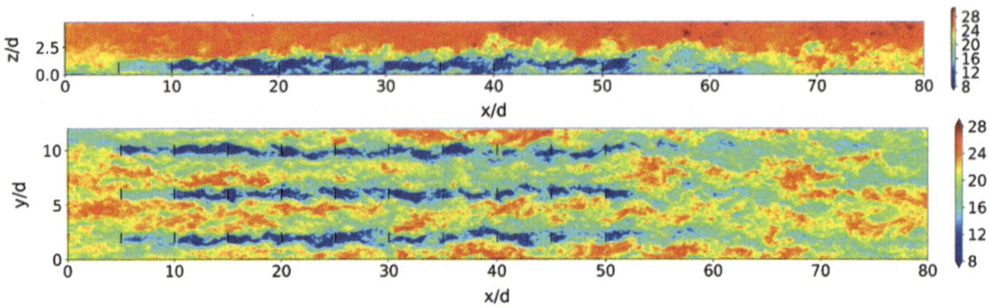

Fig. 4.7 Contours of the streamwise velocity at a representative time instant in a wind farm comprised of 30 turbines in a 10 × 3 aligned arrangement. Black vertical lines denote the turbine rotors. The top panel shows the side-view along the spanwise-centerline of one of the turbine columns. The bottom panel shows the plan view at the hub-height. Wake interactions within the farm, meandering of wakes and the presence of a large range of scales are seen in the flow

require specification of the wake growth rate (k^*) and the initial wake width (σ_0) for each turbine row. To focus on the wake-merging strategy, we extract k^* directly from the LES results and use those wake growth rate values for each turbine row. We further assume that the initial wake width is a tunable parameter and evaluate the model predictions for different values of σ_0/D ranging from 0.34 to 0.44 in steps of 0.01.

The model predictions show sensitivity to the isolated turbine model type (Gaussian vs. corrected-Gaussian), the wake-merging methodology as well as the value of the tunable parameter σ_0. The linear wake-merging methodology tends to generally over-predict the velocity deficits as seen in Fig. 4.8a, while the quadratic wake-merging methodology under-predicts the velocity deficits as seen in Fig. 4.8b. The

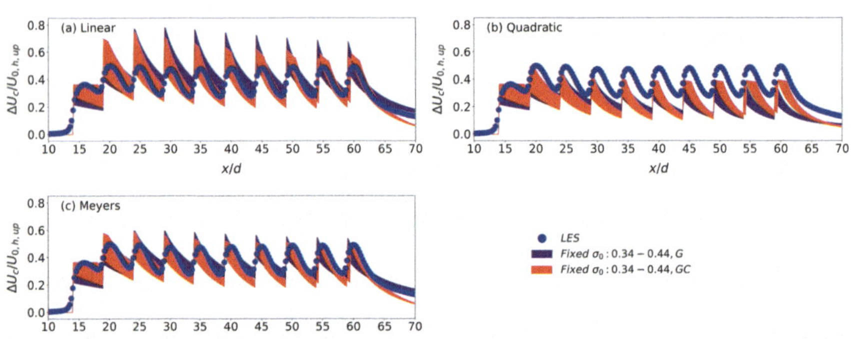

Fig. 4.8 Axial variation of the centerline velocity deficit in an extended wind farm. LES results are compared to predictions of the different bottom-up wake model combinations. 'G' denotes the Gaussian wake model of Bastankhah and Porté-Agel (2014). 'GC' denotes the corrected-Gaussian model of Kethavath et al. (2022). Results are shown assuming **a** linear wake-merging, **b** quadratic wake-merging and **c** wake-merging of Lanzilao and Meyers (2022)

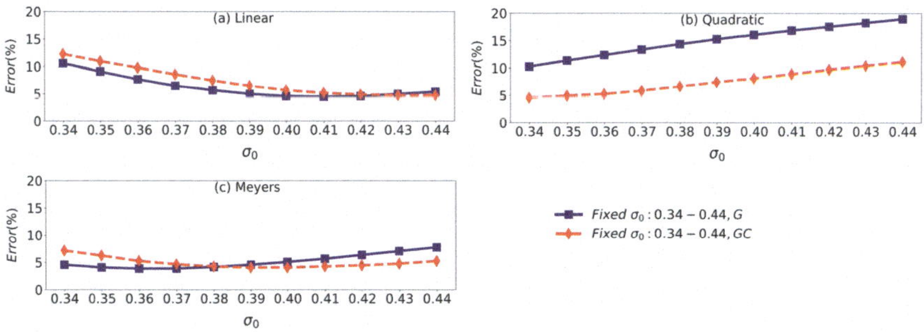

Fig. 4.9 Errors between the LES results and model predictions of the centerline velocity deficit in an extended wind farm as a function of the tunable parameter σ_0. 'G' denotes the Gaussian wake model of Bastankhah and Porté-Agel (2014). 'GC' denotes the corrected-Gaussian model of Kethavath et al. (2022). Results are shown assuming **a** linear wake-merging, **b** quadratic wake-merging and **c** wake-merging of Lanzilao and Meyers (2022)

model predictions are in closest agreement with the LES results on using the wake-merging methodology of Lanzilao and Meyers (2022). In each case, the velocity deficit predictions of the first few (up to 4) turbines are very sensitive to the σ_0 value. The sensitivity to σ_0 reduces as distance into the wind farm increases. The errors plotted in Fig. 4.9 show that the results of linear and quadratic wake-merging strategies are very sensitive to σ_0. The results of the Lanzilao and Meyers (2022) wake-merging methodology are least sensitive to the value of σ_0. Finally, the corrected-Gaussian model with Lanzilao and Meyers (2022) wake-merging and a $\sigma_0 = 0.42$ gives errors less than 5% which is much better than other wake-merging and isolated-turbine model combinations.

4.4 Concluding Remarks

Due to the exorbitant computational cost of high-fidelity simulations, such as large-eddy simulations, the efficient design of wind farms depends critically on reduced analytical wake models. Two broad wake-modelling frameworks have been described in the above chapter and a few variants of each are tested against high-fidelity LES data. All wake models invoke simplifying assumptions which are often encoded into tunable parameters. The sensitivity of the results to the tunable parameters has also been demonstrated in the above chapter. Each of the models discussed above provides a prediction in at most a few seconds for a given combination of parameter values. Thus, large parametric sweeps, which are essential for design studies, can be performed at a fraction of the cost associated with execution of a single high-fidelity calculation.

Development of such wake models continues to spawn further studies focused on (a) calibrating the model parameters either as static coefficients or through dynamic procedures, (b) quantifying uncertainties introduced by the tunable parameters, (c) incorporation of further complicating features such as yaw-mismatch, or atmospheric stability and (d) employing the wake models into layout optimization and/or wind farm control routines.

Acknowledgements We would like to thank the DST National Supercomputing Mission Grant No. DST/NSM/R&D_HPC_Applications/2021/28 and DST SERB Grant No. CRG/2022/0006735 for providing HPC resources on Param Brahma cluster at IISER Pune and Param Seva cluster at IIT Hyderabad.

References

Abkar M, Porté-Agel F (2015, Mar) Influence of atmospheric stability on wind-turbine wakes: a large-eddy simulation study. Phys Fluids 27(3). https://doi.org/10.1063/1.4913695

Barthelmie RJ, Jensen LE (2010) Evaluation of wind farm efficiency and wind turbine wakes at the Nysted offshore wind farm. Wind Energy 13(6):573–586. https://doi.org/10.1002/we.408

Barthelmie RJ, Larsen GC, Frandsen ST, Folkerts L, Rados K, Pryor SC, Lange B, Schepers G (2006) Comparison of wake model simulations with offshore wind turbine wake profiles measured by Sodar. J Atmos Ocean Technol 23(7):888–901. https://doi.org/10.1175/JTECH1886.1

Barthelmie RJ, Frandsen ST, Nielsen MN, Pryor SC, Rethore PE, Jørgensen HE (2007) Modelling and measurements of power losses and turbulence intensity in wind turbine wakes at Middelgrunden offshore wind farm. Wind Energy Int J Prog Appl Wind Power Convers Technol 10(6):517–528. https://doi.org/10.1002/we.238

Bastankhah M, Porté-Agel F (2014, Oct) A new analytical model for wind-turbine wakes. Renew Energy 70:116–23. https://doi.org/10.1016/j.renene.2014.01.002

Betz A (1926) Wind-energie und ihre ausnutzung durch windmühlen. Vandenhoeck & Ruprecht

Calaf M, Meneveau C, Meyers J (2010) Large eddy simulation study of fully developed wind-turbine array boundary layers. Phys Fluids 22(1). https://doi.org/10.1063/1.3291077

Chamorro LP, Porté-Agel F (2009) A wind-tunnel investigation of wind-turbine wakes: boundary-layer turbulence effects. Bound-Layer Meteorol 132:129–149. https://doi.org/10.1007/s10546-009-9380-8

Cheng WC, Porté-Agel F (2018, Oct) A simple physically-based model for wind-turbine wake growth in a turbulent boundary layer. Bound-Layer Meteorol 169:1–0. https://doi.org/10.1007/s10546-018-0366-2

Cleve J, Greiner M, Enevoldsen P, Birkemose B, Jensen L (2009) Model-based analysis of wake-flow data in the Nysted offshore wind farm. Wind Energy Int J Prog Appl Wind Power Convers Technol 12(2):125–135. https://doi.org/10.1002/we.314

Frandsen S (1992) On the wind speed reduction in the center of large clusters of wind turbines. J Wind Eng Ind Aerodyn 39(1–3):251–265. https://doi.org/10.1016/0167-6105(92)90551-K

Frandsen S, Barthelmie R, Pryor S, Rathmann O, Larsen S, Højstrup J, Thøgersen M (2006) Analytical modelling of wind speed deficit in large offshore wind farms. Wind Energy Int J Prog Appl Wind Power Convers Technol 9(1–2):39–53. https://doi.org/10.1002/we.189

Gaumond M, Réthoré PE, Ott S, Pena A, Bechmann A, Hansen KS (2014) Evaluation of the wind direction uncertainty and its impact on wake modeling at the Horns Rev offshore wind farm. Wind Energy 17(8):1169–1178. https://doi.org/10.1002/we.1625

Ghaisas N, Mridula KV, Ghate A (2024) Large-eddy simulation and empirical top-down models for single-layer and two-layer very large wind farms. Submitted

Jensen NO (1984) Note on wind generator interaction (Citeseer, 1983), vol 2411

Katic I, Højstrup J, Jensen NO (1987) A simple model for cluster efficiency. In: Raguzzi A (ed) European wind energy association conference and exhibition, pp 407–410

Kethavath NN, Ghaisas NS (2024, May) Effect of an abrupt rough-to-smooth surface roughness transition on wind farm wakes: an LES and analytical modeling study. J Renew Sustain Energy 16(3). https://doi.org/10.1063/5.0202733
Kethavath NN, Mondal K, Ghaisas NS (2022, Dec) Large-eddy simulation and analytical modeling study of the wake of a wind turbine behind an abrupt rough-to-smooth surface roughness transition. Phys Fluids 34(12). https://doi.org/10.1063/5.0129022
Lanzilao L, Meyers J (2022) A new wake-merging method for wind-farm power prediction in the presence of heterogeneous background velocity fields. Wind Energy 25(2):237–259. https://doi.org/10.1002/we.2669
Markfort CD, Zhang W, Porté-Agel F (2012) Turbulent flow and scalar transport through and over aligned and staggered wind farms. J Turbul 13(1):N33. https://doi.org/10.1080/14685248.2012.709635
Medici D, Alfredsson PH (2006) Measurements on a wind turbine wake: 3D effects and bluff body vortex shedding. Wind Energy Int J Prog Appl Wind Power Convers Technol 9(3):219–236. https://doi.org/10.1002/we.156
Meyers J, Meneveau C (2012) Optimal turbine spacing in fully developed wind farm boundary layers. Wind Energy 15(2):305–317. https://doi.org/10.1002/we.469
Mondal K, Kethavath NN, Abhinay K, Ghaisas NS (2023, Aug) Large eddy simulation study of atmospheric boundary layer flow over an abrupt rough-to-smooth surface roughness transition. Bound-Layer Meteorol 188(2):229–57. https://doi.org/10.1007/s10546-023-00811-3
Nygaard NG, Jensen L, Downey R, Méchali M (2013) Construction and validation of a new offshore wake model. In: Proceedings of the international conference on aerodynamics of offshore wind energy systems and wakes (ICOWES) conference, Lyngby, Denmark, June 2013, vol 1719
Porté-Agel F, Wu YT, Lu H, Conzemius RJ (2011) Large-eddy simulation of atmospheric boundary layer flow through wind turbines and wind farms. J Wind Eng Ind Aerodyn 99(4):154–168. https://doi.org/10.1016/j.jweia.2011.01.011
Stevens RJ, Gayme DF, Meneveau C (2015) Coupled wake boundary layer model of wind-farms. J Renew Sustain Energy 7(2). https://doi.org/10.1063/1.4915287
Stevens RJ, Gayme DF, Meneveau C (2016, Feb) Effects of turbine spacing on the power output of extended wind-farms. Wind Energy 19(2):359–70. https://doi.org/10.1002/we.1835
Stevens RJ, Meneveau C (2014, Jul) Temporal structure of aggregate power fluctuations in large-eddy simulations of extended wind-farms. J Renew Sustain Energy 6(4). https://doi.org/10.1063/1.4885114
Vahidi D, Porté-Agel F (2022) A physics-based model for wind turbine wake expansion in the atmospheric boundary layer. J Fluid Mech 943:A49. https://doi.org/10.1017/jfm.2022.443
Xie S, Archer C (2015, Oct) Self-similarity and turbulence characteristics of wind turbine wakes via large-eddy simulation. Wind Energy 18(10):1815–38. https://doi.org/10.1002/we.1792
Xie S, Archer CL, Ghaisas N, Meneveau C (2017) Benefits of collocating vertical-axis and horizontal-axis wind turbines in large wind farms. Wind Energy 20(1):45–62. https://doi.org/10.1002/we.1990
Zhang W, Markfort CD, Porté-Agel F (2013) Wind-turbine wakes in a convective boundary layer: a wind-tunnel study. Bound-Layer Meteorol 146:161–179. https://doi.org/10.1007/s10546-012-9751-4
Zhang H, Ge M, Liu Y, Yang XI (2021) A new coupled model for the equivalent roughness heights of wind farms. Renew Energy 171:34–46. https://doi.org/10.1016/j.renene.2021.02.076

Chapter 5
Wind Farm Performance Assessment Using CFD: Key Factors, SOWFA, and Yaw Misalignment Case Study

Utkarsh Deep Tiwari, Niranjan S. Ghaisas, and Kishalay Mitra

Abstract The wind industry sector has emerged as a leading contender in addressing contemporary clean energy demands. This industry is experiencing rapid growth, driven by concerted endeavors to enhance the efficacy of wind farms. A critical aspect in this pursuit revolves around mitigating wake losses, which significantly impact overall power generation. This article delineates the significance of key factors such as inter-turbine distance, yaw misalignment, atmospheric thermal stability, and wind direction in influencing wake dynamics within wind farms. The discourse also highlights the importance of employing high-fidelity computational fluid dynamics (CFD) simulations to understand the intricate physics governing the flow patterns within the wind farms. Furthermore, it includes a brief introduction to the Simulator for Wind Farm Applications (SOWFA) package, a sophisticated tool instrumental in simulating flow dynamics within wind farm environments. Meanwhile, a case study investigating the influence of yaw misalignment on a medium-sized wind farm is also presented. The study unveils a discernible dependency of overall power output and time-averaged turbulence intensity on the angle of yaw misalignment. Notably, the findings underscore the potential benefits of strategic turbine yawing, particularly in scenarios characterized by elevated wake losses.

Keywords Wind farm · Yaw misalignment · Wind flow direction · Wake dynamics

U. D. Tiwari · K. Mitra
Department of Chemical Engineering, Indian Institute of Technology Hyderabad, Telangana 502284, India

N. S. Ghaisas (✉)
Department of Mechanical and Aerospace Engineering, Indian Institute of Technology Hyderabad, Telangana 502284, India
e-mail: nghaisas@mae.iith.ac.in

N. S. Ghaisas · K. Mitra
Department of Climate Change, Indian Institute of Technology Hyderabad, Telangana 502284, India

K. Mitra
Department of Artificial Intelligence, Indian Institute of Technology Hyderabad, Telangana 502284, India

K. Mitra et al. (eds.), *Optimization, Uncertainty and Machine Learning in Wind Energy Conversion Systems*, Engineering Optimization: Methods and Applications,
https://doi.org/10.1007/978-981-97-7909-3_5

5.1 Introduction

Wind energy is one of the most rapidly growing renewable energy resources, offering a viable alternative to conventional electricity generation methods. In India's comprehensive strategy aimed at reducing carbon emissions within its energy sector, the wind power industry holds a crucial position. As of July 2023, India has an installed wind power capacity of approximately 42.8 gigawatts (GW), positioning it as the world's fourth largest wind power market (Invest India 2023). Forecasts indicate that India's installed wind energy capacity is poised to reach an estimated 99.895 GW by the conclusion of the fiscal year 2029–2030 (Central Electricity Authority 2023).

Although wind energy is a green source of energy and has the potential to reduce carbon footprints, wind farms also face several problems that can affect their efficiency, safety, and environmental impact. Firstly, one of the primary challenges faced by wind farm operators is the intermittent nature (Albadi and El-Saadany 2010) of wind power generation. Wind speed and direction can be highly variable and unpredictable, which causes fluctuations in the amount of power generated by the turbines. This can make it difficult for grid operators to balance the supply and demand of electricity, which can lead to grid instability and the need for additional backup power sources (Das et al. 2018). Second, wind turbines can also be affected by extreme weather conditions such as high winds (Tavner et al. 2006), lightning strikes, and ice build-up (Katsaprakakis, Papadakis, and Ntintakis 2021). These events can cause damage to the turbines, reducing their efficiency and leading to costly repairs and downtime. Third, wind farms can have negative impacts on local wildlife, particularly birds and bats (Smallwood and Bell 2020), which can collide with the turbine blades. This can have significant ecological implications, as many bird and bat species are already under threat from habitat loss and other human activities (Wang, Wang, and Smith 2015). Fourthly, wind farms can also be a source of noise pollution (Knopper et al. 2014), which can affect nearby communities and wildlife (Kumara et al. 2022). While modern wind turbines are designed to be relatively quiet (Oerlemans 2021), the sound they produce can still be a concern for some people living in close proximity. Lastly, wind farms also face challenges related to their installation and maintenance (Mustafa and Al-Mahadin 2018), including logistical issues such as transportation and access to remote sites (Cotrell et al. 2014), as well as the need for skilled technicians to perform repairs and upgrades. These challenges can add to the cost and complexity of wind farm projects, making them less attractive to investors and developers.

In addition to the aforementioned challenges, another significant problem that notably impacts the efficiency of wind farms is the occurrence of wake losses. A fundamental working principle of a wind turbine is that it extracts only a part of the kinetic energy of the incoming wind and converts it into rotational kinetic energy of the rotor (Koch et al. 2005), which is in turn converted to electrical energy by a generator. Consequently, the wind downstream of the turbine has lower speed, larger shear and is more turbulent than upstream of the turbine. The region of reduced speed and enhanced turbulence levels is called the wind turbine wake region (González-Longatt,

Wall, and Terzija 2012). With increasing distance from the turbine in the downwind direction, the wake expands radially, the velocity deficit reduces in magnitude, and the conditions in the wake gradually regain their upstream conditions by entraining faster-moving fluid from the surroundings (see Fig. 5.1). This phenomenon is called wake recovery. Depending upon the freestream turbulence levels and the atmospheric stability conditions, it may require up to twenty times the rotor diameter for the wake to recover substantially (Sanderse, Pijl, and Koren 2011). Turbines situated closer than this distance are influenced by the wake of upstream turbines. Consequently, downstream turbines experience reduced wind speeds and are capable of producing only a fraction of their rated power. In wind farm scenarios, such wake interactions occur extensively, leading to a decrease in overall power output compared to the theoretical linear increase in power with the number of turbines, as would be expected without wake interactions. The discrepancy between the realized power output of a wind farm and its theoretical maximum output is quantified as wake losses.

Apart from physical characteristics of each turbine (such as hub height, rotor diameter, blade pitch, aerofoil shapes, speed of rotation), factors such as the atmospheric boundary layer height, incoming turbulence levels, and thermal stratification and stability affect wake losses. Furthermore, the positioning of turbines in a wind farm, also termed as wind turbine micro-siting, plays a crucial role in determining the overall initial cost, operation, and maintenance costs of the plant, as well as its annual energy production (Kusiak and Song 2010). For designing a wind farm layout, key factors of utmost importance include the spacing between the turbines, yaw misalignment of individual turbines, direction of prevailing wind flow, and the atmospheric thermal stability. Previous work and key insights obtained on some of these issues are described in more detail below.

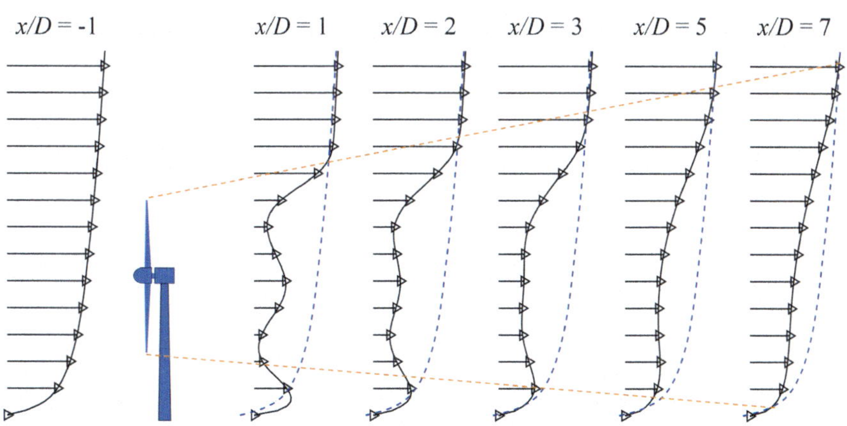

Fig. 5.1 Illustration of the wake recovery process past the wind turbine in the vertical direction

5.1.1 Effect of Inter-turbine Distance

The inter-turbine distance is the most obvious design parameter in wind farm layout that affects wake losses. To reduce the costs of land acquisition, cabling, other power harvesting infrastructure, and monitoring and civil works, wind turbines are often closely grouped together in a wind farm. If the inter-turbine distance is too large, all of the above-mentioned costs increase, at different rates. For example, land purchase/lease costs increase with the square of the horizontal footprint of a wind farm, while cabling costs increase linearly with the length of a wind farm (Guirguis, Romero, and Amon 2016; Stevens et al. 2017; Diógenes et al. 2020).

If the inter-turbine distance is too less, the wakes of individual turbines do not recover to a sufficient level before a downstream turbine is encountered (Wu and Porté-Agel 2011; Xie and Archer 2015). The wake effects become prominent, and the power output of the downstream turbines is significantly reduced because of the large velocity deficit. A compact arrangement of turbines also induces higher fatigue loads on the blades of downstream wind turbines, reducing their service life (Choi et al. 2014). A consequence of the above two competing effects is that an optimal inter-turbine spacing needs to be arrived at so as to maximize the power output and the lifetime of the farm (Meyers and Meneveau 2012; Choi et al. 2013; Stevens et al. 2014) and minimize the capital and operation and maintenance costs.

5.1.2 Effect of Yaw Misalignment

Yaw misalignment is the angle between the wind flow direction and normal to the area swept by wind turbine blades. A schematic of the wake behind a yaw-misaligned turbine is shown in Fig. 5.2. Yawing the turbine leads to two advantageous effects. First, the swept area of a yawed wind turbine is not perpendicular to the wind flow direction, which reduces the energy extracted by the turbine. This leads to a reduction in the velocity deficit within the wake, thereby resulting in higher wind velocities experienced by the downstream turbine. Secondly, yawing the turbine in a certain direction causes the wake to deflect in the opposite direction. Depending upon geometrical parameters, such as the inter-turbine distance and the angle of deflection of the wake, a wind turbine placed directly behind a yawed upstream turbine may lie completely outside the wake of the upstream turbine. These factors combine to reduce the wake effects of upstream turbines on the downstream turbines and increase the power production by the latter.

An effective yaw control mechanism can lead to high overall power production by a wind farm by degrading the power output of a few yawed wind turbines. Numerous research initiatives have been pursued to cultivate a comprehensive understanding of the impact of yaw misalignment on wind farm performance. Fleming et al. (2014) conducted simulations to elucidate the wake redirection resulting from yaw misalignment and its ramifications for overall power production. In a contrasting

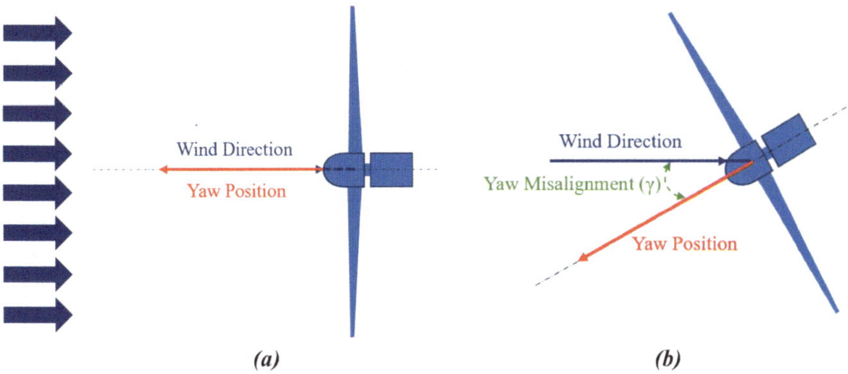

Fig. 5.2 Schematic diagram depicting top view of wind turbine in **a** non-yawed configuration **b** yawed or yaw-misaligned configuration

approach, Howland et al. (2020) systematically assessed the influence of different yaw control algorithms on power generation within wind farms. In parallel, analytical models have also emerged as valuable tools for predicting the dynamics of deflected wakes. Jiménez et al. (2009) formulated a straightforward analytical model for predicting wake deflection and trajectories past the yawed turbines. Subsequently, Bastankhah and Porté-Agel (2015) conducted controlled experiments involving miniature turbines within a wind tunnel, leveraging the obtained data to construct an Bastankhah and Porte-Agel (2016) analytical representation of wake characteristics behind isolated yawed wind turbines. Collectively, the strategic deployment of deliberate yaw misalignment as an effective means to optimize wind farm power generation has garnered substantial recognition in recent years.

5.1.3 Effect of Wind Flow Direction

In the atmospheric environment, wind direction exhibits variability rather than remaining constant. The distribution of wind flow probabilities across different directions (ϕ) is graphically represented using a "wind rose." Figure 5.3 illustrates a wind rose for a specific location, revealing distinct peaks at angles $\phi = 225°$ and $270°$. This analysis underscores the diverse wind flow patterns encountered by wind farms in operational settings.

Figures 5.4a and b display time-averaged velocity contours derived from large-eddy simulations of a 24-turbine wind farm under two distinct wind directions, namely $\phi = 270°$ and $210°$. The figures distinctly illustrate how alterations in wind direction profoundly influence the wake dynamics throughout the wind farm. These changes manifest in various aspects, such as the number of turbines exposed to

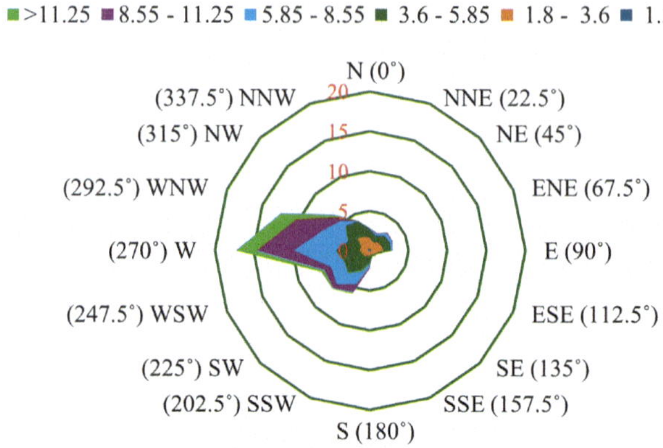

Fig. 5.3 Wind rose diagram for Kern County, California, United States of America, adapted from the Midwestern Regional Climate Center, Purdue University (Mojave Wind Rose, 1986–2024)

undisturbed wind, the delineation of upstream and downstream turbines, and the consequent adjustments in effective spacing between them. Such alterations consequently lead to significant shifts in the wind farm's performance metrics. Noteworthy studies exploring the impact of wind flow direction on wind farm dynamics include those conducted by Stevens et al. (2014), Ghaisas and Archer (2016) and Porté-Agel et al. (2020). The interaction between the effect of wind direction and other parameters such as wind farm length, atmospheric stability, and horizontal layout has been investigated in these studies. In particular, it is now clear that when the wind flows in a direction that is parallel to the columns of the wind turbine array, the wake effects (velocity deficit, turbulence intensity) and, consequently, the power losses are the highest (Wu and Porté-Agel 2015). A small, up to 15°, misalignment between the wind flow direction and the columns of the wind turbine array leads to a very large decrease in wake losses (Porte-Agel et al., 2013; Ghaisas et al., 2017). Furthermore, due to the highly turbulent nature of turbine wakes, variations in wake dynamics resulting from changes in wind direction significantly impact the fluctuating loads experienced by turbines, thereby influencing their structural integrity and service life (Kazacoks et al., 2019; Uchida and Gagnon 2022). In summary, the location and orientation of wind turbines within a wind farm need to be carefully planned to ensure that the turbines are positioned to capture the most wind and generate the most power.

5.1.4 Effect of Atmospheric Thermal Stability

Apart from the aforementioned parameters, thermal stratification or the local atmospheric stability conditions play a crucial role in determining the vertical motions, the turbulent mixing, and wake losses in wind farms. Atmospheric stability can be

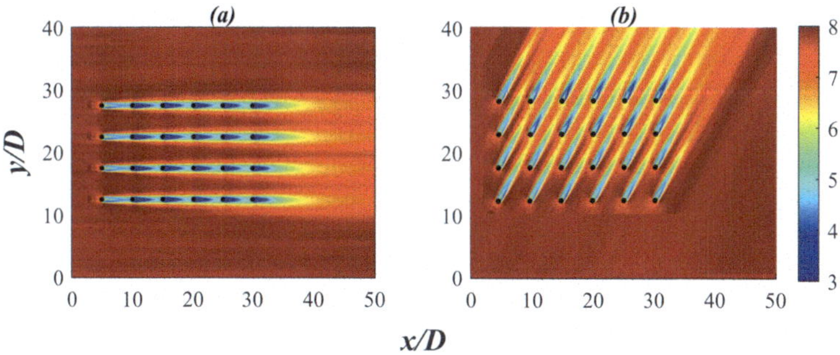

Fig. 5.4 Time-averaged velocity contours, $\overline{\overline{V}}_j$ (m/s), at hub height corresponding to wind direction **a** $\phi = 270°$ and **b** $\phi = 210°$

classified into three categories, namely stable, neutral, and unstable. Unstable and stable conditions generally occur during the daytime and nighttime, respectively, for onshore wind farms. Neutral stability is a good approximation for certain parts of the day for onshore wind farms, and for most of the time for offshore farms.

Neutral stability refers to adiabatic conditions inside the domain, wherein the buoyancy force is considered unimportant in the momentum equations. Under neutral conditions, the turbulent kinetic energy generation in the flow is primarily because of the vertical shear caused by the rough planetary surface to the atmospheric wind. Unstable conditions exist in the atmospheric boundary layer when there is a persistent positive heat flux to the domain at the lower boundary. The resulting temperature profile decreases with height very close to the surface and is followed by a well-mixed layer. This unstable temperature profile, along with the vertical buoyancy force, promotes turbulent mixing, and, as a result, the level of turbulent kinetic energy is higher under unstable conditions. Stable atmospheric conditions are set up when heat is lost by the air within the boundary layer to the ground surface below it. Contrary to unstable conditions, the effect of buoyancy dampens the vertical motions, and the turbulent mixing levels are generally lower for stable conditions as compared to neutral and unstable conditions.

The effect of stability on isolated wind turbine and wind farm wakes has been studied extensively in the past few years (Bhaganagar and Debnath 2014; Machefaux et al. 2016; Xie and Archer 2017; Du et al. 2021; Liu and Stevens 2021). For example, Wharton and Lundquist (2012a, 2012b) noted that, in the case of isolated turbines, stable atmospheric conditions tend to enhance power generation when compared to unstable and neutral conditions. In contrast, an LES study performed by Abkar and Porteagel (2015) showed that for an isolated wind turbine, the wake recovers faster under convective conditions compared to neutral and stable conditions, when subjected to same wind speed at hub height. Furthermore, Barthelmie and Jensen (2010) found that, for an entire wind farm, convective or unstable atmospheric conditions result in more favorable power output when compared to conditions characterized by low turbulence levels. It's noteworthy that Vanderwende and Lundquist (2012) observed that under stable conditions, higher wind speeds correlate with increased power output. However, in situations characterized by low wind speeds, convective conditions yield greater power generation compared to the other two conditions. Several studies have also pointed out that the effect of atmospheric stability on wind farm wakes, and consequently on overall power generation, is strongly modulated by factors such as wind speed, level of shear, geostrophic forcing, turbulence intensity, and wind farm layout (staggered or aligned) (Wagner et al. 2009; Ghaisas et al. 2017; Gomez and Lundquist (2020); Kim et al. (2021); Stieren et al. 2022).

5.1.5 Need for High-Fidelity Simulations

As alluded to above, a large number of phenomena, operating at multiple length and time scales, affect the performance of a wind farm installed in the atmospheric boundary layer. These include atmospheric turbulence, vertical wind shear, Earth's rotation, atmospheric stability, ground surface properties and processes, in addition to design and operational features of the individual wind turbines, and the layout and arrangement in a wind farm. At the fundamental level, the equations of fluid motion, governing conservation of mass, momentum, and energy govern the physical processes occurring in a wind farm. First principle calculations using the methods of computational fluid dynamics (CFD) are theoretically capable of simulating all these flow processes and can provide insights into flow physics. The primary challenge with CFD methods is that of huge computational expense due to the large length and time scales in the problem, primarily on account of flow turbulence. With advancements in computer hardware (high-performance computing on multi-CPU and multi-GPU clusters) as well as software (numerical libraries, open-source solver packages), high-fidelity CFD calculations have become an attractive and feasible tool for obtaining fundamental insights into the physics of flows around wind turbines and wind farms. In particular, high-fidelity simulations have an important role to play in the development of accurate, reliable and fast reduced-order (ODE-based or analytic) models that can serve as design tools in the future.

In this chapter, we discuss the methodology used for high-fidelity CFD simulations of wind turbines and wind farms. The governing equations, closure models using the large-eddy simulation (LES) technique, and reduced representations of wind turbines are discussed next. An open-source software package, named SOWFA (Churchfield et al. 2014), is described briefly. In the last part of this chapter, we describe results from sample calculations involving moderately sized wind farms with the wind turbines subjected to yaw misalignment and to different wind flow directions.

5.2 Governing Equations and Methodology

5.2.1 Large-Eddy Simulation Formulation

The equations governing conservation of mass, momentum, and energy, assuming that the flow is almost incompressible, with a constant density ρ_0, and that the temperature fluctuations are sufficiently small around a mean value for the Boussinesq approximation to be valid, are

$$\frac{\partial \overline{V}_j}{\partial x_j} = 0 \quad (5.2.1)$$

$$\frac{\partial \overline{V}_j}{\partial t} + \frac{\partial\left(\overline{V}_j\overline{V}_i\right)}{\partial x_i} = -2\epsilon_{j3k}\Omega_3\bar{V}_k - \frac{\partial \tilde{P}}{\partial x_j} + \frac{1}{\rho_0}\frac{\partial P_0}{\partial x_j} - \frac{\partial \tau_{ij}^D}{\partial x_i} + g\left(\frac{\bar{\theta}-\theta_0}{\theta_0}\right)\delta_{j3} + \frac{F_j^B}{\rho_0} \quad (5.2.2)$$

$$\frac{\partial \overline{\theta}}{\partial t} + \frac{\partial\left(\overline{V}_j\overline{\theta}\right)}{\partial x_j} = -\frac{\partial q_j}{\partial x_j} \quad (5.2.3)$$

where $\overline{V}_j$ stands for the spatially filtered resolved scale velocity and is the primary quantity that is to be solved for in the large-eddy simulation (LES) technique. The instantaneous full velocity field (V_j) is related to the instantaneous filtered velocity field ($\overline{V}_j$) by the sub-filter scale velocity, v_j' $\left(\overline{V}_j = V_j - v_j'\right)$. The LES filtering operation removes these sub-filter scales, v', from the simulations. The first term on the right-hand side of Eq. 5.2.2 indicates the Coriolis force due to Earth's rotation, with ϵ_{ijk} being the alternating tensor and Ω_j the rotation rate vector defined as $\Omega = \omega[0,\ \cos\varphi,\ \sin\varphi]$. Here, $\omega = 7.27 \times 10^{-5}$ rad/s and φ is the latitude of the location of the study. The second term in the momentum equation represents the gradient of density normalized pressure variable ($\tilde{P}$). It is obtained by adding the trace of sub-filter stress tensor ($\tau_{kk}/3$) and the difference between the time-averaged pressure (P_0) and the instantaneous resolved scale pressure field, normalized by the density. The third term is the driving pressure gradient (P_0), which mimics the effect of a geostrophic forcing in the atmospheric boundary layer. The fourth term is the divergence of the deviatoric part of the sub-filter scale stress tensor, τ_{ij}^D. The effects of buoyancy due to temperature differences inside the domain are taken into account by the fifth term. In these equations, $\overline{\theta}$ and $g = 9.81$ m/s^2 denote the filtered potential temperature and gravitational acceleration, respectively. δ_{ij} stands for the Kronecker delta, which equals 1 for $i = j$ and equals 0 otherwise. The last term (F_i^B) expresses body forces in the domain. The effect of the wind turbines is introduced through body forces in accordance with a variety of actuator-type models. The energy equation takes the form of Eq. 2.3 when the Boussinesq approximation is valid and when the specific heat of air is constant. The term on the right-hand side of Eq. 5.2.3

signifies the divergence of the thermal fluxes due to viscous and sub-filter scale effects.

5.2.2 Sub-filter Scale Models

Flows in the atmospheric boundary layer and, consequently, in wind farms, are characterized by very high Reynolds numbers. At such high Reynolds numbers, the flow is fully turbulent and there exists a wide range of length and time scales in the flow. The LES filter removes scales smaller than its characteristic width, which is most often equal to the size of a typical control volume. The effect of scales smaller than the LES filter width on the larger, resolved, scales is introduced in the governing equations through the terms involving τ_{ij}^{D} and q_j. Several sub-filter scale (SFS) closure models have been developed for these terms, starting from the earliest Smagorinsky (1963) model. A family of SFS models, called eddy viscosity models, specify the deviatoric part of the SFS stress tensor and the SFS heat flux as

$$\tau_{ij}^{D} = -2\vartheta^{sfs}\overline{S}_{ij} \tag{5.2.4}$$

$$q_j'' = -\frac{\vartheta^{sfs}}{\mathrm{Pr}_t}\frac{\partial\overline{\theta}}{\partial x_j} \tag{5.2.5}$$

where ϑ^{sfs} is called the sub-filter scale viscosity and Pr_t is the turbulent Prandtl number. Some of the widely used SFS viscosity models include the classical Smagorinsky (1963) model, and the more recent Vreman (2004), Sigma (2011), and Anisotropic Minimum Dissipation (2015) models. As an example, the classical one equation eddy viscosity model prescribes the SFS viscosity as

$$\vartheta^{\mathrm{sfs}} = C_k l\left(k_{\mathrm{sgs}}\right)^{1/2} \tag{5.2.6}$$

$$\mathrm{Pr}_t = \frac{1}{1+\frac{2l}{\Delta}} \tag{5.2.7}$$

where $\mathrm{C}_k = 0.0673$ is a constant, and l is the LES filter width, defined as

$$l = \Delta = (\Delta x\ \Delta y\ \Delta z)^{1/3} \tag{5.2.8}$$

For neutral and unstable atmospheric conditions, where Δx, Δy, and Δz represent the grid spacing in x, y, and z directions, respectively. Whereas for stable atmospheric conditions,

$$l = l_s = \min\left(\Delta, 0.76(k_{\text{sgs}})^{1/2}\left(\frac{g}{\theta_0}\frac{\partial\overline{\theta}}{\partial z}\right)^{-1/2}\right) \tag{5.2.9}$$

In all of the above, $\overline{S}_{ij}$ is the strain-rate tensor,

$$\overline{S}_{ij} = \frac{1}{2}\left(\frac{\partial \overline{\boldsymbol{V}}_j}{\partial x_i} + \frac{\partial \overline{\boldsymbol{V}}_i}{\partial x_j}\right) \tag{5.2.10}$$

The term k_{sgs} in Eq. 5.2.6 is sub-grid scale kinetic energy which is computed through the solution of an auxiliary equation, as

$$\frac{\partial k_{\text{sgs}}}{\partial t} + \overline{V}_j\frac{\partial k_{\text{sgs}}}{\partial x_j} = \tau_{ij}^{D}\overline{S}_{ij} + \frac{g}{\theta_o}q_j'' + C_f\frac{(k_{\text{sgs}})^{3/2}}{l} + \frac{\partial}{\partial x_j}\left(2\vartheta^{\text{sfs}}\frac{\partial k_{\text{sgs}}}{\partial x_j}\right) \tag{5.2.11}$$

With,

$$C_f = 0.19 + 0.74\frac{l}{\Delta} \tag{5.2.12}$$

The value of C_f is calibrated to 3.9 at the lowest level, as recommended by Deardorff (1980), to account for wall effects. Typically, the turbulent Prandtl number is typically given the same value as the molecular Prandtl number. Several dynamic SFS models (Moin et al. 1991; Meneveau et al. 1996; You and Moin 2007; Lee et al. 2010), where the coefficients c_s {Smagorinsky constant} and $\Pr_t$ are not specified a-priori but are calculated using the instantaneous resolved flow fields as part of the solution process, have also been developed and are found to be much more accurate for a wide variety of flow problems including atmospheric boundary layer and wind farm flows.

5.2.3 *The SOWFA Package*

5.2.3.1 Introduction

SOWFA (Simulator for Wind Farm Applications) is an open-source numerical tool that is used to simulate and analyze the flow physics in wind farm (Churchfield et al. 2014). SOWFA has been developed by NREL (National Renewable Energy Laboratory) in the USA in 2013. It is based on the Open Field Operation and Manipulation (OpenFOAM) software, which is a widely used open-source CFD tool. SOWFA is a modular code, which means it is divided into several smaller components or modules that can be used independently or in combination. This modular design allows for flexibility and ease of use, as users can choose to run only the specific modules that are relevant to their simulation and can easily add or remove modules as needed. Examples of other tools that have the capability to simulate the flow inside wind farms are PadeOps (Ghate et al. 2018), Openwind (UL Solutions), and OpenWindPower (Bentley System), each with its advantages and disadvantages.

SOWFA uses the finite volume method which offers flexibility in handling complex geometries. The most widely used numerical discretization methodology is one that is second-order accurate in space and time. It has an underlying LES solver which is specifically designed to simulate the flow in the atmospheric boundary layer and in wind farms. SOWFA, by default, contains five sub-filter scale models with the flexibility to the user of choosing any one out of (1) Static Smagorinsky model (Smagorinsky 1963), (2) Dynamic Smagorinsky model (Meneveau et al. 1996), (3) Dynamic Smagorinsky model with bounds (Churchfield et al. 2014), (4) One equation eddy viscosity model (Moeng 1984), and (5) Kosovic and Curry's one equation model (Kosović and Curry 2000). Additional features of the SOWFA package are discussed below.

5.2.3.2 Wall Treatment and Rotor Models

The typical velocity in the atmospheric boundary layer (10 m/s) and its height (1000 m) yield a Reynolds number of the order of 10^8. At such high Reynolds numbers, the velocity gradient becomes very high near the ground surface in an exceedingly thin viscous sublayer. The direct effects of viscosity are important only inside this thin layer and are negligible over the rest of the domain. One way to capture the fluid flow dynamics near the wall is to refine the mesh near the wall that fully resolves the small motions close to the wall. However, this imposes a very high computational cost as well as giving rise to the need for large storage space. An efficient alternative to fully resolving the near-wall motions is to use wall models. SOWFA has an inbuilt Schumann wall model (Schumann 1975). Input to this model is von-Karman constant, κ; surface roughness z_0; Monin-Obukhov wall shear stress model constants, β and γ; as well as the preferred type of averaging.

The purpose of SOWFA is to simulate fluid flow processes at the wind farm scale rather than at the wind turbine blade scale. It is known that the details of the flow close to the wind turbine blades, tower, and nacelle are localized to regions of the wind turbine and die down very quickly downstream of the turbine. Fully resolving flows close to the turbine would require a code with the capability of handling complex geometries due to the complicated shapes of the tower, nacelle, and blades. To overcome all the above-mentioned issues, several reduced representations of the turbine rotors are utilized. Popular among them are actuator line models (ALM) (Sørensen and Shen 2002), actuator disk model (ADM) (Jens N. Sørensen and Myken 1992), and actuator surface models (Yang and Sotiropoulos 2018). These rotor models parametrize the interaction of wind turbine blades with the incoming wind flow without actually resolving the boundary layer flow over turbine blades. Two of these models have been implemented in SOWFA.

Actuator Line Model: Actuator line model (ALM) considers the wind turbine blade as a rotating line which is discretized into several points. These points represent discrete sections of the blade respective to sectional area and characteristics. Forces acting at these locations are evaluated using blade element method (Hansen 2015) combined with an airfoil table that contains the values of lift and drag coefficients at each point. These blade forces are then projected onto the flow field through the term F_i^B in the momentum equations. The ALM model allows for explicitly simulating effects of the rotating blades. Features such as root vortices and tip vortices can be simulated accurately. However, these tip and root vortices can be resolved only if the so-called tip-CFL criterion is satisfied, i.e., the tip of the blade moves by a distance that is smaller than one grid point in one time step of the unsteady simulation. As a result, a simulation involving ALM is generally slower than one involving other actuator models described below.

Actuator Disk Model: Actuator disk models (ADM) consider the effect of rotating wind turbine blades as a circular permeable disk inside the flow field which is characterized by the swept area of the rotor. The effect of incoming wind flow on the wind turbine is considered by calculating force at different points on the circular disk using blade element momentum theory (Masson et al. 2001) in conjunction with tabulated values for lift and drag coefficient of the aerofoils. The projection of the blade forces onto the flow field is carried out in the same manner as for ALM. Other features of the turbine wakes, such as growth and recovery, largest velocity deficit, and turbulence intensity, are generated very accurately using ADM models for representing wind turbines.

Actuator Surface Model: While the use of ALM provides a tangible representation of wind turbine blade characteristics, it is important to note that ALM inherently features a single point along the aerofoil section of the blade. This two-dimensional modeling approach poses challenges in accurately depicting the flow patterns past an aerofoil. To address this limitation, three-dimensional actuator surface models (ASMs) were developed to effectively simulate the airflow around horizontal axis wind turbines (Shen et al. 2007). Within the ASM framework, the body force that originally represents the two-dimensional aerofoil section in ALM, denoted as f_{2D}, is enhanced by multiplying it with the term $f_d(x)$. This multiplication accounts for the three-dimensional rotational effects and results in an updated $f_{2D}(x)$. Notably, the function $f_d(x)$ indirectly depends on various factors, including the angle of attack, aerofoil thickness, and camber. Subsequently, the updated function $f_{2D}(x)$ is subjected to integration across both the radius and chord length, following a regularization process. The outcome of this integration is denoted as F_i^B signifying the manifestation of blade forces on to the flow field, as accounted for in the momentum equations.

The ADM and ALM have been implemented in SOWFA for representing wind turbine rotors. There are conflicting results in the literature about the accuracy of ALM and ADM among different researchers. The ADM has demonstrated its proficiency in analyzing the far wake region (Kalvig et al. 2014). It may serve as an appropriate approach for turbine representation in LES of very large wind farms, particularly when the focus lies on the characterization of main flow structures (Stevens et al. 2018). On the contrary, modeling the blade as a line of point forces in the ALM enables it to represent flow structures associated with the turbine blades that are missing from the ADM approach, where the actual movement of blades is not properly represented (Martinez et al. 2012). Capturing flow structures such as root and tip vortices depends upon the mesh resolution near the rotor (Jha et al. 2013), rendering the ALM computationally more expensive than the ADM (Sanderse et al. 2011). On a relatively coarser grid resolution, the accuracy of ADM resembles with that of ALM (Troldborg et al. 2015; Stevens et al. 2018).

5.2.3.3 Case Setup in SOWFA

The SOWFA package provides different example cases containing the complete setup for a simulation. These example cases serve as good starting points and can be modified by users according to their desired application. The geometry of the flow field and the default number of finite volumes in each direction have to be specified in the "setUp" dictionary. This dictionary contains information about the initial conditions for the velocity, pressure, and temperature, as well as the direction and magnitude of freestream velocity. Information about the surface conditions, and sub-filter scale model is also specified in this dictionary. These values are incorporated in the appropriate ".C" source files via the compiler directive '#include "./setUp"' in those files.

Details of the turbine locations and operating conditions of the wind turbines, such as yaw angle, rotor speed etc., are to be specified inside the file "constant/turbineArrayProperties." The speed of rotation of the turbines can be determined dynamically as a response to the wind conditions using different control algorithms generator torque controller defined as "GenTorqueControllerType" in the latter dictionary (Jonkman et al. 2009). The turbine dimensions and design parameters have to be specified in the "constant/turbineProperties/$turbineType" dictionary. Additionally, it has information about the mechanical features responsible for converting the rotational kinetic energy of the rotor to electrical energy, such as the gearbox ratio, drivetrain system, type of generator, etc. Each wind turbine blade is marked by a distribution of aerofoil cross-sections at different locations along its length. The details of the lift and drag features of each of these aerofoils are provided in the "constant/airfoilProperties/" directory. Details, such as blade profiles, airfoil lift and drag lookup tables, generator and rotor properties, corresponding to two turbine models are included by default in the SOWFA package. These turbine models are the NREL 5 MW reference turbine that has been developed for the purpose of standardizing research studies and the NREL Phase VI turbine.

5.2.3.4 Simulation Methodology

A typical simulation of a wind farm using SOWFA consists of two components, namely (a) precursor run and (b) windplant run. These are illustrated in Fig. 5.5. The

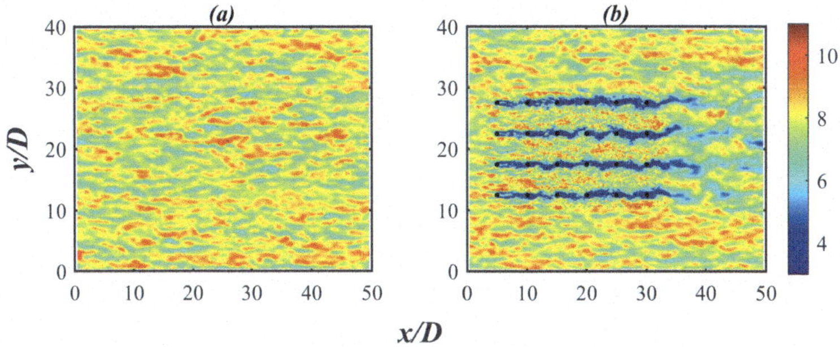

Fig. 5.5 Instantaneous velocity contours, $\overline{V}_j$ (m/s), at hub height in a statistically stationary state for **a** the precursor run and **b** the windplant run

precursor run simulates the flow field in the absence of wind turbines (Fig. 5.5a). The domain is assumed to be horizontally homogeneous. Cyclic (or periodic) boundary conditions are applied on all horizontal boundaries to mimic the characteristics of a fully developed boundary layer on a horizontally homogeneous, infinitely large, ground surface. The purpose of the precursor is to provide physically realistic initial conditions and inflow conditions to the subsequent windplant run. Running a precursor case helps ensure that the simulation results of the windplant run are more accurate and reliable, as the initial flow field and the inflow throughout the simulation have spatial and temporal correlations that are fully representative of the actual conditions in the atmospheric boundary layer.

The precursor run is further divided into two stages. In the first stage, the simulation is run for a sufficient period of time to ensure that the statically stationary flow field is achieved. This flow field is then used as the initial condition for the windplant run (Fig. 5.5b). In the second stage of the precursor run, the flow fields at one or more planes, depending upon the direction of wind flow, are stored to disk. These stored flow fields serve as boundary conditions at those respective planes for the windplant run. This is achieved by adding the "sampling/boundaryDataPre" function in the "system/controlDict" file, which creates a "boundaryDataPre" folder inside the "postProcessing" directory. By default, the "boundaryDataPre" utility records the instantaneous velocity in the "vectorField" directory and the instantaneous temperature, along with sub-grid scale kinetic energy in the "scalarField" directory at each time step. The user has the flexibility to change the write interval and to add more flow variables as needed. To allow the utilization of the boundary data stored using the "boundaryDataPre" dictionary, the "makeBoundaryDataFiles.$face.sh" script needs to be executed. This generates a "boundaryData" directory which contains the same data as in the "boundaryDataPre" folder, but in a format that can be readily used by the windplant solver. The "boundaryData/" folder can be linked to inside the "constant/" directory of the windplant run.

Similar to the precursor run, the windplant run is performed in two stages. The first, spin-up stage allows the flow to attain a statistically stationary state. Following this, statistical averaging is performed every few time steps of the windplant run. This averaging stage is carried out for a sufficiently long period of time for the statistics to converge to an acceptable level.

The sample scripts provided along with the SOWFA package serve as useful starting points for illustrating the process of conducting simulations of flows inside wind farms with realistic inflow and boundary conditions.

5.3 Sample Results

5.3.1 Cases Simulated and Boundary Conditions

A numerical study is conducted on a wind farm consisting of 24 NREL 5 MW wind turbines. Each turbine, with a diameter (D) of 126 m and a hub height (H) of 90 m, is situated within a cuboid-shaped wind farm. This farm spans 50 D in the X direction, 40 D in the Y direction, and rises to a height of 8D along the Z direction, as illustrated in Fig. 5.6. The turbines are organized in a grid pattern of 4 columns and 6 rows, with adjacent columns and rows separated by a distance of 5 D. We assume neutral stability conditions by setting the ground surface heat flux equal to zero. The temperature is constant at 300 K up to a height of 700 m from the bottom surface and varies linearly with a slope of 0.05 K/m from 700 m to 800 m. This leads to a strong capping inversion at the top of the atmospheric boundary layer. Above 800 m to the top of the domain, the temperature increases at the rate of 0.003 K/m, where additional forcing is employed which acts as a numerical damping layer to prevent reflection of waves from the top boundary. The driving pressure gradient is adjusted such that the wind speed at the height of 100 m attains a value of 8.0 m/s.

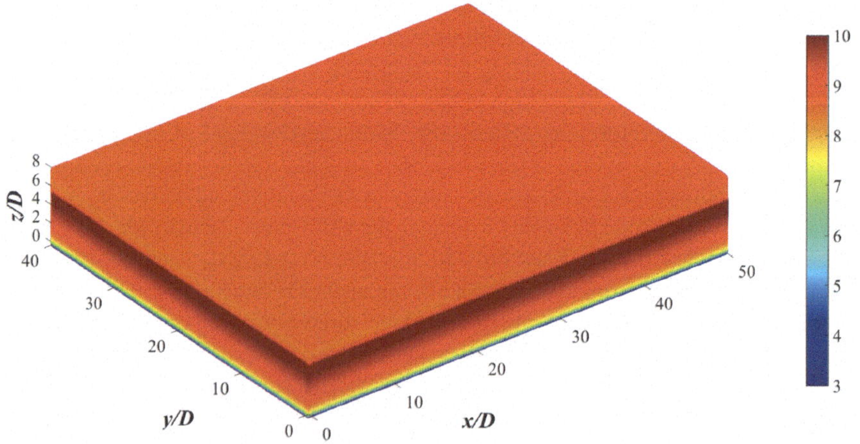

Fig. 5.6 Three-dimensional representation of the domain employed in the current investigation, featuring time-averaged velocity contours, $\overline{\overline{V}}_j$ (m/s)

In the current scenario, the Y–Z plane located at the front ($x/D = 0$) acts as the inflow boundary for the wind farm, while the opposite plane ($x/D = 50$) is set as the outflow boundary. The lateral boundaries ($y/D = 0$ and 40) are defined as cyclic boundary conditions. Additionally, large-scale turnover times of 125 units are chosen to facilitate both the spin-up and averaging stages of the wind plant simulation. Here, each time unit corresponds to D/u_*, with D and u_* representing the rotor diameter and friction velocity, respectively.

The SOWFA solver has been well-validated over the years for a single isolated turbine (Fleming et al. 2018; Zhang and Zhao 2020; Kale et al. 2022) and for large but finite-sized wind farms (Archer et al. 2013; Yan and Archer 2018; Pan and Archer 2018), leading to the fully developed wind turbine array boundary layer, in neutral, unstable, and stable conditions of differing strengths. In this section, we examine the results obtained from a numerical study conducted on a wind farm comprising 24 turbines. The study investigates the effects of four distinct yawed conditions, denoted by yaw angles (γ) of 0°, 15°, 30°, and 45°, on the performance of the wind farm.

5.3.2 Grid Sensitivity and Validation

Figure 5.7 depicts the normalized power output characteristics of a single NREL 5 MW turbine across a range of yaw misalignment angles (γ), specifically $\gamma = 0°$, 10°, 20°, 30°, and 40°. Normalization involves the division of the turbine's power output under varied yaw conditions by that of the reference non-yawed condition. These normalized results are juxtaposed with empirical data from Bastankhah and Porté-Agel (2015), outcomes derived from the large-eddy simulation (LES) study conducted by Ciri et al. (2018), and predictions based on the cosine law of power decay (Adaramola and Krogstad 2011). Our analysis reveals a substantial agreement between the outcomes derived from our numerical simulations and the existing literature, emphasizing the credibility and accuracy of the presented findings.

Sensitivity to the size of the computational grid is conducted. Simulations are performed on three different meshes, labeled M1, M2, and M3, with details provided in Table 5.1. Here, N_{cell} denotes the number of control volumes in the simulation domain, whereas $\Delta x_{\min}$, $\Delta y_{\min}$, and $\Delta z_{\min}$ refer to the minimum cell sizes in x, y, and z directions, respectively. The grid sensitivity is quantified by computing the normalized percentage difference (δU_{diff}) between streamwise profiles of the centerline velocity on successive grids. For example, the percentage change between grids M1 and M2 is

$$\delta U_{\text{diff}} = \frac{U_{M1} - U_{M2}}{U_{M1} + U_{M2} + U_{M3}} \tag{5.2.8}$$

And similarly, for between $M2$ and $M3$. Figure 5.7b shows that the velocity profile changes by about ±2.5 % as the mesh changes from $M2$ to $M3$. Comparing the

Table 5.1 Details of different block-structured meshes, along with minimum grid size in each direction.

Mesh	N_{cell}	Δx_{min}	Δy_{min}	Δz_{min}	Computational time (CPU-hrs)
M1	7,511,719	0.120 D	0.103 D	0.020 D	13,500
M2	14,134,074	0.098 D	0.084 D	0.016 D	30,000
M3	42,667,076	0.078 D	0.0067 D	0.013 D	71,000

CPU time is measured on a cluster with Intel Xeon "Cascadelake" processors with a clock speed of 2.9 GHz and with a 100 Gbps Infiniband interconnect.

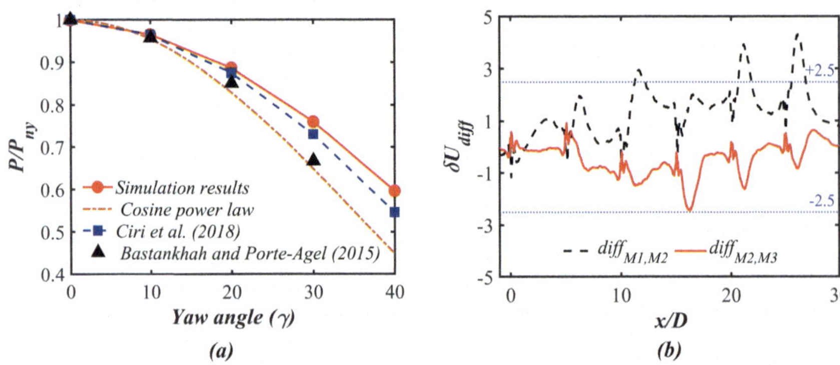

Fig. 5.7 **a** Variation of normalized power output of an isolated turbine with yaw misalignment; **b** Differences in streamwise velocity along the streamwise direction at hub height across various mesh configurations

number of control volumes, and consequently, the computational cost in CPU-hours, it is evident that Mesh $M2$ is the optimum mesh with respect to grid convergence as well as computational cost. We select this mesh resolution for subsequent computations.

5.3.3 *Effect of Yaw Misalignment*

The performance of the 6 × 4 wind turbine array wind farm with four different yaw misalignment angles ($\gamma = 0°, 15°, 30°$ and $45°$) is compared in this section. Figure 5.8 shows the contours of the time-averaged velocity field ($\overline{\overline{V}}_j$). For $\gamma = 0°$, the normal to the rotor swept area aligns with the wind flow direction. Each upstream turbine is generating power as per its design conditions. However, the wake losses are very high in this case, because the wake region of each upwind turbine fully envelops the respective downstream turbines. The downstream turbines are under what can be termed as "full wake" conditions (Fig. 5.8a). The reduced velocity incident on the

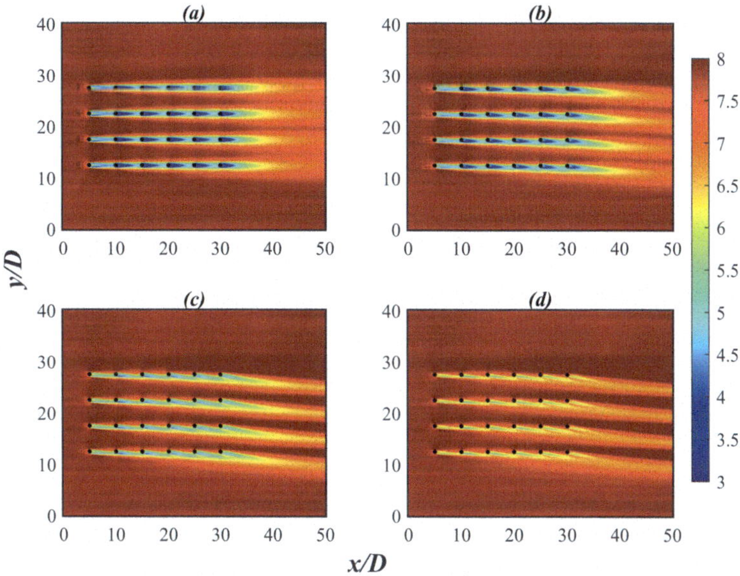

Fig. 5.8 Time-averaged velocity, $\overline{\overline{V}}_j$ (m/s), contours at the hub height with wind flow directions $\phi = 0°$ for different yaw misalignment angles, i.e., **a** $\gamma = 0°$, **b** $\gamma = 15°$, **c** $\gamma = 30°$, and **d** $\gamma = 45°$

fully waked turbines reduces the power output of these downstream turbines. Since there are large numbers of fully waked turbines, the overall power output of the wind farm is reduced significantly.

Yawing the wind turbines reduces the power output of the upstream turbines. This is because only that portion of the incoming wind velocity that is along the normal to the rotor plane can be harnessed. The component of the wind velocity in the rotor plane does not get converted to electricity. However, as stated earlier, a consequence of introducing a yaw misalignment in a turbine leads to the wake getting deflected in the direction opposite to the direction of the yaw. In these cases, only a portion of each downwind turbine is covered by the wake region of the upstream turbine. This situation is termed as "partial wake" condition. Figure 5.8(b, c, d) show that several wind turbines are partially waked when $\gamma = 15°$, 30° and 45°. Due to the wake deflection, the wake effects of the upstream turbines on the downstream turbines are reduced.

Figure 5.9 shows the variation of streamwise velocity and turbulence intensity in downstream of the turbine along a representative column in each wind farm. At each turbine location (dotted vertical lines), there is a sharp reduction in wind velocity because of the extraction of kinetic energy from the freestream flow by the turbine (Fig. 5.9a). At the same time, the wind becomes more turbulent (Fig. 5.9b). As the streamwise distance from a turbine increases, the wind speed increases and the turbulence intensity reduces as the wake recovers because of continuous entrainment and interaction with the surrounding atmospheric wind. This wake recovery continues

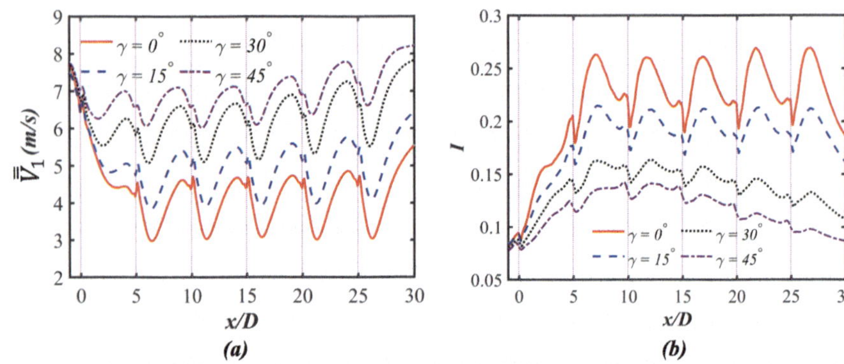

Fig. 5.9 Variation of time-averaged **a** streamwise velocity component, $\overline{\overline{V}}_1$ **b** turbulence intensity, I in the downstream of wind turbine for different values of yaw misalignment

till the next downstream turbine is encountered, which leads to the formation of a new combined wake. This pattern is observed repeatedly in the downstream direction. With a change in yaw misalignment from $\gamma = 0°–45°$, the streamwise velocity curves monotonically move up while the turbulence intensity curves all move down. This indicates that the wakes become weaker, and the turbulence levels reduce as the yaw misalignment is increased.

Figure 5.10a shows the average power output of the wind farm for different yaw angles. It can be observed that for a closely packed 6×4 arrangement with mean wind direction parallel to the column of turbines, the average power output of the wind farm increases with an increase in the yaw misalignment by 15° and remains relatively stable for $\gamma = 30°$. This is explained by the fact that due to the spanwise deflections of the turbine wakes, wake losses of the downstream turbines are reduced and the overall power output of the wind farm as a whole is increased. However, for $\gamma = 45°$, there is a sharp decrease in the power output of the wind farm, even though the wake losses at downstream turbines are smaller even than those for $\gamma = 30°$. This is because the increased power generation of all the downwind turbines is insufficient to compensate for the reduced power generation of the upwind turbines.

The averaged flow turbulence intensity inside the domain, shown in Fig. 5.10b, keeps on decreasing until $\gamma = 45°$, which indicates that the fatigue load on turbines is significantly reduced with an increase in yaw misalignment, ensuring the long service life of the turbine. Figure 5.10a and b collectively point to the compromise that must be made between the expected power output and the expected lifetime of the wind farm. For example, for the configurations studied, selecting $\gamma = 30°$ appears beneficial because it yields the highest power output but with a much lower turbulence intensity than for the default, non-yawed wind farm.

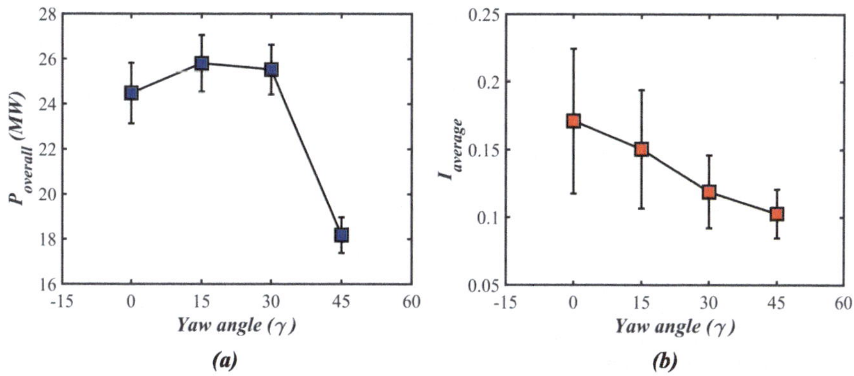

Fig. 5.10 Plot for **a** Overall power output, P_{overall} **b** average turbulence intensity, I_{average} in the wind farm when wind turbines are placed under different yawed conditions

5.4 Discussion and Conclusions

This chapter discussed some of the factors influencing the efficiency of wind farms. A significant portion of energy loss within farms can be attributed to wake losses, highlighting the importance of understanding and mitigating these phenomena. Our analysis indicates that increasing the inter-turbine distance can enhance wind farm performance. However, the practical implementation of these adjustments is constrained by both financial and physical limitations. Additionally, a change in wind direction changes the number of turbines exposed to fresh and undisturbed wind. This significantly affects the flow field inside the wind farm and, thereby, the overall power production.

This chapter has demonstrated the use of the CFD package SOWFA for studying the effect of yaw misalignment on the wind farm performance. In particular, LES studies of a medium-sized wind farm containing 24 turbines in a 6 × 4 arrangement are discussed. It has been observed that an increase in yaw misalignment (γ) deflects the wind turbine wake in a direction opposite to the yaw, thereby reducing the wake effects of upstream turbines on downstream turbines. It helps in improving the power output of downwind turbines, thereby increasing their contribution to the overall power output of the wind farm. Higher values of yaw misalignment reduce the turbulence intensity and velocity deficit across the wind turbine. An increase in yaw misalignment up to 30 degrees reduces the power of the upwind turbines but promotes the overall power output of the wind farms. It also reduces the average turbulence inside the domain, thereby reducing fatigue loads on downstream turbines and ensuring longer their service life. However, with a further increase in the yaw misalignment ($\gamma = 45°$), the power output of downstream turbines is significantly reduced, which results in lower power output of the wind farm.

Numerical investigation of wind farms can provide detailed information about the flow dynamics inside the domain. However, the computational cost involved with each CFD simulation is exorbitant, which constrains wind farm designers from performing a large number of simulations and precludes the use of high-fidelity CFD calculations as the primary design tool. The result of these CFD simulations can, however, be used to develop simple but accurate analytical and machine learning models, which are suitable as design tools.

Acknowledgments Utkarsh Deep Tiwari kindly acknowledges the support provided by the Prime Minister's Research Fellowship (PMRF), Ministry of Education, Government of India. Niranjan S. Ghaisas thanks the Science and Engineering Research Board (SERB) Core Research Grant (CRG/2022/006735). Kishalay Mitra would like to thank the Department of Science and Technology (DST), Government of India, for the support provided by the project (DST/NSM/R&D_HPC_Applications/2021/23) under the National Supercomputing Mission (NSM). The authors thank computational resources on ParamSeva at IIT Hyderabad.

References

Abkar M, Porté-Agel F (2015) Influence of atmospheric stability on wind-turbine wakes: a large-eddy simulation study. Phys Fluids 27. https://doi.org/10.1063/1.4913695

Adaramola MS, Krogstad PÅ (2011) Experimental investigation of wake effects on wind turbine performance. Renew Energy 36(8):2078–2086. https://doi.org/10.1016/j.renene.2011.01.024

Aditya G, Subramaniam A, Niranjan Ghaisas S (2018) Michael Howland, and Jacob West. 'PadeOps'

Albadi MH, El-Saadany EF (2010) Overview of wind power intermittency impacts on power systems. Elect Power Syst Res 80(6):627–632. https://doi.org/10.1016/j.epsr.2009.10.035

Archer CL, Mirzaeisefat S, Lee S (2013) Quantifying the sensitivity of wind farm performance to array layout options using large-eddy simulation. Geophys Res Lett 40(18):4963–4970. https://doi.org/10.1002/grl.50911

Bentley System. n.d. 'OpenWindPower'

Barthelmie RJ, Jensen LE (2010) Evaluation of wind farm efficiency and wind turbine wakes at the Nysted offshore wind Farm. Wind Energy 13(6):573–586. https://doi.org/10.1002/we.408

Bastankhah M, Porté-Agel F (2015) A wind-tunnel investigation of wind-turbine wakes in yawed conditions. J Phys Conf Series 625(June):012014. https://doi.org/10.1088/1742-6596/625/1/012014

Bastankhah M, Porte-Agel F (2016) Experimental and theoretical study of wind turbine wakes in yawed conditions. J Fluid Mech 806(Nov):506–541. https://doi.org/10.1017/jfm.2016.595

Bhaganagar K, Debnath M (2014) Implications of stably stratified atmospheric boundary layer turbulence on the near-wake structure of wind turbines. Energies 7(9):5740–5763. https://doi.org/10.3390/en7095740

Central Electricity Authority (2023) 'Report on optimal generation capacity mix for 2029-30'. https://cea.nic.in/wp-content/uploads/irp/2023/05/Optimal_mix_report__2029_30_Version_2.0__For_Uploading.pdf

Choi NJ, Nam SH, Jeong JH, Kim KC (2013) Numerical study on the horizontal axis turbines arrangement in a wind farm: effect of separation distance on the turbine aerodynamic power output. J Wind Eng Ind Aerodyn 117(June):11–17. https://doi.org/10.1016/j.jweia.2013.04.005

Choi N, Nam S, Jeong J, Kim K (2014) CFD Study on aerodynamic power output changes with inter-turbine spacing variation for a 6 MW offshore wind farm. Energies 7(11):7483–7498. https://doi.org/10.3390/en7117483

Churchfield MJ, Lee S, Moriarty PJ (2014) Adding complex terrain and stable atmospheric condition capability to the OpenFOAM-based flow solver of the simulator for on/offshore wind farm applications (SOWFA). ITM Web Conf 2(February):02001. https://doi.org/10.1051/itmconf/20140202001

Ciri U, Rotea MA, Leonardi S (2018) Effect of the turbine scale on yaw control. Wind Energy 21(12):1395–1405. https://doi.org/10.1002/we.2262

Cotrell J, Stehly T, Johnson J, Roberts JO, Parker Z, Scott G, Heimiller D (2014) Analysis of transportation and logistics challenges affecting the deployment of larger wind turbines: summary of results. Golden CO United States. https://doi.org/10.2172/1123207

Das P, Mathur J, Bhakar R, Kanudia A (2018) Implications of short-term renewable energy resource intermittency in long-term power system planning. Energy Strategy Rev 22(November):1–15. https://doi.org/10.1016/j.esr.2018.06.005

Deardorff JW (1980) Stratocumulus-capped mixed layers derived from a three-dimensional model. Boundary-Layer Meteorol 18(4):495–527. https://doi.org/10.1007/BF00119502

Diógenes JR, Farkat JC, Rodrigues JC, Loureiro MV (2020) Barriers to onshore wind energy implementation: a systematic review. Energy Res Soc Sci 60(February):101337. https://doi.org/10.1016/j.erss.2019.101337

Du B, Ge M, Zeng C, Cui G, Liu Y (2021) Influence of atmospheric stability on wind-turbine wakes with a certain hub-height turbulence intensity. Phys Fluids 33(5):055111. https://doi.org/10.1063/5.0050861

Fleming PA, Gebraad PMO, Lee S, van Wingerden J-W, Johnson K, Churchfield M, Michalakes J, Spalart P, Moriarty P (2014) Evaluating techniques for redirecting turbine wakes using SOWFA. Renew Energy 70:211–218. https://doi.org/10.1016/j.renene.2014.02.015

Fleming P, Annoni J, Churchfield M, Martinez-Tossas LA, Gruchalla K, Lawson M, Moriarty P (2018) A simulation study demonstrating the importance of large-scale trailing vortices in wake steering. Wind Energy Sci 3(1):243–255. https://doi.org/10.5194/wes-3-243-2018

Ghaisas NS, Archer CL (2016) Geometry-based models for studying the effects of wind farm layout. J Atmos Ocean Technol 33(3):481–501. https://doi.org/10.1175/JTECH-D-14-00199.1

Ghaisas NS, Archer CL, Xie S, Sicheng W, Maguire E (2017) Evaluation of layout and atmospheric stability effects in wind farms using large-eddy simulation. Wind Energy 20(7):1227–1240. https://doi.org/10.1002/we.2091

González-Longatt F, Wall P, Terzija V (2012) Wake effect in wind farm performance: steady-state and dynamic behavior. Renew Energy 39(1):329–338. https://doi.org/10.1016/j.renene.2011.08.053

Guirguis D, Romero DA, Amon CH (2016) 'Efficient wind turbine micrositing in large-scale wind farms'. In: American society of mechanical engineers. https://doi.org/10.1115/ES2016-59594

Hansen M (2015) Aerodynamics of wind turbines. Routledge. https://doi.org/10.4324/9781315769981

Howland MF, González CM, Martínez JJP, Quesada JB, Larrañaga FP, Yadav NK, Chawla JS, Dabiri JO (2020) Influence of atmospheric conditions on the power production of utility-scale wind turbines in yaw misalignment. J Renew Sustain Energy 12:063307. https://doi.org/10.1063/5.0023746

Invest India (2023) 'Renewable Energy'. National investment promotion & facilitation agency. Accessed September 25. https://www.investindia.gov.in/sector/renewable-energy

Jha P, Churchfield M, Moriarty P, Schmitz S (2013) 'Accuracy of state-of-the-art actuator-line modeling for wind turbine wakes'. In 51st AIAA Aerospace sciences meeting including the new horizons forum and aerospace exposition. Reston, Virigina: American Institute of Aeronautics and Astronautics. https://doi.org/10.2514/6.2013-608

Jiménez Á, Crespo A, Migoya E (2009) Application of a LES technique to characterize the wake deflection of a wind turbine in yaw. Wind Energy 13:559–572. https://doi.org/10.1002/we.380

Jonkman J, Butterfield S, Musial W, Scott G (2009) Definition of a 5-MW reference wind turbine for offshore system development. Golden CO. https://doi.org/10.2172/947422

Kale B, Buckingham S, van Beeck J, Cuerva-Tejero A (2022) Implementation of a generalized actuator disk model into WRF v4.3: a validation study for a real-scale wind turbine. Renew Energy 197(September):810–827. https://doi.org/10.1016/j.renene.2022.07.119

Kalvig S, Manger E, Hjertager B (2014) Comparing different CFD wind turbine modelling approaches with wind tunnel measurements. J Phys Conf Series 555(December):012056. https://doi.org/10.1088/1742-6596/555/1/012056

Katsaprakakis DAl, Papadakis N, Ntintakis I (2021) A comprehensive analysis of wind turbine blade damage. Energies 14(18):5974. https://doi.org/10.3390/en14185974

Kazacoks R, Amos L, Leithead PW (2019) Effect of wind flow direction on the loads at wind farm. J Phys Conf Series 1356(1):012005. https://doi.org/10.1088/1742-6596/1356/1/012005

Kim DY, Kim YH, Kim BS (2021) Changes in wind turbine power characteristics and annual energy production due to atmospheric stability, turbulence intensity, and wind shear. Energy 214(January):119051. https://doi.org/10.1016/j.energy.2020.119051

Knopper LD, Ollson CA, McCallum LC, Whitfield Aslund ML, Berger RG, Souweine K, McDaniel M (2014) Wind turbines and human health. Front Public Health. https://doi.org/10.3389/fpubh.2014.00063

Koch F, Gresch M, Shewarega F, Erlich I, Bachmann U (2005) 'Consideration of wind farm wake effect in power system dynamic simulation'. In: 2005 IEEE Russia Power Tech, IEEE. pp 1–7. https://doi.org/10.1109/PTC.2005.4524572

Kosović B, Curry JA (2000) A large eddy simulation study of a quasi-steady, stably stratified atmospheric boundary layer. J Atmos Sci 57(8):1052–1068

Kumara HN, Babu S, Babu Rao G, Mahato S, Bhattacharya M, Rao NVR, Tamiliniyan D et al (2022) Responses of birds and mammals to long-established wind farms in India. Sci Rep 12(1):1339. https://doi.org/10.1038/s41598-022-05159-1

Kusiak A, Song Z (2010) Design of wind farm layout for maximum wind energy capture. Renew Energy 35(3):685–694. https://doi.org/10.1016/j.renene.2009.08.019

Lee J, Choi H, Park N (2010) Dynamic global model for large eddy simulation of transient flow. Phys Fluids. https://doi.org/10.1063/1.3459156

Liu L, Stevens RJ (2021) Effects of atmospheric stability on the performance of a wind turbine located behind a three-dimensional hill. Renew Energy 1(175):926–935. https://doi.org/10.1016/j.renene.2021.05.035

Machefaux E, Larsen GC, Koblitz T, Troldborg N, Kelly MC, Chougule A, Hansen KS, Rodrigo JS (2016) An experimental and numerical study of the atmospheric stability impact on wind turbine wakes. Wind Energy 19(10):1785–1805. https://doi.org/10.1002/we.1950

Martinez L, Leonardi S, Churchfield M, Moriarty PA (2012) 'A Comparison of actuator disk and actuator line wind turbine models and best practices for their use'. In: 50th AIAA Aerospace sciences meeting including the new horizons forum and aerospace exposition. Reston, Virigina: American Institute of Aeronautics and Astronautics. https://doi.org/10.2514/6.2012-900

Masson C, Smaïli A, Leclerc C (2001) Aerodynamic analysis of HAWTs operating in unsteady conditions. Wind Energy 4(1):1–22. https://doi.org/10.1002/we.43

Meneveau C, Lund TS, Cabot WH (1996) A lagrangian dynamic subgrid-scale model of turbulence. J Fluid Mech 319(1):353. https://doi.org/10.1017/S0022112096007379

Meyers J, Meneveau C (2012) Optimal turbine spacing in fully developed wind farm boundary layers. Wind Energy 15(2):305–317. https://doi.org/10.1002/we.469

Moeng CH (1984) A large-eddy-simulation model for the study of planetary boundary-layer turbulence. J Atmos Sci 41(13):2052–2062

Moin P, Squires K, Cabot W, Lee S (1991) A dynamic subgrid-scale model for compressible turbulence and scalar transport. Phys Fluids A Fluid Dyn 3(11):2746–2757. https://doi.org/10.1063/1.858164

'Mojave Wind Rose (1986-2024)'. n.d. Midwestern Regional Climate Center (1986-2024)

Mustafa AM, Al-Mahadin A (2018) 'Risk assessment of hazards due to the installation and maintenance of onshore wind turbines'. In: 2018 Advances in science and engineering technology international conferences (ASET), IEEE. pp 1–7. https://doi.org/10.1109/ICASET.2018.8376789

Nicoud F, Toda HB, Cabrit O, Bose S, Lee J (2011) Using singular values to build a subgrid-scale model for large eddy simulations. Phys Fluids. https://doi.org/10.1063/1.3623274

Oerlemans S (2021) 'Wind turbine noise mitigation'. In: Handbook of wind energy aerodynamics, Cham: Springer International Publishing. pp 1–16, https://doi.org/10.1007/978-3-030-05455-7_73-1

Pan Y, Archer CL (2018) A hybrid wind-farm parametrization for mesoscale and climate models. Boundary-Layer Meteorol 168(3):469–495. https://doi.org/10.1007/s10546-018-0351-9

Porté-Agel F, Wu Y-T, Chen C-H (2013) A Numerical Study of the Effects of Wind Direction on Turbine Wakes and Power Losses in a Large Wind Farm. Energies (Basel) 6:5297–5313. https://doi.org/10.3390/en6105297

Porté-Agel F, Bastankhah M, Shamsoddin S (2020) Wind-turbine and wind-farm flows: a review. Boundary-Layer Meteorol 174(1):1–59. https://doi.org/10.1007/s10546-019-00473-0

Rozema W, Bae HJ, Moin P, Verstappen R (2015) Minimum-dissipation models for large-eddy simulation. Phys Fluids. https://doi.org/10.1063/1.4928700

Sanchez Gomez M, Lundquist JK (2020) The effect of wind direction shear on turbine performance in a wind farm in central IOWA. Wind Energy Sci 5(1):125–139. https://doi.org/10.5194/wes-5-125-2020

Sanderse B, Pijl SP, Koren B (2011) Review of computational fluid dynamics for wind turbine wake aerodynamics. Wind Energy 14(7):799–819. https://doi.org/10.1002/we.458

Schumann U (1975) Subgrid scale model for finite difference simulations of turbulent flows in plane channels and annuli. J Comput Phys 18(4):376–404. https://doi.org/10.1016/0021-9991(75)90093-5

Shen WZ, Sørensen JN, Zhang J (2007). 'Actuator surface model for wind turbine flow computations'. In: Proceedings of European wind energy conference. https://core.ac.uk/download/pdf/13720989.pdf

Smagorinsky J (1963) General circulation experiments with the primitive equations. Monthly Weather Rev 91(3):99–164

Smallwood KS, Bell DA (2020) Effects of wind turbine curtailment on bird and bat fatalities. J Wildlife Manag 84(4):685–696. https://doi.org/10.1002/jwmg.21844

UL Solutions. n.d. 'Openwind'

Sørensen JN, Myken A (1992) Unsteady actuator disc model for horizontal axis wind turbines. J Wind Eng Ind Aerodyn 39(1–3):139–149. https://doi.org/10.1016/0167-6105(92)90540-Q

Sørensen JN, Shen WZ (2002) Numerical modeling of wind turbine wakes. J Fluids Eng 124(2):393–399. https://doi.org/10.1115/1.1471361

Stevens RJAM, Gayme DF, Meneveau C (2014) Large eddy simulation studies of the effects of alignment and wind farm length. J Renew Sustain Energy. https://doi.org/10.1063/1.4869568

Stevens RJAM, Hobbs BF, Ramos A, Meneveau C (2017) Combining economic and fluid dynamic models to determine the optimal spacing in very large wind farms. Wind Energy 20(3):465–477. https://doi.org/10.1002/we.2016

Stevens RJAM, Martínez-Tossas LA, Meneveau C (2018) Comparison of wind farm large eddy simulations using actuator disk and actuator line models with wind tunnel experiments. Renew Energy 116(February):470–478. https://doi.org/10.1016/j.renene.2017.08.072

Stieren A, Kasper JH, Gadde SN, Stevens RJAM (2022) Impact of negative geostrophic wind shear on wind farm performance. PRX Energy 1(2):023007. https://doi.org/10.1103/PRXEnergy.1.023007

Tavner P, Edwards C, Brinkman A, Spinato F (2006) Influence of wind speed on wind turbine reliability. Wind Eng 30(1):55–72. https://doi.org/10.1260/030952406777641441

Troldborg N, Zahle F, Réthoré PE, Sørensen NN (2015) Comparison of wind turbine wake properties in non-sheared inflow predicted by different computational fluid dynamics rotor models. Wind Energy 18(7):1239–1250. https://doi.org/10.1002/we.1757

Uchida T, Gagnon Y (2022) Effects of continuously changing inlet wind direction on near-to-far wake characteristics behind wind turbines over flat terrain. J Wind Eng Ind Aerodyn 220(January):104869. https://doi.org/10.1016/j.jweia.2021.104869

Vanderwende BJ, Lundquist JK (2012) The modification of wind turbine performance by statistically distinct atmospheric regimes. Environ Res Lett 7(3):034035. https://doi.org/10.1088/1748-9326/7/3/034035

Vreman AW (2004) An eddy-viscosity subgrid-scale model for turbulent shear flow: algebraic theory and applications. Phys Fluids 16(10):3670–3681. https://doi.org/10.1063/1.1785131

Wagner R, Antoniou I, Pedersen SM, Courtney MS, Jørgensen HE (2009) The influence of the wind speed profile on wind turbine performance measurements. Wind Energy 12(4):348–362. https://doi.org/10.1002/we.297

Wang S, Wang S, Smith P (2015) Ecological impacts of wind farms on birds: questions, hypotheses, and research needs. Renew Sustain Energy Rev 44(April):599–607. https://doi.org/10.1016/j.rser.2015.01.031

Wharton S, Lundquist JK (2012a) Atmospheric stability affects wind turbine power collection. Environ Res Lett 7(1):014005. https://doi.org/10.1088/1748-9326/7/1/014005

Wharton S, Lundquist JK (2012b) Assessing atmospheric stability and its impacts on rotor-disk wind characteristics at an onshore wind farm. Wind Energy 15(4):525–546. https://doi.org/10.1002/we.483

Wu YT, Porté-Agel F (2011) Large-eddy simulation of wind-turbine wakes: evaluation of turbine parametrisations. Boundary-Layer Meteorol 138(3):345–366. https://doi.org/10.1007/s10546-010-9569-x

Wu YT, Porté-Agel F (2015) Modeling turbine wakes and power losses within a wind farm using LES: an application to the horns rev offshore wind farm. Renew Energy 75(March):945–955. https://doi.org/10.1016/j.renene.2014.06.019

Xie S, Archer C (2015) Self-similarity and turbulence characteristics of wind turbine wakes via large-eddy simulation. Wind Energy 18(10):1815–1838. https://doi.org/10.1002/we.1792

Xie S, Archer CL (2017) A numerical study of wind-turbine wakes for three atmospheric stability conditions. Boundary-Layer Meteorol 165(1):87–112. https://doi.org/10.1007/s10546-017-0259-9

Yan C, Archer CL (2018) Assessing compressibility effects on the performance of large horizontal-axis wind turbines. Appl Energy 212(February):33–45. https://doi.org/10.1016/j.apenergy.2017.12.020

Yang X, Sotiropoulos F (2018) A new class of actuator surface models for wind turbines. Wind Energy 21(5):285–302. https://doi.org/10.1002/we.2162

You D, Moin P (2007) A dynamic global-coefficient subgrid-scale eddy-viscosity model for large-eddy simulation in complex geometries. Phys Fluids. https://doi.org/10.1063/1.2739419

Zhang J, Zhao X (2020) A novel dynamic wind farm wake model based on deep learning. Appl Energy 277(November):115552. https://doi.org/10.1016/j.apenergy.2020.115552

Chapter 6
Synergizing Machine Learning with Physics-Based Modelling: A Unified Approach for Characterizing Wake Effects

NagaSree Keerthi Pujari, Srinivas Soumitri Miriyala, and Kishalay Mitra

Abstract Effects of wake from turbines significantly influencing the power production in a wind farm are crucial part in wind farm design. Several wake models starting from the analytical to the Computational Fluid Dynamics (CFD) are developed in the literature. The analytical models are simple but not accurate, whereas large eddy simulations (LES) models are accurate but time-consuming. The major concern of analytical models is that they consider linear wake growth rate which might not be true. To address this issue, a data-driven wake model under the machine learning framework is proposed in this work to reconcile the limitations of analytical wake models. Further, the data-driven wake model is compared with the power produced through analytical models provided in the literature such as Jensen model to validate its efficiency. The proposed model has shown better accuracy when compared with other models, justifying the significance of nonlinear growth rate of wake and its impact on power production through wind farm.

Keywords Nonlinear wake expansion · Machine learning · Artificial neural networks · Jensen wake model

N. K. Pujari · S. S. Miriyala · K. Mitra (✉)
Department of Chemical Engineering, Indian Institute of Technology Hyderabad, Hyderabad, Telangana 502284, India
e-mail: kishalay@che.iith.ac.in

N. K. Pujari
e-mail: ch20resch11006@iith.ac.in

K. Mitra
Department of Climate Change, Indian Institute of Technology Hyderabad, Hyderabad, Telangana 502284, India

Department of Artificial Intelligence, Indian Institute of Technology Hyderabad, Hyderabad, Telangana 502284, India

K. Mitra et al. (eds.), *Optimization, Uncertainty and Machine Learning in Wind Energy Conversion Systems*, Engineering Optimization: Methods and Applications,
https://doi.org/10.1007/978-981-97-7909-3_6

6.1 Introduction

The climate change and energy shortage caused by the conventional fossil fuels drove the industrialists and researchers towards the renewable energy sources. Wind energy is experiencing rapid growth among all renewable energy sources, attributed to its abundance and capacity for clean energy generation. Wind energy offers significant potential in reducing long-term and short-term greenhouse gas emissions and reduces the usage of fossil fuels. With the new addition of 94 GW, the total global wind power capacity has reached 837 GW which avoids 1.2 billion tonnes of CO_2 annually (GWEC 2023). To make use of the full potential of the wind energy, many turbines have been installed to extract the wind energy and the size of wind farms is growing worldwide from the past few decades. However, in any wind farm, many of the turbines operate under the influence of upwind turbines, thus exposing to lower incoming wind velocities. The aerodynamic interactions between the wind turbines lead to significant reduction in power production of downstream turbines, thus leading to lower power production from a wind farm. The downstream of the turbine exhibits turbulence and reduced velocity as the energy is extracted and is referred as wake (Archer et al. 2018). Since the wake flow contains an accumulated turbulence from upwind turbines as well as atmosphere, it will impose structural, and fatigue loads on downwind turbines. The wake effects account for a power loss of 46% and are identified through experimental investigations, demonstrating the significance of wake characterizations in wind farm design (Adaramola and Krogstad 2011). Since the turbines are always prone to wake effects in the large wind farms, it will lead to 10–20% power reduction from a wind farm (Barthelmie et al. 2007). One solution to minimize these wake effects is to increase the spacing between the turbines so that the wake recovers back to the free stream wind speeds, but it leads to complicated and costlier maintenance due to longer lines for transmission and large space. The strategic design of a wind farm helps in optimal energy generation, and therefore, the research community has proposed many strategies to minimize wake effect, which includes the design and control of wind farm. In wind farm design studies or control studies, the key challenge is to leverage good models to study the wake interactions for efficient energy production from a wind farm. Therefore, the research of wake interactions has been presented in the existing literature at different levels of prediction granularity (analytical to LES) (Sedaghatizadeh et al. 2018; Tao et al. 2020; Kaldellis et al. 2021).

The analytical models are simple and computationally efficient and are widely used by the wind energy community. Though the analytical models are not precise when compared to high fidelity models, they will provide the fundamental insights of physics as they rely on the basic equations which govern the conservation of flow properties (Porté-Agel et al. 2019). Many studies focussing on developing analytical wake models are present in the literature. Jensen model, proposed by Jensen (1983), assumes a top-hat distribution for the velocity deficit with linear growth rate of wake. But the Jensen model considers only mass conservation and the top-hat velocity deficit profile led to inaccuracies. Another analytical model was proposed by

Frandsen based on conservation of mass and momentum around the control volume of turbine assuming a top-hat profile of velocity deficits (Frandsen et al. 2006). The assumption of top-hat profile for velocity deficit underestimates the velocity deficit and leads to substantial errors. A self-similar Gaussian profile can be more appropriate and can provide a better representation of velocity deficit in far wake (Chamorro and Porté-Agel 2009). Therefore, another analytical model was proposed by Bastankhah and Porté-Agel based on conservation of mass and momentum and Gaussian profile for velocity deficit (Bastankhah and Porté-Agel 2014). Though the model doesn't underestimate the velocity profile, an estimate for the wake growth rate needs to be determined based on inflow turbulence characteristics. The streamwise turbulence intensity and wake growth rate modelled with an empirical relation based on LES have been proposed by Niayifar and Porté-Agel (2016). A three-dimensional wake model was introduced utilizing law of conservation of flow flux and accounting the wind direction variations with height while improving the accuracy (Sun and Yang 2018). However, analytical models still impose inaccuracies as they have simplified rotor-flow interaction mechanics, absence of turbulence modelling, dependence on empirical constants (Wu and Porté-Agel 2015; Niayifar and Porté-Agel 2016).

Another method for efficiently characterizing wakes involves employing CFD, which encompasses turbulence modelling, boundary modelling and rotor-flow interactions. To improve the flow resolutions, CFD techniques such as LES, Reynolds-Averaged Navier Stokes equations (RANS) have been used to model the wake characteristics (Verbrugge et al. 2016; Sedaghatizadeh et al. 2018). A CFD model was proposed based on actuator disc model (ADM) and approximating the surface boundary layer while simulating the wind turbines and ambient conditions. Then, the results are processed with a Gaussian weighted accounting for uncertainty in wind direction and are compared with the experimental data to prove the efficacy of the proposed model (Antonini et al. 2019). Six widely used wake modelling approaches which include analytical and CFD wake models were presented and the subcomponents are analysed using the data obtained from onshore and offshore wind farms (Göçmen et al. 2016). A very fine mesh is required to resolve the high turbulence flow which makes LES models to be computationally expensive. RANS simulations, which focus on time-averaged characterization of fluids, result in lower computational costs. To reduce the computational costs while capturing the wake flows, a RANS solver is developed with parabolic and boundary layer approximations for axisymmetric wake flows (Iungo et al. 2018). Boussinesq hypothesis and a mixing length model were used for capturing the turbulence effects of time-averaged wake velocity field. The results are validated with NREL 5 MW turbines proving that the developed solver has good accuracy and lower computational costs (Iungo et al. 2018). Since RANS model can result in fast recovery of wake by overestimating the turbulence viscosity, an ADM is integrated to describe the wake characteristics near to the wake region (Li et al. 2020). Though CFD models are accurate for wake characterization, it is computationally expensive and sophisticated for industrial problems such as designing an optimal layout of wind farm.

The utilization of machine learning tools has been increased over the recent years in several disciplines to solve complex problems. The local inflows and wake expansion are mapped using Random Forest by developing a data-based wake model (Nai-Zhi et al. 2022). Artificial Neural networks (ANN) are utilized to model the inflows and wake flows (Biswas et al. 2016). Since the wake recovery is influenced by turbulence inflows, ANN model is developed by integrating turbulence model with the velocity model which is validated with LES dataset proving the efficiency in predicting the wake losses (Biswas et al. 2016). Machine learning models are utilized to estimate the power accurately considering the wake losses (Göçmen and Giebel 2018). A Convolutional Neural Network is used for building a relationship between the flow field and a pressure around a bluff body (Jin et al. 2018). To summarize, it is necessary to establish a sophisticated relation between rotor parameters, inflow patterns and spatial patterns of wake fields while predicting turbine wakes and power from a wind farm. A significant interest has been shown by industrialists in analytical wake models for characterizing the wake because of their cost-effectiveness. However, conventional analytical wake models primarily concentrate on the wake growth rate and its interactions, making a fundamental presumption regarding linear wake growth rate which is not accurate. The CFD models involve rotor-flow interactions and turbulence modelling with governing equations of flow, thus providing higher accuracy at the cost of higher computations for large-scale wind farms. Although machine learning tools have shown superiority over traditional wake models, insufficient data in near-wake regions could result in inaccuracies. Therefore, machine learning techniques could be integrated with physics-based models while modelling turbulence and expansion rates concerning inflows to enhance accuracy and reduce the computational costs. This study introduces a new data-driven analytical wake model leveraging machine learning tools which creates a balance between computational cost and accuracy. The major drawback in analytical models is the assumption of linear wake expansion. Therefore, the authors have used ANNs to bring in the nonlinear wake characteristics for analytical models to improve the accuracy of the wake characterization. In this work, a pioneering analytical wake model, Jensen model, is used as a foundation for understanding wake dynamics and ANNs are then integrated with it to capture the nonlinear wake growth rate. The proposed model is validated on a real dataset which accounts for wake effects. The unique contributions of this research are outlined as follows, and the salient features are presented in Table 6.1:

- A novel data-based analytical wake model including the nonlinear wake growth rates.
- ANNs are utilized to characterize the nonlinear wake growth rate in the wake models.
- A novel wake model was proposed by superimposing ANNs on physics-based wake model to create the balance between accuracy and computational costs.
- The proposed wake model was validated with the single wake effect dataset.

Table 6.1 Salient features of the proposed Jensen–ANN model

Jensen–ANN: integration of physics-based model with data-based models	
Engineering wake models assume linear expansion of wake $r_w = r_0 + \alpha x$	ANN wake models consider nonlinear expansion of wake (w-weights of ANN) $r_w = f(x, \theta)$; f is ANN
Jensen model	*Jensen–ANN model*
Near wake radius: $r_0 = r_r\sqrt{\frac{1-a}{1-2a}}$	Near wake radius: $r_0 = r_r\sqrt{\frac{1-a}{1-2a}}$
Velocity: $u_w = \left[1 - \frac{2a}{\left(\frac{r_0+\alpha x}{r_0}\right)^2}\right]u_0$	Velocity: $u_w = \left[1 - \frac{2a}{\left(\frac{f(x,\theta)}{r_0}\right)^2}\right]u_0$

In the rest of the chapter, a brief description on dataset, details of the proposed novel wake model along with the descriptions of ANNs and Jensen wake model are presented in Sect. 6.2. Section 6.3 presents the results of the proposed Jensen–ANN wake model. Section 6.4 outlines the summary of the proposed work.

6.2 Formulation

In this section, a brief description about the data and the physics-based wake model along with ANNs are presented.

6.2.1 Data

Wind characteristics data over a span of one year with 10 min resolution was collected for wind speed and direction (Ding 2019). The dataset includes 12 turbines and 3 met masts. The turbines are placed in pairs, and each pair has been chosen such that none of the turbines are placed within 10 times rotor diameter of the turbine pair. The turbines are arranged in such a way to replicate single wake effect, i.e. only the turbine pair has wake effect while the pair of turbines are free from the wake effect of other turbines. The dataset includes wind power, speed and direction. The wind power and speed are measured on the turbine, and the direction is measured at the met masts. In the collected dataset, 70% was used for training the model and 30% was used to test the proposed model.

6.2.2 Jensen Wake Model

As stated earlier, the sophisticated CFD wake models are computationally expensive and the use of CFD models are less practical for commercial wind farm operations. In this study, Jensen wake model is used which is one of the simplest engineering wake models and was proposed by Jensen (1983) with an assumption of linear wake growth rate. The major assumptions of the Jensen wake model are described as follows:

- Velocity deficits follows top-hat distribution.
- Linear wake growth rate along the downstream of the turbine.

The governing equations of Jensen model were based on the conservation of mass around the control volume which is considered immediately downwind of the turbine. The schematic of control volume and the velocity deficit profile of Jensen model are presented in Fig. 6.1. The right cross-sectional area of control volume is equal to the area of wake, A_{w}, and the left cross-sectional area is equal to the area swept by blades, A_0. u_0 represents the free stream velocity entering the control volume through lateral side. Using conservation of mass around the control volume, the mass flow rate of incoming wind through the lateral side $\dot{m}_2$ is described as follows:

$$\dot{m}_2 = \rho u_{\mathrm{w}} A_{\mathrm{w}} - \rho u_{\mathrm{r}} A_0 \tag{6.1}$$

Equation (6.1) can be modified as

$$\rho u_0 (A_{\mathrm{w}} - A_0) = \rho u_{\mathrm{w}} A_{\mathrm{w}} - \rho u_{\mathrm{r}} A_0 \tag{6.2}$$

$$\pi r_0^2 u_{\mathrm{r}} + \pi \left(r_{\mathrm{w}}^2 - r_0^2\right) u_0 = \pi r_{\mathrm{w}}^2 u_{\mathrm{w}} \tag{6.3}$$

In Eqs. (6.2) and (6.3), u_{w} is the wake velocity at a downstream distance of x from the rotor, u_{r} is the wind velocity behind the turbine rotor of radius r_0, and r_{w} is the wake radius at downstream distance of x from the rotor. Behind the rotor, wind

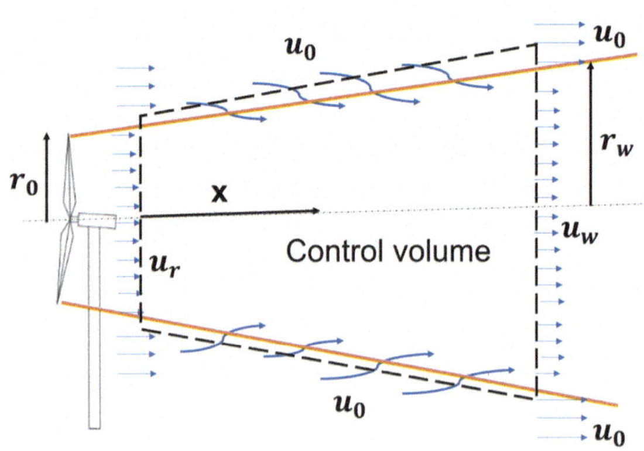

Fig. 6.1 The control volume and velocity deficit profile for Jensen model

velocity is defined from the actuator disc theory, and is defined as a function of free stream velocity and axial induction factor a as shown in Eq. (6.4)

$$u_{\mathrm{r}} = u_0(1 - 2a) \tag{6.4}$$

Since, linear wake growth rate is assumed in Jensen model, the wake radius at a distance x from the turbine is presented in Eq. (6.5)

$$r_{\mathrm{w}} = r_0 + \alpha x \tag{6.5}$$

where α is decay constant. From Eqs. (6.3) to (6.5), the deficit velocity is represented as:

$$\frac{\Delta u}{u_0} = \frac{2a}{\left(\frac{r_0+\alpha x}{r_0}\right)^2} \tag{6.6}$$

6.2.3 ANNs

ANNs draw inspiration from their biological counterparts and employ interconnected processing units, known as nodes and hidden layers, to capture the inherent nonlinearities in the given data (Bishop 2006). The data is segregated linearly as a hyperplane through the hidden layer, necessitating several hidden layers while modelling the nonlinearities. The parameters of ANN, weights and biases, are employed to establish connections between nodes across the layers. The outputs of ANN are derived by processing the weighted sum of inputs and biases through the activation function. Consider the input data as X where M is its dimension and N is the length of data. A multilayered ANN is presented in Fig. 6.2 and Eqs. (6.7)–(6.9).

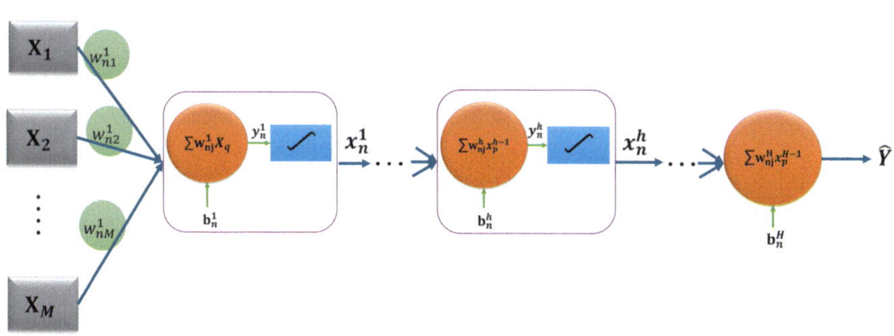

Fig. 6.2 ANN network representation

For 1st hidden layer, the output of nth node activated through activation function, φ, x_n^1 is evaluated as:

$$y_n^1 = \sum_{q=1}^{M} w_{nj}^1 X_q + b_n^1, x_n^1 = \varphi(y_n^1) \quad (6.7)$$

For h^{th} hidden layer, the output of nth node activated through activation function, φ, x_n^h is evaluated as:

$$y_n^h = \sum_{p=1}^{N^{h-1}} w_{nj}^h x_p^{h-1} + b_n^h, x_n^h = \varphi(y_n^h) \quad (6.8)$$

The evaluated output of the network is as follows:

$$\hat{Y} = \sum_{p=1}^{n^{H-1}} w_{nj}^H x_p^{H-1} + b_n^H \quad (6.9)$$

where M represents the input size, H denotes number of layers, and N^h signifies number of nodes in layer h. The network parameters, weights, w_{nj}^h and biases, b_n^h, are adjusted to determine the estimate, $\hat{Y}$. The error between the ground truth Y and the estimate $\hat{Y}$, known as loss function, L is determined. Subsequently, the network parameters undergo fine-tuning aimed at loss function minimization and this process is referred to as Training. As neural network mappings are continuous, and information of the gradient is also available, gradient-based optimization techniques can be used to address the loss function minimization. The back propagation algorithm is used to determine the loss function gradients with respect to parameters. The formulation for optimizing parameters of ANNs is presented in Eq. (6.10).

$\hat{Y} = F(X_q|_{1 \text{ to } M}, \theta)$, as presented in Eqs. (6.7)–(6.9), θ represents the network parameters, i.e. weights and biases.

$L = \sum_{i=1}^{T} \left(Y_i - \widehat{Y_i}\right)^2$, T denotes the size of training data.

$$\theta^* = \arg\min(L) \quad (6.10)$$

6.2.4 Jensen–ANN Wake Model

Though Jensen model is widely used wake model in the literature, the major concerns of Jensen wake model are linear wake growth rate, top-hat distribution profile of wake deficits and conservation of mass. Several analytical wake models in the literature were developed to improvise the accuracy, but all the analytical models consider the linear wake growth rate while modelling the wake characteristics. In this study, ANNs are utilized to model the nonlinear wake growth rate and are superposed with Jensen model to improve the accuracy of modelling the wake deficits in a wind farm. The wake growth rate is modelled as a nonlinear function of downstream distance from the turbine. Consequently, we modify the Eq. 6.5 to Eq. 6.11 in our proposed Jensen-ANN model

$$\widehat{r_w} = F(x, \theta) \tag{6.11}$$

where x is the downstream distance, F is the mapping function and in this work, ANN is considered as mapping function and θ are the parameters of ANN. Since r_w data is not accessible, the loss function is defined on the available effective velocity data, u_w. Hence $\widehat{u_w}$ is determined through Jensen model using the estimated wake growth rate, $\widehat{r_w}$ as shown below:

$$\frac{\Delta u}{u_0} = \frac{2a}{\left(\frac{\widehat{r_w}}{r_0}\right)^2}$$

$$\widehat{u_w} = u_0\left(1 - \frac{2a}{\left(\frac{\widehat{r_w}}{r_0}\right)^2}\right). \tag{6.12}$$

Though ANNs are used to model the wake growth rate, the estimated growth rate is utilized to determine the effective velocity, $\widehat{u_w}$ and thus the loss function is defined on velocities (see Eq. 6.13).

$$L = \sum_{i=1}^{T}\left(u_{wi} - \widehat{u_{wi}}\right)^2$$

$$\theta^* = \arg\min(L) \tag{6.13}$$

The back propagation algorithm is used to tune the network parameters, represented as θ. Given that the loss function is based on effective velocities but ANNs are employed to map the wake growth rate r_w and downstream distance, x, the following chain rule of differentiation is formulated to determine the gradients for back propagation:

$$\frac{\partial L}{\partial w} = \frac{\partial L}{\partial r_w} \cdot \frac{\partial r_w}{\partial w} \tag{6.14}$$

The derivative $\frac{\partial r_w}{\partial w}$ is derived from ANNs using the chain rule, whereas $\frac{\partial L}{\partial r_w}$ is derived through Eqs. (6.12) and (6.13) and is outlined in Eq. (6.15)

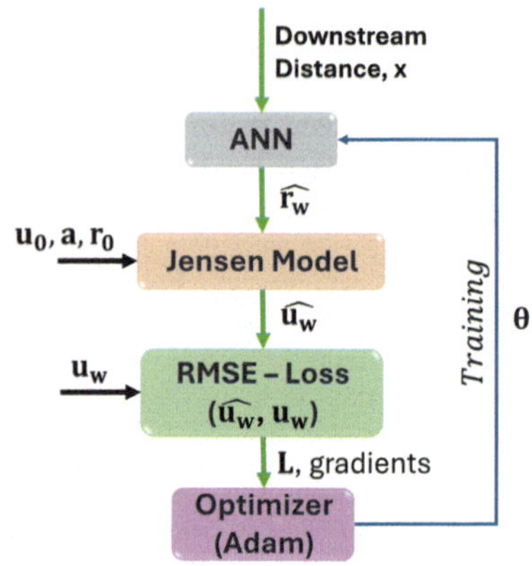

Fig. 6.3 Summary of proposed Jensen–ANN model

$$\frac{\partial L}{\partial r_{\mathrm{w}}} = \sum_{i=1}^{T} \left[-2\left(u_i - \widehat{u_i}\right) \right] \frac{4 a u_0 r_0^2}{r_{\mathrm{w}}^3} \tag{6.15}$$

With the gradients defined in Eqs. (6.14) and (6.15), the ANN parameters are adjusted using Adam optimizer (Kingma and Ba 2014) and Root Mean Square Error (RMSE) is used as loss function. The proposed methodology of Jensen–ANN wake model is summarized in Fig. 6.3.

6.3 Results

6.3.1 Results of Jensen–ANN Wake Model

A dataset of wind speed and direction over a time horizon of one year from 6 pairs of turbines is collected for this study. The turbines are arranged in pairs so that they have single wake effects, i.e. wake effects are present within the pair but not with the other pair of turbines. With the available dataset, the novel ANN–Jensen model was trained with 70% data and the remaining 30% was used for testing the model. The conventional Jensen model was integrated with ANN as described in Sect. 6.2.4 to estimate the wake growth rate, and the integrated model was trained with the collected dataset. The input for the developed wake model is the downstream distance of turbines, and the wake radius is estimated through the ANN. The approximated wake radius through ANN is used in Jensen model to estimate the velocity deficits. Since the available dataset is on velocities, loss function is defined on the estimated velocities. Later Adam optimizer is used to update the parameters of ANN with the gradients obtained through Eqs. (6.14)–(6.15). The architecture [1–2–2–1] was employed to train the ANN model, and reported the training loss, RMSE as 0.08. Subsequently, the validation of the trained model is reported with RMSE of 0.06 and an accuracy of 93%.

The efficacy of the model is presented by the comparison of the proposed novel wake model with the conventional Jensen wake model. The test data which is extracted from the available data is used to estimate the velocities through the conventional Jensen model. The accuracy is reported as 86% with an RMSE of 0.09. The comparison plots and error plots considering conventional and the proposed model are shown in Figs. 6.4, 6.5 and 6.6. From the plots, we can observe the superiority of Jensen–ANN model with the conventional Jensen model and the same reflects in RMSE and accuracy values. The results have shown the importance of considering nonlinear wake growth rate while characterizing the wake effects in a wind farm. Since the sophisticated CFD wake models are computationally expensive, the industrialists incorporate the analytical wake models for practical wind farm operations; hence, there is a need to improvise the accuracy of the analytical wake models. Therefore, in this work, ANNs are integrated with conventional techniques to model the nonlinear wake characteristics more efficiently, which have influence on wind farm operations.

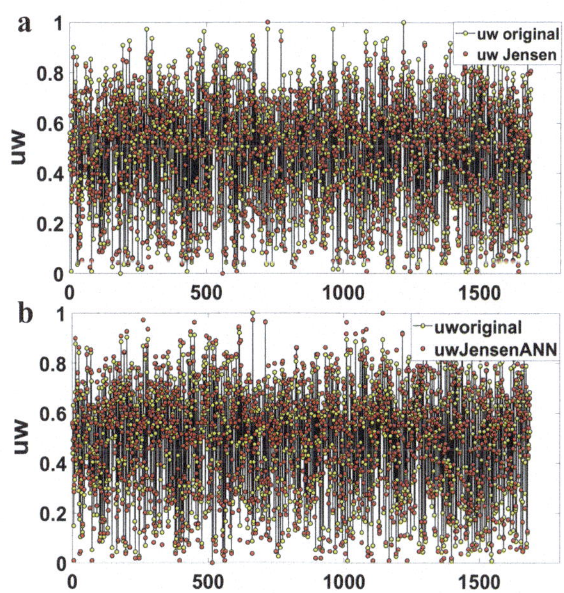

Fig. 6.4 Plots for the reduced velocities obtained through models **a** conventional and **b** proposed

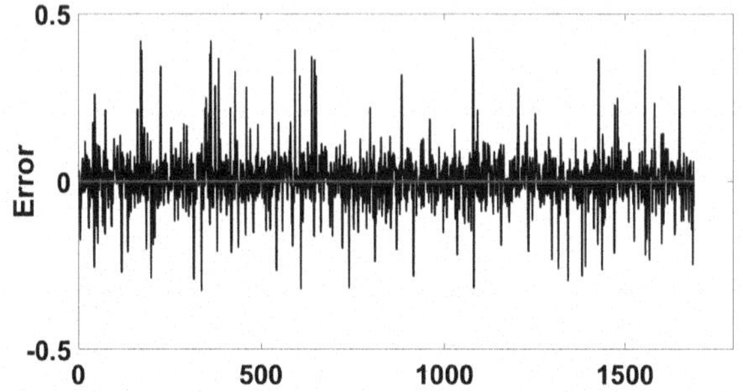

Fig. 6.5 Error plot (error between original and predicted velocity deficit) for the reduced velocities obtained through conventional model

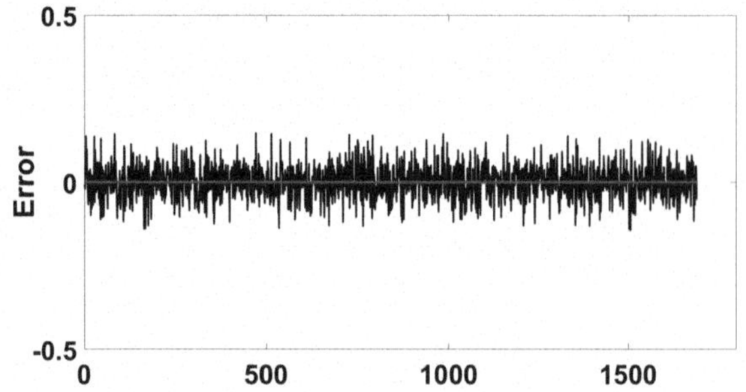

Fig. 6.6 Error plot (error between original and predicted velocity deficit) for the reduced velocities obtained through proposed model

6.3.2 Significance of Wake Modelling in Wind Farm Studies

The focus of any wake model is to determine the reduced wind speeds due to wake effects. To quantify the power losses through wake effects or to quantify the energy production from a wind farm, the power is calculated from every turbine and the overall power production is the summation of power from individual turbines. Therefore, in this study, the importance of wake modelling is shown through the power calculation with two different wake models: conventional and the proposed model, where accuracy in wake calculations can be captured. In this work, we have considered a 12-turbine layout, with 6 pairs of turbines in which each pair has a wake effect within and devoid of wake effects from other turbines. Calculating the expected power, P, for all the turbines ($N_t = 12$ for this study) considering data for wind speed and direction for 1 year, we have

$$P = \sum_{i=1}^{N_t} \sum_{d=1}^{12} \sum_{s=1}^{16} p_c(u_d(u_s, \theta_d, N_t)) \times F(u_s, \theta_d) \tag{6.16}$$

u_s, θ_d are the values of wind speed and wind direction at s^{th} and d^{th} intervals, u_d is effective speed of wind from a turbine after wake characterization, $p_c(u_d(u_s, \theta_d, N_t))$

is the power curve obtained from turbine manufacturer and is a function of effective wind speed, u_d. F is the wind frequency which is a joint probability mass function of two random variables: wind speed and wind direction.

In this work, the significance of wake characterization is examined through estimating the power from the 12 turbines with different methods of modelling the velocity deficits. In the first case study, the velocity deficits are modelled using the conventional model and the power is evaluated using Eq. (6.16). Later the velocity deficits are estimated using the proposed model and the power is evaluated. These two cases are compared with the power estimated from the velocity deficits collected in the field. The power values are obtained as 3865.23 kW for the conventional Jensen model, 3712.23 kW for the Jensen–ANN model and 3597.22 for the data obtained from field. The comparison of all these 3 cases has proven that the power estimated from the proposed model is more reliable and revealing that the conventional Jensen model tends to overestimate the velocities by assuming a linear wake growth rate. Thus, this study emphasizes the significance of correct consideration of nonlinear growth rate of wake while estimating the accurate values of wake deficits leading to more accurate estimation of power.

6.4 Conclusion

One of the important aspects in wind farm design and wind farm control studies is the modelling of wake characteristics. The importance behind the wake characterization is the reduced wind speeds of downwind turbines which are caused by upwind turbines due to the increased turbulence and leading to a significant decrease in power generation from a wind farm. Therefore, numerous wake models are emerging in the literature to characterize the wakes of wind turbines. Despite their superiority in accuracy, the LES models are not utilized for practical operations by the industrialists as they are time-consuming. On the other side, the engineering wake models which are fast have been utilized prominently while modelling wind energy conversion systems. Though the analytical models are fast, the accuracy is not good due to the underlying assumptions of wake model which do not consider the turbulence effects and rotor-flow interactions. To mitigate these issues, ANNs are superposed with the analytical models, thereby proposing a novel wake model which creates a balance between accuracy and computational costs. The significance of modelling wake characteristics is presented using a study on power production from a layout. The work is summarized below:

- A nonlinear wake growth rate is incorporated by modelling the wake radius as a function of streamwise distance between the turbines using ANNs.
- The trained ANNs are embedded in Jensen model to determine the wake radius and in turn velocity deficits.
- Velocity deficits are used to define the Loss function, and the gradients used for tuning the parameters of ANN are reformulated using the Jensen models.

- The proposed Jensen–ANN model is compared with the conventional Jensen model, and the accuracies are reported as 93% and 86%, respectively.
- An enhanced model for wake characterization enables precise estimation of power production therefore assisting the wind farm owners to mitigate the potential losses.

Acknowledgements We would like to acknowledge the support of SPARC project (SPARC/2018-2019/P1084/SL) sponsored by the Ministry of Education (MoE), Government of India and National Supercomputing Mission (NSM) for the project (DST/NSM/R&D_HPC_Applications/2021/23) sponsored by the Department of Science and Technology (DST), Government of India.

References

Adaramola MS, Krogstad PÅ (2011) Experimental investigation of wake effects on wind turbine performance. Renew Energy 36:2078–2086. https://doi.org/10.1016/J.RENENE.2011.01.024

Antonini EGA, Romero DA, Amon CH (2019) Improving CFD wind farm simulations incorporating wind direction uncertainty. Renew Energy 133:1011–1023. https://doi.org/10.1016/J.RENENE.2018.10.084

Archer CL, Vasel-Be-Hagh A, Yan C et al (2018) Review and evaluation of wake loss models for wind energy applications. Appl Energy 226:1187–1207. https://doi.org/10.1016/J.APENERGY.2018.05.085

Barthelmie RJ, Schepers JG, Van Der Pijl SP et al (2007) Flow and wakes in complex terrain and offshore. Model development and verification in UPWIND. In: European wind energy conference & exhibition (EWEC 2007), European wind energy conference & exhibition (EWEC 2007), Milan, Italy, 07–10 May 2007

Bastankhah M, Porté-Agel F (2014) A new analytical model for wind-turbine wakes. Renew Energy 70:116–123. https://doi.org/10.1016/J.RENENE.2014.01.002

Bishop CM (2006) PRML

Biswas A, Sarkar S, Gupta R (2016) Application of artificial neural network for performance evaluation of vertical axis wind turbine rotor. Int J Ambient Energy 37:209–218. https://doi.org/10.1080/01430750.2014.915889

Chamorro LP, Porté-Agel F (2009) A wind-tunnel investigation of wind-turbine wakes: boundary-layer turbulence effects. Boundary Layer Meteorol 132:129–149. https://doi.org/10.1007/S10546-009-9380-8/METRICS

Ding Y (2019) Data science for wind energy. Google Books. CRC Press. Accessed 4 Mar 2024. https://books.google.com.au/books?hl=en&lr=&id=nwadDwAAQBAJ&oi=fnd&pg=PP1&dq=Ding+Y.+Data+science+for+wind+energy.+CRC+Press%3B+2019+Jun+4.+&ots=59tV2K_87b&sig=UDyWM3cvRS8NRe7dAXUwfaz6pN8#v=onepage&q=Ding%20Y.%20Data%20science%20for%20wind%20energy.%20CRC%20Press%3B%202019%20Jun%204.&f=false

Frandsen S, Barthelmie R, Pryor S et al (2006) Analytical modelling of wind speed deficit in large offshore wind farms. Wind Energy 9:39–53. https://doi.org/10.1002/WE.189

Göçmen T, Giebel G (2018) Data-driven wake modelling for reduced uncertainties in short-term possible power estimation. J Phys Conf Ser 1037:072002. https://doi.org/10.1088/1742-6596/1037/7/072002

Göçmen T, Van Der LP, Réthoré PE et al (2016) Wind turbine wake models developed at the technical university of Denmark: a review. Renew Sustain Energy Rev 60:752–769. https://doi.org/10.1016/J.RSER.2016.01.113

GWEC (2023) Global wind report 2023. Global Wind Energy Council. Accessed 4 Mar 2024. https://gwec.net/globalwindreport2023/

Iungo GV, Santhanagopalan V, Ciri U et al (2018) Parabolic RANS solver for low-computational-cost simulations of wind turbine wakes. Wind Energy 21:184–197. https://doi.org/10.1002/WE.2154

Jensen NO (1983) A note on wind generator interaction

Jin X, Cheng P, Chen WL, Li H (2018) Prediction model of velocity field around circular cylinder over various Reynolds numbers by fusion convolutional neural networks based on pressure on the cylinder. Phys Fluids 30:47105. https://doi.org/10.1063/1.5024595/260354

Kaldellis JK, Triantafyllou P, Stinis P (2021) Critical evaluation of wind turbines' analytical wake models. Renew Sustain Energy Rev 144:110991. https://doi.org/10.1016/J.RSER.2021.110991

Kingma DP, Ba JL (2014) Adam: a method for stochastic optimization. In: 3rd international conference on learning representations, ICLR 2015—conference track proceedings

Li N, Liu Y, Li L et al (2020) Numerical simulation of wind turbine wake based on extended k-epsilon turbulence model coupling with actuator disc considering nacelle and tower. IET Renew Power Gener 14:3834–3842. https://doi.org/10.1049/IET-RPG.2020.0416

Nai-Zhi G, Ming-Ming Z, Bo L (2022) A data-driven analytical model for wind turbine wakes using machine learning method. Energy Convers Manag 252:115130. https://doi.org/10.1016/J.ENCONMAN.2021.115130

Niayifar A, Porté-Agel F (2016) Analytical modeling of wind farms: a new approach for power prediction. Energies 9:741. https://doi.org/10.3390/EN9090741

Porté-Agel F, Bastankhah M, Shamsoddin S (2019) Wind-turbine and wind-farm flows: a review. Boundary-Layer Meteorol 174:1–59. https://doi.org/10.1007/S10546-019-00473-0

Sedaghatizadeh N, Arjomandi M, Kelso R et al (2018) Modelling of wind turbine wake using large eddy simulation. Renew Energy 115:1166–1176. https://doi.org/10.1016/J.RENENE.2017.09.017

Sun H, Yang H (2018) Study on an innovative three-dimensional wind turbine wake model. Appl Energy 226:483–493. https://doi.org/10.1016/J.APENERGY.2018.06.027

Tao S, Xu Q, Feijóo A et al (2020) Wind farm layout optimization with a three-dimensional Gaussian wake model. Renew Energy 159:553–569. https://doi.org/10.1016/J.RENENE.2020.06.003

Verbrugge MW, Baker DR, Xian HX et al (2016) CFD modelling approaches against single wind turbine wake measurements using RANS. J Phys Conf Ser 753:032062. https://doi.org/10.1088/1742-6596/753/3/032062

Wu YT, Porté-Agel F (2015) Modeling turbine wakes and power losses within a wind farm using LES: an application to the Horns Rev offshore wind farm. Renew Energy 75:945–955. https://doi.org/10.1016/J.RENENE.2014.06.019

Chapter 7
Automated LSTMs for Wind Behaviour Predictions

NagaSree Keerthi Pujari, Srinivas Soumitri Miriyala, and Kishalay Mitra

Abstract Wind being one of the most utilized renewable energy sources, the researchers aimed towards development of new techniques to harness it efficiently. The complex nature of wind characteristics presents hurdle for its accurate prediction. As wind speed has a close link with the power production capability of a wind farm, accurate prediction of it is the need-of-the-hour for realistic studies. In this work, long short-term memory networks (LSTMs) have been utilized by the authors for accurately forecasting wind behaviour characterized by its speed and direction. The proposed work brings novelty in the evolutionary approach for the automated design of LSTMs, which enables the estimation of the hyper-parameters in them, including their architecture, activation function, and backpropagation length, without the intervention of heuristics. The optimally designed LSTMs are trained using real wind data with very high accuracy. The forecasted data with optimal LSTMs are validated with the available original data, which shows the credibility of the proposed methodology.

Keywords Wind characteristics forecasting · Deep learning · LSTMs · Evolutionary optimization · Effective wind farm design

---

N. K. Pujari · S. S. Miriyala · K. Mitra (✉)
Department of Chemical Engineering, Indian Institute of Technology Hyderabad, Hyderabad, Telangana 502284, India
e-mail: kishalay@che.iith.ac.in

N. K. Pujari
e-mail: ch20resch11006@iith.ac.in

K. Mitra
Department of Climate Change, Indian Institute of Technology Hyderabad, Hyderabad, Telangana 502284, India

Department of Artificial Intelligence, Indian Institute of Technology Hyderabad, Hyderabad, Telangana 502284, India

K. Mitra et al. (eds.), *Optimization, Uncertainty and Machine Learning in Wind Energy Conversion Systems*, Engineering Optimization: Methods and Applications,
https://doi.org/10.1007/978-981-97-7909-3_7

7.1 Introduction

Increasing energy demand with the depleting fossil fuels and rising concerns of climate change have shifted the global attention towards renewable energy resources as sustainable alternative. There has been a significant stride in renewable energy development over the past decade, driven by several government initiatives, private sector investments. As of 2024, India has achieved substantial growth in its installed renewable energy capacity standing over 160 GW with an addition of 65 GW in the last 5 years which leads to its contribution as the fourth largest in the world (MNRE 2023a). Wind energy plays a pivotal role towards sustainable energy future with its clean and economically viable proposal providing numerous environmental, technological, and socio-economic benefits. The wind energy sector in India holds significant potential with around 42 GW installed capacity by 2022 making it fourth highest in the world and further expecting to reach 56 GW with ongoing projects (MNRE 2023b). Despite the extensive use of wind energy for power generation, the variability and unpredictability of wind pose significant challenges to the reliable operation of wind farms. Therefore, developing an accurate wind characteristics forecasting model is essential to achieve consistent wind energy production. Research in Hu et al. (2018) indicated that prediction accuracy that increases by 1% may lead to 1.38 billion USD savings for one year. Therefore, many studies reported several models in the literature aiming at enhancing the forecasting accuracy.

The approaches of deterministic wind characteristics forecasting can be classified into three categories: physical, statistical, and artificial intelligence models. The physical models make use of meteorological and geographical factors to capture the physical processes (Ratjiranukool and Ratjiranukool 2017; Zhao et al. 2021). As an example, numerical weather prediction method uses meteorological factors such as temperature, pressure, surface roughness, etc., to build a mathematical model of the phenomena occurring in atmosphere (Wang et al. 2019). The difficulties in physics-based models such as the complexities with first principles for weather modelling, maintenance of crucial data while validating, and high computational costs turned researchers towards data-driven techniques. The traditional statistical models such as Auto-regressive models (Lydia et al. 2016), probabilistic methods (Heng et al. 2022), exponential smoothing methods (Yang et al. 2015) are commonly used for modelling and forecasting wind characteristics. In Prasetyowati et al. (2017), nonlinear auto-regressive neural network model along with wavelet decomposition is utilized for wind power forecasting. A robust short-term forecasting model with wavelet-based neural network with significant accuracy was presented in Abhinav et al. (2017) to predict wind power generation. Similarly, forecasting wind speed in short-term using support vector machine is presented in Salcedo-Sanz et al. (2011).

With the ability to model complex relationships in the data, deep learning techniques such as recurrent neural networks (RNNs), gated recurrent units, and convolutional networks have proven their prominence in recent times. (S)ARIMA models are compared with LSTMs models in Kumar Dubey et al. (2021), where LSTMs are proven to be efficient. Convolutional neural networks were proposed in Trebing and Mehrkanoon (2020) and a hybrid model with Autoencoders and LSTMs was presented in Wu et al. (2019) for predicting speed of wind. In this study, stacked denoising autoencoders and batch normalization are utilized for extracting the features, and then LSTMs are used for accurate wind speed forecasting (Wu et al. 2019). The applicability of system identification and deep learning tools has shown a tremendous rise recently using open-source packages. But the heuristic way of hyperparameter selection made it tedious task. Many recent works have reported reinforcement learning (Dong et al. 2021), Bayesian optimization (Cho et al. 2020) and evolutionary optimization (Han et al. 2020) frameworks to tackle the trial-and-error approaches of fixing the hyperparameters in data-driven models. The novelty lies in developing an optimal algorithm to tune the hyperparameters with a multiobjective formulation which balances accuracy and model complexity.

The uncertain nature of wind characteristics is one of the most neglected but practical issues in wind farm designing due to which the farm faces variability in production of power. Conventionally, wind behaviour is modelled by capturing a probability mass function (PMF) using two random variables: wind direction and wind speed while designing a wind farm layout or wind farm control techniques. Due to data scarcity, PMF construction may not be accurate and thus leading to unrealistic results. Therefore, the current work aims to develop a technique which forecasts the wind data accurately by learning the nonlinear hidden patterns and transient behaviour of wind characteristics. In this work, first the realistic wind data are analysed thoroughly using time series analysis to decide on the appropriate modelling tool and the data are decomposed to separate out the repetitive seasonality component from it. Since the wind time series data are extremely nonlinear and have long-term dependencies, deep learning technique LSTM is utilized for modelling the data. The heuristics associated with the open-source software were addressed by an evolutionary multiobjective optimization framework with two conflicting objectives of accuracy maximization and model complexity minimization. The designed models were validated with the four-year real wind characteristics data obtained from a wind farm in France. The contributions of this work are as follows:

- Hyper-parameters optimization using evolutionary optimization framework.
- Validating the optimally designed models with the real wind characteristics data.

In the rest of the chapter, a brief description on time series analysis techniques and details of the proposed algorithm for tuning the hyperparameters of LSTM are presented in Sect. 7.2. Section 7.3 presents the results of the described models. Section 7.4 outlines the summary of the proposed work.

7.2 Formulation

In this section, data description and its analysis are presented. Followed with the techniques of data analysis, the proposed algorithm for the hyperparameter tuning of LSTMs is described.

7.2.1 About Data

Wind behaviour (speed and direction) over a span of four years with six-hour interval was considered from an open repository (ENGIE 2020). Sensors were used to collect data with four turbines placed at corners around 9 km^2 in the farm. As the wind turbines are placed a large distance apart, wake effects are assumed to be negligible. As the data are collected through sensors, the moving average is used to filter the measurement noise. Figure 7.1 presents stacked bar chart representing wind speed and direction data used for this study.

In this work, a univariate time series approach has been used to model data, which results in two such time series, one for speed and the other for direction. Therefore, the time series analysis and time series decomposition were performed on speed and direction data. Nonlinearity, stationarity, and long-range dependencies are examined in time series analysis whereas the time series decomposition helps in deciphering hidden and sequential patterns.

The nonlinear nature of input–output mapping with the irregular temporal behaviour can be identified using Brock–Dechert–Scheinkman (BDS) test (Akintunde 2015). Stationary nature of data, i.e., not changing statistical properties, such as mean and variance with time, is determined by a statistical Augmented–Dickey–Fuller (ADF) test (Dickey and Fuller 1979). The long-term dependencies in the data, dependency of a current time step on the previous time steps, are determined using Hurst exponent (Kalo et al. 2019). Several literature studies have shown that the time series decomposition can model the data accurately by capturing different patterns and sequential factors. There are several techniques that decompose the data into trend, seasonal and random components and seasonal and trend decomposition using LOESS (STL) is found to be robust and efficient among them. Here, STL has been used to decompose the data and the results were presented in Sect. 7.3.

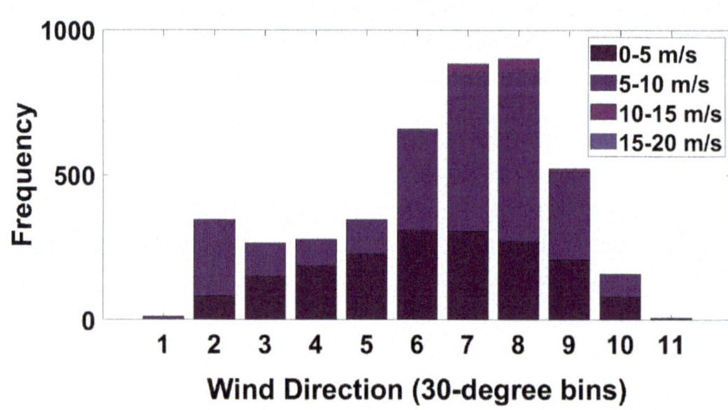

Fig. 7.1 Stacked bar chart representing wind speed and wind direction with 4 speed bins and 12 direction sectors

7.2.2 *Modelling Nonlinear Time Series Wind Data Using LSTMs*

A univariate time series data at time instance, t is represented as $X^t = [X_1^t X_2^t X_3^t \ldots X_M^t]$, $\forall\, t = 1{:}\ T$, where M represents data dimension and T represents data length. Here, $M = 1$ since the data is modelled as univariate time series considering wind speed and direction one at a time. A univariate time series data is modelled by estimating the target value at time step t, $\widehat{X^t}$, as a function of previous data points, b_{T}, and a set of tunable parameters, W. The tunable parameters, W, are optimized by minimizing the loss, L between the ground truth variable, X^t and estimate $\widehat{X^t}$. In the current work, we have considered LSTMs as input–output functional mapping with weights and biases as tunable parameters. LSTMs regress on hidden states which models time series data.

LSTMs

LSTMs are an offshoot of RNNs which regress on hidden states. A hidden state is an activated output of a node and a feedback loop on this node allows the network to regress on hidden states, and many such hidden states together process information simultaneously by culminating into a single hidden layer. Several such layers are connected sequentially giving rise to a deep neural network, and the RNNs are trained by unrolling them. The unrolled RNN is shown in Fig. 7.2. In LSTMs, the parameters of regression are functions of time and the contextual memory in the network signifies hidden state regression. The unique design of LSTMs in which the output is determined through a hypothetical cell state devoid of nonlinear activation function, therefore, allows to capture the long-term dependencies by overcoming the problem of vanishing gradients (Sherstinsky 2020). Acting as a memory unit of LSTM, the hypothetical cell state controls the previous as well as current cell state information using three gates, e.g., forget, input, and intermittent cell value. The forget gate estimates how much information is needed from the previous cell state, i.e., it regulates the information that needs to be forgotten from previous state. The intermittent cell value is the current information. The input gate regulates the information that needs to be carried forward from the intermittent cell value. Further, the output from hypothetical cell state is regulated using the output gate, which determines how much information is carried from the hypothetical cell state and results in current hidden state. Figure 7.3 presents the difference between RNN and LSTM units and Eq. (7.1) presents the LSTM.

$$F^t = \log\mathrm{sig}\left(w_{\mathrm{F}x}X^t + w_{\mathrm{FF}}a^{t-1} + b_{\mathrm{F}}\right) \tag{7.1}$$

$$I^t = \log\mathrm{sig}\left(w_{\mathrm{I}x}X^t + w_{\mathrm{II}}a^{t-1} + b_{\mathrm{I}}\right)$$

$$\widetilde{C^t} = \mathcal{F}_{\text{non-linear}}\left(w_{\mathrm{C}x}X^t + w_{\mathrm{CC}}a^{t-1} + b_{\mathrm{C}}\right)$$

$$C^t = F^t C^{t-1} + I^t \widetilde{C}^t$$

$$\Theta^t = \log \mathrm{sig}\left(w_{ox} X^t + w_{oo} a^{t-1} + b_o\right)$$

$$a^t = \Theta^t \mathcal{F}_{\text{non-linear}}\left(C^t\right)$$

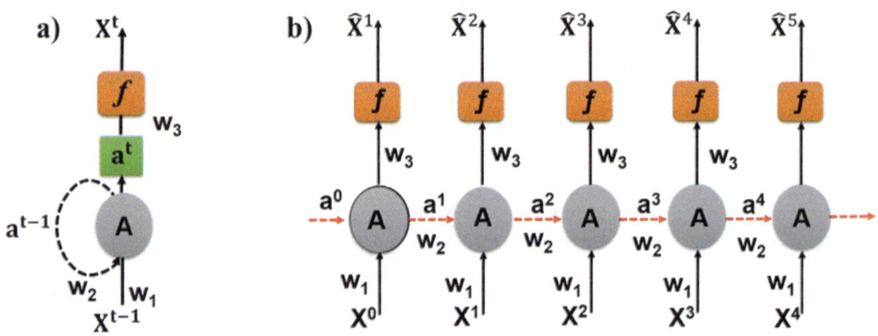

Fig. 7.2 Recurrent neural network. **a** Represents the RNN unit. **b** Represents the unrolled network of RNN

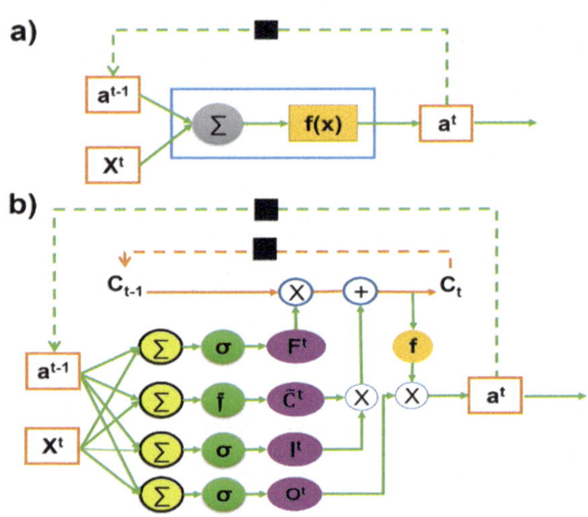

Fig. 7.3 RNN and LSTM networks. **a** RNN unit. **b** LSTM unit

7.2.3 Hyperparameter Optimization of LSTMs

In this study, stacked LSTMs are utilized for the wind characteristics modelling. While modelling LSTMs, certain parameters such as number of hidden layers and the number of nodes in them, activation function to be used and the unrolling length which governs the models need to be fixed in priori and these are called hyperparameters. In all the hyperparameters, the architectural parameters, i.e., hidden layers and nodes in hidden layers as well as learning rate are considered as most important (Greff et al. 2017). To enable the auto-tuning of learning rate, Adam optimizer, a variant of steepest descent is used for this study (Kingma and Ba 2014). The traditional approach is to fix the other hyperparameters heuristically which leads to several inaccuracies and limited potential of the LSTM network. Therefore, in this work, a novel algorithm has been proposed by the authors in machine learning framework to create a balance between accuracy and overfitting by exploring different designs of the model. A Pareto optimization is proposed in this algorithm by utilizing the compromise between accuracy of prediction and complexity in model defined by number of parameters and unrolling length. The optimization problem has been formulated with maximizing accuracy and minimizing number of parameters and unrolling length as objectives. The hyperparameters to be tuned during this exercise are the number of hidden layers and nodes in each of them, activation function and unrolling length. The prediction accuracy on unseen dataset can be determined by the statistical metric called correlation coefficient, R^2. The integral decision variables and nonlinear objectives frame the formulation as integer nonlinear programming problem (INLP) and can be solved using the efficient evolutionary algorithm, non-dominated sorting genetic algorithm (NSGA-II) (Deb 2011). The formulation for INLP is presented in Eq. (7.2), and the algorithm is presented below:

$$\underset{b_{\mathrm{T}},\left\{\mathbb{N}^h:h=1:\mathbb{H}^{\mathrm{ub}}\right\}\text{ and }A}{\text{minimize}} \quad -R^2,\ \mathbb{N}^{\mathrm{P}} \text{ and } b_{\mathrm{T}} \tag{7.2}$$

where $R^2 = \left(\frac{\mathrm{cov}(X,\hat{X})}{\sqrt{\mathrm{var}(X)\mathrm{var}(\hat{X})}}\right)^2$

$$\mathrm{cov}(X,\hat{X}) = \mathbb{T}\sum_{t=0}^{\mathbb{T}} X^t\widehat{X^t} - \sum_{t=0}^{\mathbb{T}} X^t \sum_{t=0}^{\mathbb{T}} \widehat{X^t}$$

$$\mathrm{var}(X) = \mathbb{T}\sum_{t=0}^{\mathbb{T}} X^{t^2} - \left(\sum_{t=0}^{\mathbb{T}} X^t\right)^2$$

such that, $b_{\mathrm{T}}^{\mathrm{lb}} \le b_{\mathrm{T}} \le b_{\mathrm{T}}^{\mathrm{ub}}$

$$\mathbb{L}^h \le \mathbb{N}^h \le \eta^h \text{ and } \mathbb{L}^h = \begin{cases} 1, & \text{if } h = 1 \\ 0, & \text{if } h > 1 \end{cases}$$

$$A \in \{1, 2\} | \text{if } A = \begin{cases} 1, & \text{activation is using tan sigmoid} \\ 2, & \text{activation is using log sigmoid} \end{cases}$$

$$\{b_{\mathrm{T}}^{\mathrm{lb}}, b_{\mathrm{T}}^{\mathrm{ub}}, \mathbb{H}^{\mathrm{ub}}, \eta^{h}\} \in \mathbb{Z}_{+}$$

$\mathbb{H}$	is the number of hidden layers;
$\mathbb{H}^{\mathrm{ub}}$	is the upper bound on number of hidden layers;
$\mathbb{N}^{h}$	is the number of nodes in hidden layer, h;
$\mathbb{L}^{h}$ and η^{h}	are the lower and upper bounds on $\mathbb{N}^{h}$;
b_{T}	is the sequence length;
$b_{\mathrm{T}}^{\mathrm{lb}}$ and $b_{\mathrm{T}}^{\mathrm{ub}}$	are the lower and upper bounds on b_{T};
$\mathbb{N}^{\mathrm{P}}$	is the number of parameters in the network;
A	is the choice of activation function;
$\mathbb{T}$	is the number of test data samples.

Algorithm: Optimal LSTMs

1: Initialize variables: Binary variables—$\mathbb{H}^{\mathrm{ub}} + 2$; Real variables—0.
2: Set NSGA-II parameters: Population size and generation count. Start the algorithm.
3: For population size, use first $\mathbb{H}^{\mathrm{ub}}$ decision variables.
    Run a loop: $h = 1$ to $\mathbb{H}$.
    If $h = 1$: $\mathbb{N}^{h} = h^{th}$ decision variable; $h = h + 1$.
        If h^{th} decision variable = zero: $h = h - 1$ and go to 4.
    Else $\mathbb{N}^{h} = h^{th}$ decision variable; $h = h + 1$.
4: Store architecture [1, $\{\mathbb{N}^{h}$: $h = 1$: $\mathbb{H}\}$, 1].
5: Determine A and b_{T} using the final 2 decision variables.
6: If the database consists of architecture, b_{T} and A the objective function values are returned else go to step 7.
7: Train and validate LSTM using t-BPTT and ADAM.
8: Test LSTM and determine the objective functions.
9: Repeat 3–8 for all the population of NSGA II.
10: Get the new generation and repeat 3–10 till convergence.

7.3 Results

7.3.1 Time Series Analysis and Decomposition

Since the wind time series data are collected from anemometers, a 5-point moving average method is used to filter noise. The results of analysis are presented below followed by time series decomposition.

A statistical hypothesis test, BDS test, can be used to determine the nonlinearity in the data in which the data are considered as linear in the null hypothesis (H_0) and nonlinear in the alternate hypothesis (H_1). The p values are determined by considering 5% significance level, and the p-value less than 0.05 signifies the rejection of null hypothesis. BDS test results presented in Table 7.1 have shown that the p values are less than 0.05, inferring the data under consideration are nonlinear. The other important characteristics of time series data, stationarity, is determined through hypothesis test, ADF test. The null hypothesis (H_0) states that the data are non-stationary, i.e., have a unit root and stationary in the alternate hypothesis (H_1). The results are presented in Table 7.1 has shown that the p values are less than 0.05, inferring the wind characteristics data is stationary. Any time series data can have long-term dependencies which play a major role while modeling the data. The long-term dependencies in the data can be explored through a rescaled range analysis on several time spans of data: Hurst exponent, H. The H values in the range of 0.5–1 signify the presence of long-term dependencies in the data. The results in Table 7.1 represent the presence of long-term dependencies in the data.

The sequential time series can be highly impacted by several structures and patterns hidden in the data. Hence the time series can be decomposed to explore the patterns in data such as Trend-cycle, seasonality, and reminder components. The dynamics of the data is influenced by seasonal components and can imply inaccuracies while modelling the data. Hence, seasonality components should be filtered using decomposition techniques. In this work, the seasonality component is filtered using STL decomposition, and the results are presented in Figs. 7.4 and 7.5, respectively. After filtering, the remaining components, showing components of remainder and trend, are modelled using the appropriate modelling techniques. Since the data are nonlinear, stationary, and have long-term dependencies, LSTMs are used. After

Table 7.1 Results of data analysis for wind speed and wind direction time series data

Wind speed	
Test for nonlinearity	BDS test with null hypothesis saying data is linear
	p-value: 0.001
Test for stationarity	ADF test with null hypothesis saying unit root is present in the data
	p-value: 0.008
Test for long-term dependency	Hurst exponent
	H-value: 0.83
Wind direction	
Test for nonlinearity	BDS test with null hypothesis saying data is linear
	p-value: 0.003
Test for stationarity	ADF test with null hypothesis saying unit root is present in the data
	p-value: 0.01
Test for long-term dependency	Hurst exponent
	H-value: 0.79

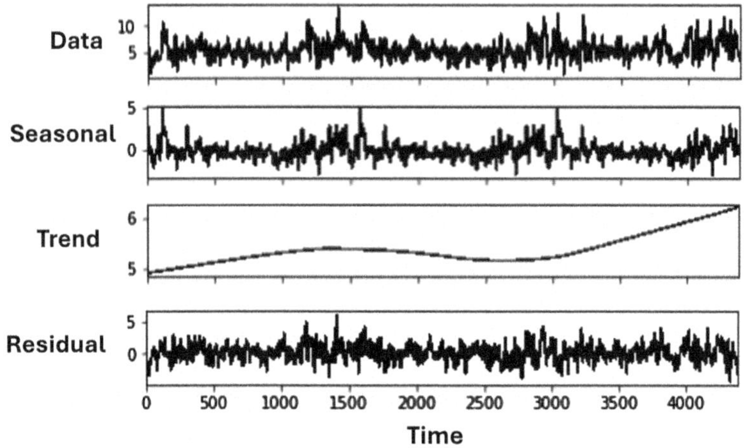

Fig. 7.4 Filtering seasonality through STL decomposition (wind speed)

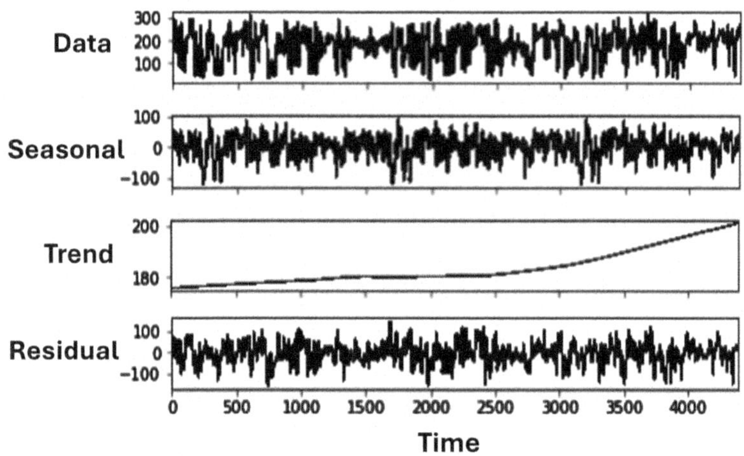

Fig. 7.5 Filtering seasonality through STL decomposition (wind direction)

modelling was performed to the desired accuracy, the seasonality component was again added with the other components to obtain the overall picture.

7.3.2 LSTMs with Optimal Design Parameters

The wind speed and direction data of four years were collected and preprocessed to remove the noise in the data. The seasonality is filtered from the data using STL decomposition to model the data accurately. Then the data are divided into two sets:

(i) the first three-year data and (ii) fourth-year data. Training the LSTMs with the first dataset, prediction of the fourth-year data is performed and validated with the fourth-year data (second dataset). While training the model, 70% of the first dataset is used for training and 30% for validation. Table 7.2 presents the decision variable bounds and NSGA-II parameters for the two simulations performed for forecasting speed and direction of wind. It is not practical to find optimal architecture using heuristics as Table 7.2 shows the possibility of $16 \times 15 \times 15 \times 64 \times 2$ alternatives for architectures. The proposed algorithm finds the best LSTM architecture by evaluating only 521 alternatives.

The solutions were obtained within the first eight generations, presented as three-dimensional Pareto Fronts. Though the results were obtained within eight generations, NSGA-II was run for longer generations to observe convergence. The Pareto fronts obtained for two optimization problems are presented in Figs. 7.6 and 7.7, respectively. Each point in the Pareto front represents the decision variables and is equally important. To select one solution from the Pareto set, a nontrivial task, a separate criterion can be utilized. One of such selection techniques, Bayesian Information Criterion (BIC), is used here for selecting a point from the Pareto set. As each of these Pareto points is a LSTM model with different architecture, BIC penalizes a model by considering the number of parameters as well as the sample size, ensuring the candidate selection with less complexity (Kashyap 1977). BIC is calculated through Eq. (7.3) where S is sample size, N_{P} is Number of parameters in LSTM, RMSE is root mean square, and the LSTM model giving least BIC value is selected as a model of choice.

$$\mathrm{BIC} = S * \log\left(\mathrm{RMSE}^2\right) + \log(S) * N_{\mathrm{P}} \tag{7.3}$$

While modelling wind speed, the optimal architecture has been attained as [1–4–1] with tan sigmoid as activation function choice. Unrolling length has been attained as 14 with a total of 101 parameters present in the network. The validation accuracy has been obtained as 0.9949 and 0.076 for R^2 and RMSE, respectively. On the other hand, for wind direction, the optimal architecture has been attained as [1–4–1] with tan sigmoid as activation function choice. Unrolling length has been attained as 16 taking

Table 7.2 Parameters of NSGA II

S. No.	Parameter	Value
1	Binary and real variables	5 and 0
2	Generations and population count	200 and 100
3	Probability of mutation and crossover	0.01 and 0.9
4	$\mathbb{H}^{\mathrm{lb}}$, $\mathbb{H}^{\mathrm{ub}}$: upper and lower bounds of hidden layers	1 and 3
5	$\mathbb{L}^h$ and η^h: upper and lower bounds of nodes in each hidden layer	{1, 0, 0} and {16, 15, 15}
6	$b_{\mathrm{T}}^{\mathrm{lb}}$ and $b_{\mathrm{T}}^{\mathrm{ub}}$: upper and lower bounds on b_{T}	2 and 65
7	$\mathbb{T}$: number of test data samples	1200

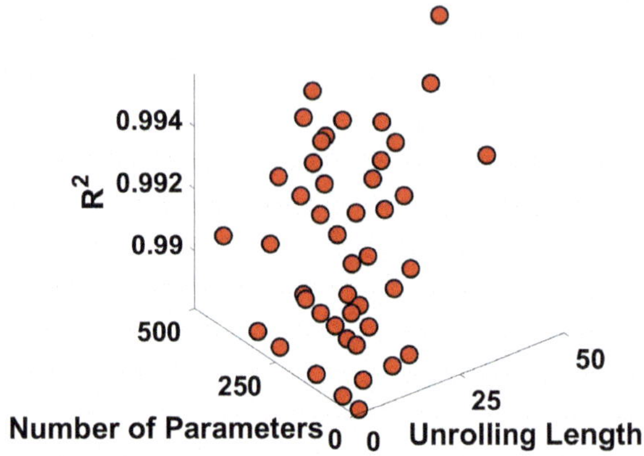

Fig. 7.6 Pareto front obtained from the proposed algorithm for wind speed data

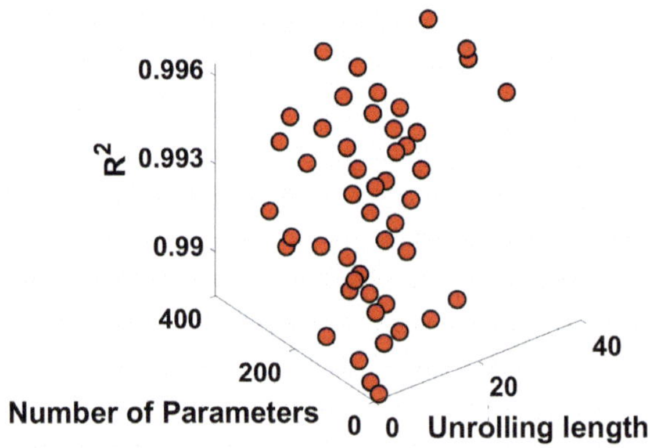

Fig. 7.7 Pareto front obtained from the proposed algorithm for wind direction data

a total of 101 parameters present in the network. The validation accuracy has been obtained as 0.9956 and 0.068 for R^2 and RMSE, respectively. Error plots in Figs. 7.8 and 7.9 show the LSTM performances on wind speed and direction, respectively. The existence of long-term dependencies in data, proven with Hurst exponent analysis, supports the usage of LSTMs. By hosting four gates, i.e., forget, input, intermittent cell value, and output, LSTM handles such dependencies resulting in large number of parameters. But better performance of LSTMs justifies this surge in parameters. Thus, LSTMs have been proven to be useful for forecasting wind characteristics efficiently. The forecasted wind characteristics can be utilized in designing wind farms with efficient energy production and in wind farm control studies.

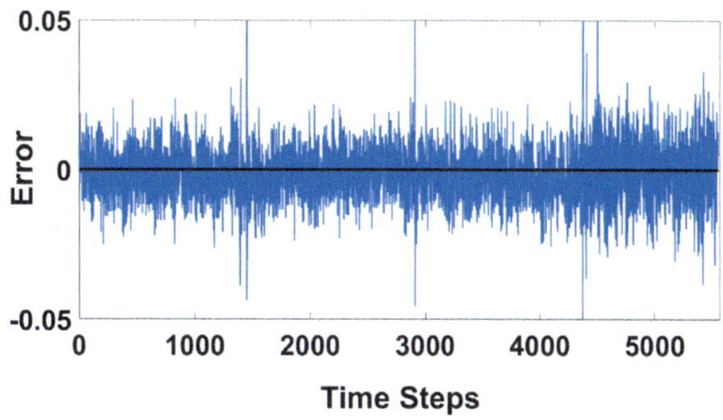

Fig. 7.8 Error plot showing LSTM performance for wind speed data

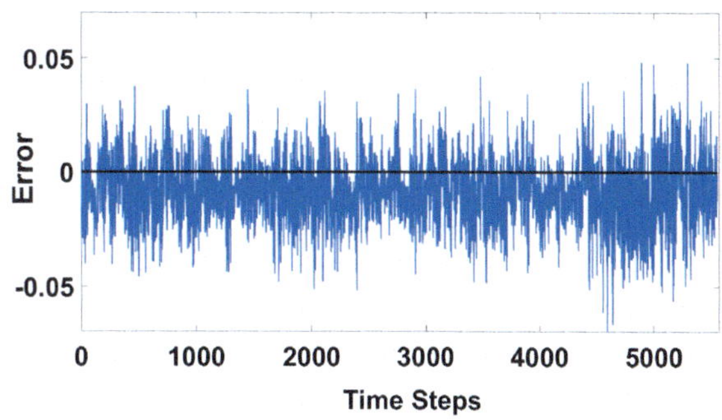

Fig. 7.9 Error plot showing LSTM performance for wind direction data

7.4 Conclusion

The necessity to bring down dependency on nonrenewable fuels and meet demands on energy surge, the research on renewable energy has been increased focusing on wind energy. As two important factors contributing towards energy extraction from wind are its speed and direction, forecasting these factors helps in designing the wind farm efficiently. Analysis of time series and decomposition was utilized to examine the wind characteristics for capturing the patterns and transient trends in its behaviour effectively. To model the transient patterns and long-term dependencies in the data, LSTMs are used. To resolve the difficulties in hyperparameter estimation for LSTMs, often performed in heuristic manner, an automated strategy has been proposed. The proposed algorithm determines the optimal hyperparameters of the LSTM using the evolutionary algorithm NSGA-II with the objectives of maximizing accuracy of prediction and minimizing the extent of overfitting. The work is summarized below:

- The hyperparameters will highly influence the efficiency of the models.
- A novel algorithm is proposed with multiobjective optimization framework to prevent the heuristic estimation involved in the LSTMs.

- The training and validation accuracy for both wind speed and wind direction is obtained as 99% using LSTMs.
- The advantage of LSTMs in forecasting wind characteristics provides a future scope in designing wind farms efficiently.

Acknowledgements We would like to acknowledge the support of SPARC project (SPARC/2018-2019/P1084/SL) sponsored by the Ministry of Education (MoE), Government of India and National Supercomputing Mission (NSM) for the project (DST/NSM/R&D_HPC_Applications/2021/23) sponsored by the Department of Science and Technology (DST), Government of India.

References

Abhinav R, Pindoriya NM, Wu J, Long C (2017) Short-term wind power forecasting using wavelet-based neural network. Energy Procedia 142:455–460. https://doi.org/10.1016/J.EGYPRO.2017.12.071

Akintunde MO (2015) Detection of non-linearity in the time series using BDS test. Sci J Appl Math Statist 3:184. https://doi.org/10.11648/J.SJAMS.20150304.13

Cho H, Kim Y, Lee E et al (2020) Basic enhancement strategies when using Bayesian optimization for hyperparameter tuning of deep neural networks. IEEE Access 8:52588–52608. https://doi.org/10.1109/ACCESS.2020.2981072

Deb K (2011) Multi-objective optimisation using evolutionary algorithms: an introduction. Multi-objective evolutionary optimisation for product design and manufacturing. Springer, London, pp 3–34

Dickey DA, Fuller WA (1979) Distribution of the estimators for autoregressive time series with a unit root. J Am Stat Assoc 74:427–431. https://doi.org/10.1080/01621459.1979.10482531

Dong X, Shen J, Wang W et al (2021) Dynamical hyperparameter optimization via deep reinforcement learning in tracking. IEEE Trans Pattern Anal Mach Intell 43:1515–1529. https://doi.org/10.1109/TPAMI.2019.2956703

ENGIE (2020) La Haute Borne-Vaudeville-le-Haut (France)—wind farms. The Wind Power. Online access. Accessed 1 Mar 2024. https://www.thewindpower.net/windfarm_en_3354_la-haute-borne.php

Greff K, Srivastava RK, Koutnik J et al (2017) LSTM: a search space odyssey. IEEE Trans Neural Netw Learn Syst 28:2222–2232. https://doi.org/10.1109/TNNLS.2016.2582924

Han JH, Choi DJ, Park SU, Hong SK (2020) Hyperparameter optimization using a genetic algorithm considering verification time in a convolutional neural network. J Electr Eng Technol 15:721–726. https://doi.org/10.1007/S42835-020-00343-7/METRICS

Heng J, Hong Y, Hu J, Wang S (2022) Probabilistic and deterministic wind speed forecasting based on non-parametric approaches and wind characteristics information. Appl Energy 306:118029. https://doi.org/10.1016/J.APENERGY.2021.118029

Hu J, Heng J, Tang J, Guo M (2018) Research and application of a hybrid model based on meta learning strategy for wind power deterministic and probabilistic forecasting. Energy Convers Manag 173:197–209. https://doi.org/10.1016/J.ENCONMAN.2018.07.052

Kalo L, Kamalanathan P, Pant HJ et al (2019) Mixing and regime transition analysis of liquid-solid conical fluidized bed through RPT technique. Chem Eng Sci 207:702–712. https://doi.org/10.1016/J.CES.2019.07.005

Kashyap RL (1977) A Bayesian comparison of different classes of dynamic models using empirical data. IEEE Trans Autom Control 22:715–727. https://doi.org/10.1109/TAC.1977.1101594

Kingma DP, Ba JL (2014) Adam: a method for stochastic optimization. In: 3rd International Conference on Learning Representations, ICLR 2015—Conference Track Proceedings

Kumar Dubey A, Kumar A, García-Díaz V et al (2021) Study and analysis of SARIMA and LSTM in forecasting time series data. Sustain Energy Technol Assess 47:101474. https://doi.org/10.1016/J.SETA.2021.101474

Lydia M, Suresh Kumar S, Immanuel Selvakumar A, Edwin Prem Kumar G (2016) Linear and non-linear autoregressive models for short-term wind speed forecasting. Energy Convers Manag 112:115–124. https://doi.org/10.1016/J.ENCONMAN.2016.01.007

MNRE (2023a) Ministry of New and Renewable Energy | India. In: MNRE. Accessed 1 Mar 2024. https://mnre.gov.in/

MNRE (2023b) Wind overview. Ministry of New and Renewable Energy | India. Accessed 1 Mar 2024. https://mnre.gov.in/wind-overview/

Prasetyowati A, Sudibyo H, Sudiana D (2017) Wind power prediction by using wavelet decomposition mode based NARX-neural network. In: ACM international conference proceeding series, pp 275–278. https://doi.org/10.1145/3168390.3168434

Ratjiranukool P, Ratjiranukool S (2017) Evaluating wind speed by WRF model over northern Thailand. Energy Procedia 138:1171–1176. https://doi.org/10.1016/J.EGYPRO.2017.10.228

Salcedo-Sanz S, Ortiz-García EG, Pérez-Bellido ÁM et al (2011) Short term wind speed prediction based on evolutionary support vector regression algorithms. Expert Syst Appl 38:4052–4057. https://doi.org/10.1016/J.ESWA.2010.09.067

Sherstinsky A (2020) Fundamentals of recurrent neural network (RNN) and long short-term memory (LSTM) network. Physica D 404:132306. https://doi.org/10.1016/J.PHYSD.2019.132306

Trebing K, Mehrkanoon S (2020) Wind speed prediction using multidimensional convolutional neural networks. In: 2020 IEEE symposium series on computational intelligence, SSCI 2020, pp 713–720. https://doi.org/10.1109/SSCI47803.2020.9308323

Wang H, Han S, Liu Y et al (2019) Sequence transfer correction algorithm for numerical weather prediction wind speed and its application in a wind power forecasting system. Appl Energy 237:1–10. https://doi.org/10.1016/J.APENERGY.2018.12.076

Wu YX, Wu QB, Zhu JQ (2019) Data-driven wind speed forecasting using deep feature extraction and LSTM. IET Renew Power Gener 13:2062–2069. https://doi.org/10.1049/IET-RPG.2018.5917

Yang D, Sharma V, Ye Z et al (2015) Forecasting of global horizontal irradiance by exponential smoothing, using decompositions. Energy 81:111–119. https://doi.org/10.1016/J.ENERGY.2014.11.082

Zhao J, Guo Z, Guo Y et al (2021) A self-organizing forecast of day-ahead wind speed: selective ensemble strategy based on numerical weather predictions. Energy 218:119509. https://doi.org/10.1016/J.ENERGY.2020.119509

Chapter 8
Artificial Intelligence Driven Forecasting for Optimal Wind Characterization and Power Generation from Wind Farm

Ravi Kiran Inapakurthi, NagaSree Keerthi Pujari, and Kishalay Mitra

Abstract The volatility in wind characteristics like speed and direction creates difficulty in wind farm layout optimization, designing energy storage systems, fluctuations on the grid and loss to businesses due to sub-optimal planning. Accurate forecasting of wind characteristics has the potential to resolve these issues. To capture the overall trend of wind characteristics, the wind data are split into trend and seasonality using STL decomposition method. The trend and remainder part are modelled together using least square support vector regression (LSSVR) whose hyper-parameters are tuned using a recently proposed algorithm from the literature. The accuracy of the speed and direction model is 0.9752 and 0.9792, respectively. These models are then used to forecast the future speed and direction values. Using the forecasts, frequency maps are generated for the wind farm with a layout selected from the literature. The generated power from the wind farm is noted to be 11,535.97 kW when compared to the benchmark value of 11,548.02 kW.

Keywords Artificial intelligence · Renewable energy · Forecasting · Time series predictions · Evolutionary optimization

---

R. K. Inapakurthi · N. K. Pujari · K. Mitra (✉)
Department of Chemical Engineering, Indian Institute of Technology Hyderabad, Hyderabad, Telangana 502284, India
e-mail: kishalay@che.iith.ac.in

N. K. Pujari
e-mail: ch20resch11006@iith.ac.in

K. Mitra
Department of Climate Change, Indian Institute of Technology Hyderabad, Hyderabad, Telangana 502284, India

Department of Artificial Intelligence, Indian Institute of Technology Hyderabad, Hyderabad, Telangana 502284, India

K. Mitra et al. (eds.), *Optimization, Uncertainty and Machine Learning in Wind Energy Conversion Systems*, Engineering Optimization: Methods and Applications,
https://doi.org/10.1007/978-981-97-7909-3_8

8.1 Introduction

With increasing population, the global energy demand is expected to increase at an exponential rate. The same trend can be confirmed from Fig. 8.1. For a large part of human history, these energy demands are met through fossil fuel-based energy sources thereby causing stress on the environment, nature, and health of all biological forms. Such human-driven activities have created many unexpected extreme weather events like floods, cyclones, and wildfires (Zachariah et al., n.d.; Herring et al. 2014). These phenomena are not restricted to any country. For example, the precipitation in June 2013 in northern India was reported to be a century scale event (Herring et al. 2014). As per the National Centers for Environmental Information, the number of extreme weather events in the USA has risen considerably as can be seen in Fig. 8.2 (2020). Increase in technology has helped in reducing the number of human casualties by mass adoption of information dissemination mechanisms like radio and internet, though the damage due to extreme weather events has increased historically as shown in Fig. 8.3. The number of deaths has been largely fluctuating around a few thousand while the cost has gone up from less than a million dollars to a couple of million dollars.

With increase in global temperature, it has been estimated that the energy demand increases to cool the residential and commercial spaces thereby further increasing the stress on the environment (Ritchie and Roser 2023). This results in a self-reinforcing loop whose outcome would be difficult to predict and control. Another indication of growing global warming is the melting of glaciers successively for the past 33 years without any break. From the year the mass balance of glaciers was tracked, there has been a continuous decline in the glacier mass balance with some intermittent uptrends only due to reduced human activity. This trend can be visualized in Fig. 8.4 with the data collected from Blunden and Arndt (2020). The melting of glaciers results in an increase in sea level, loss of land mass, increases coastal erosion and results in intense and frequent hurricanes and typhoons and causes oceans to turn acidic (Sabine

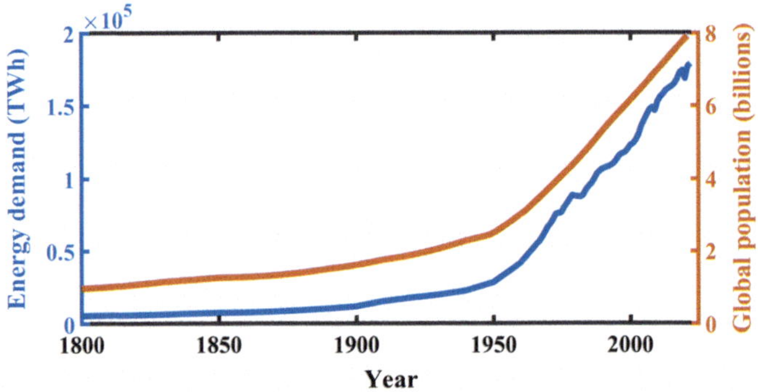

Fig. 8.1 Global total energy demand and population trends through history. As population increases, the global energy demand also increases

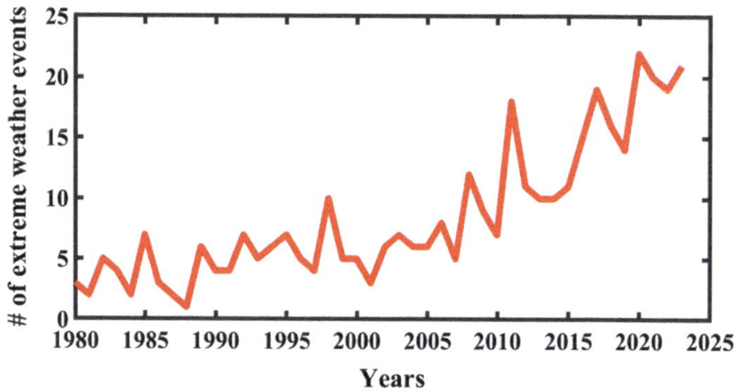

Fig. 8.2 A steady increase in the extreme weather events in the USA over the period 1980–2023

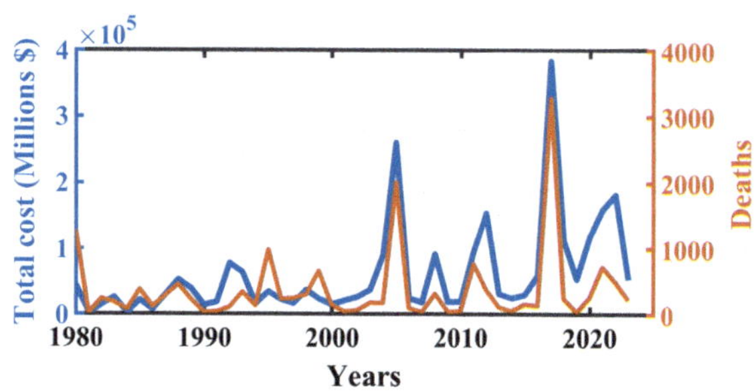

Fig. 8.3 Total cost (CPI adjusted) and deaths accrued due to extreme weather events since 1980 for the USA. The cost in million $ has been increasing

et al. 2004). As the human drive for energy continues, CO_2 emissions only increase with the years, and hence, the global temperature is going to rise inflicting further damage. For instance, the global CO_2 emissions and the corresponding change in temperature (due to the emission of greenhouse gases) compared to pre-industrial revolution period can be seen in Fig. 8.5. It is obvious that thrust must be placed on generating energy from sources which have lesser impact on the environment in terms of lesser CO_2 emissions. This helps, in a cascaded fashion, to reduce the rate of temperature rise considerably.

To meet the target of restricting the global warming of 1.5 °C compared to pre-industrial level, the share of renewable energy in the global energy demand basket needs to increase significantly. Many countries are taking bold steps in this direction by implementing policies incentivizing the generation and utilization of renewable energy (Kumar and Meena 2017) while penalizing the use of fossil fuel-based energy. With special emphasis on renewable energy sector worldwide, wind and solar energy are gaining traction. The global energy demand and the contribution of renewable energy towards the global energy needs can be understood from Fig. 8.6 (Ritchie et al. 2022). This disparity between renewable energy and total world energy demands is attributed to lack of awareness, global consensus of leading economies and industries, and volatility involved in the availability of renewable energy (like wind and solar). To a large extent, humans are responsible for this volatility, thanks to the continuous

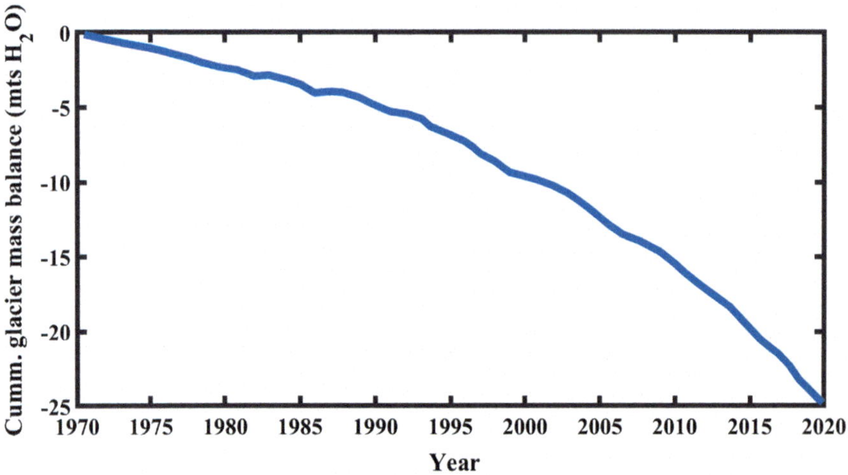

Fig. 8.4 Cumulative glacier mass balance measured in metres of water

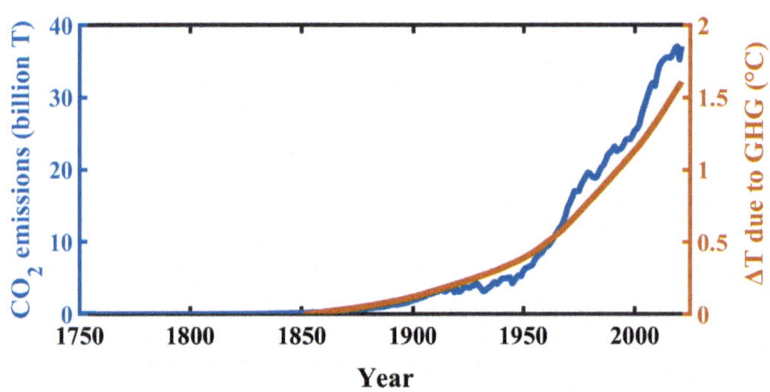

Fig. 8.5 CO_2 emissions trend and the change in temperature compared to pre-industrial revolution period. This drives the thrust for sustainable forms of energy generation

firing of fossil fuel-based energy sources and automobiles (Ritchie and Roser 2023). As per the estimates of ourworldindata.org, the contribution of energy sector to greenhouse gas emissions stands at 73.2% for the year 2016 alone (Ritchie and Roser 2023). This volatility also creates additional challenges in the design of energy storage systems (needed to store the captured renewable energy), and fluctuations in the grid. Leading industry leaders opine the dilute and intermittent availability of renewable energy across a given geographic location for not exploring renewable energy as potential sources of energy. This can be justified by looking at the wind speed and direction data collected from La Haute borne, France shown in Fig. 8.7.

With this understanding attained after thorough analysis of the global energy demands, causes for climate change and the effects thereof, it seems imperative that forecasting the availability of renewable energy would help in promoting the overall

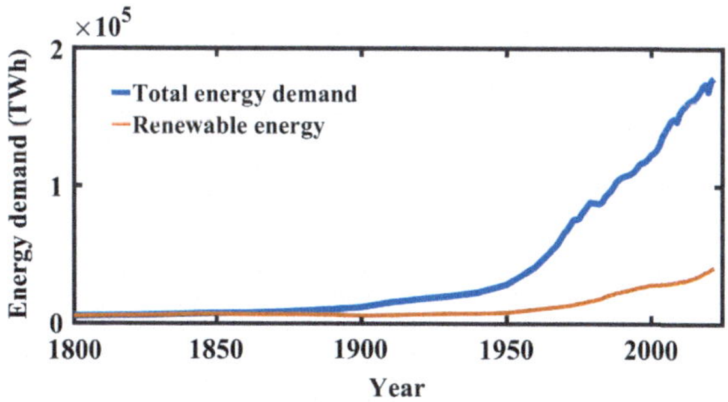

Fig. 8.6 Global total energy demand and energy demand contribution from renewable energy. The remaining energy is generated by fossil fuels

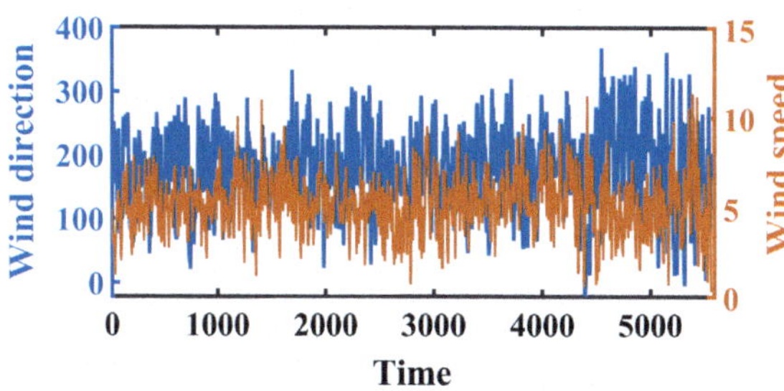

Fig. 8.7 Global intermittent nature of the wind speed and direction for the La Haute borne, France. With such huge fluctuations in the wind speed and direction data, it becomes difficult to estimate the power. The time resolution is 6 h

cause of fighting climate change. Also, it can be deciphered that the main aim of the forecasting exercise should be to provide accurate and reliable estimates of the future energy that can be extracted. This information will be useful for industries and state establishments to evaluate the techno-economic feasibility of setting up a renewable energy extraction plant like wind or solar farm in a geographic location. Modelling and forecasting of the actual wind speed and direction for the existing wind farms can also be performed in which the goal would be to integrate the control of yaw angle of the wind turbine to maximize the power output from a given farm layout. This can further facilitate robust design of energy storage systems which can help reduce the fluctuations in the grid to a considerable extent thereby tuning the dependence on fossil fuels.

With the growing popularity of artificial intelligence (AI) and big data, much research focus has been placed on the use of these techniques for forecasting wind speed and direction. Data being the basis for forecasting using AI techniques, different types of datasets are used for modelling. This includes structured time series data for diverse wind farms and unstructured image-based datasets (Islam et al. 2013; Pujari et al. 2023). It has been reported that, for time series problems with high volatility, the divide and conquer policy is more appropriate. Hence, empirical mode decomposition (EMD) was used for capturing the volatility in the wind speed in Luo et al. (2021) using multiobjective optimization (MOO) approach and the results were reportedly better than individual forecasting approach. In Lv and Wang (2022), a time series decomposition strategy is implemented to extract the linear patterns from wind speed data. Then, MOO algorithm is employed to optimize the time series decomposition parameters. Using Kruskal–Wallis test, the effectiveness of the proposed forecasting technique is established. In Wang et al. (2014), a combination of seasonal adjustment method, radial basis function and exponential smoothing function were used as ensemble method to forecast wind speed and tested on wind farm data from China. In Guo et al. (2011), the seasonal effects are eliminated from the data and modelled using back propagation neural network (BPNN) and exponential adjustments. Using the wisdom of crowd theory, the use of ensemble techniques is a great way to achieve forecasting accuracy. However, one major limitation of the above works is the use of wind speed only. If the direction of the wind changes, the net power that can be extracted would vary considerably resulting in inaccurate estimations. Also, real data collected from wind farms are seldom free of noise (missing values included) as such necessitating the use of data de-noising techniques as a precursor to forecasting stage.

As stated in Wu and Xiao (2019), data decomposition techniques such as EMD and wavelet decomposition are known to simplify the data structure (a crude example of this can be seen in Guo et al. (2011) where the seasonal effects of wind speed are eliminated). Though this can be considered as a good start, the actual volatility in the data is discarded during this process making the forecasts (or the forecasting approach, depending on how one sees) less likely to be applicable for long-time horizons. Note that the applicability of forecasts for long-time horizon would be of interest for industry leaders and policy makers to make informed decisions. In Akçay and Filik (2017), data de-trending is used to address the issue of missing values before employing a forecasting approach. Though the challenge of missing values is addressed with such mechanisms, a complete picture of the wind power remains missing as other important features are not considered during forecasting. Considering multiple dimensions for forecasting results in the curse of dimensionality. One way to mitigate this curse is to use feature dimensionality reduction techniques to identify and retain the most important features. This has additional advantages like reducing the training time as well as the chances of over-fitting among a host of other benefits. Another way to mitigate this curse is to resort to big data techniques. In Arun Jees and Gomathi (2019), k means algorithm and support vector machines (SVM) are used in Hadoop framework for load forecasting. In Galicia et al. (2018), Apache Spark framework is used for implementing random forests, and linear regression method. In Xu et al. (2020), Apache Spark is used for executing extreme learning method models for multistep ahead wind speed forecasting. The bottleneck with the big data methods is the silent assumption on the large-scale availability of computational resources, i.e., such methods can be explored by only those having resources at their dispense.

It is imperative to design and implement AI forecasting techniques which:

(a) Provides accurate forecast for a long-time horizon.
(b) Not dependent on the huge computational resources.
(c) Can take multiple dimensions like speed and direction.

From the above, the use of just one forecasting technique (instead of an ensemble of methods) is also possible as the computational overhead would be less for one technique. But this does not provide any slack to the forecasting method in terms of forecast accuracy. In fact, forecasting accuracy remains the topmost priority for the industry leaders and policy makers.

With the above requirements and as per our analysis of the literature, least square support vector regression (LSSVR) is explored in this study. There is ample literature to prove the capabilities of support vector class of techniques on diverse applications like EEG feature reduction (Yin and Zhang 2014), clustering tasks (Ben-Hur et al. 2001), market segmentation (Huang et al., n.d.), engine reliability prediction (Hong and Pai 2006), predictive modelling in steel industries (Acosta et al. 2021) and stock selection (Huang 2012), cost prediction for transportation sector (Woldemariam 2022), optimization of crystallization unit (Inapakurthi et al. 2022) among others. The novelties in the current study include the following:

(a) Using a recently proposed algorithm based on LSSVR to model the wind speed and direction for long-time horizons.
(b) With the LSSVR models, the future values of wind speed and direction are forecasted with high accuracy.
(c) Based on the forecasted values, power frequency maps are generated for a given layout.

The rest of the book chapter is organized as follows. In Sect. 8.2, the data source, implemented methodology and wind frequency map (WFM) generation approach are briefly outlined. In Sect. 8.3, the modelling results are presented. This includes demonstrating the efficacy of the modelling approach, model performance for both wind speed and direction, power frequency maps for a given layout. The conclusions are then summarized in the conclusion section.

8.2 Formulation

8.2.1 Wind Characteristics Data

The wind speed and direction data used in this study are generated by La Haute borne, France and accessed through an electric utility company ENGIE (Pujari et al. 2023). The wind data thus collected for four years was processed in the following way before modelling. Since the turbines in the wind farm are placed at the corners, it is assumed that there are no wake effects. In this work, the data are considered as two univariate time series: wind speed time series and wind direction time series. First, the data are analysed to check the nonlinearity (Akintunde et al. 2015), stationarity (Dickey and Fuller 1979) and long-term dependencies (Kalo et al. 2019) and then the model is formulated such that it accounts for long-term dependencies and nonlinearity. Since the time series data have hidden behavioural patterns which influence the sequence of the data, the time series data must be decomposed into trend, seasonal and residual components (Guignard et al. 2019). In this work, STL decomposition is used as it is proven to be efficient and robust (Pujari et al. 2023). The results of time series analysis and decomposition are presented in Sect. 7.3. To validate the proposed LSSVR model, the data is divided into two sets. Set 1 consists of 3 years of data which is used for training and validation. Set 2 consists of fourth year data which are unseen to the model. After training the model using set 1, one-year data are forecasted using LSSVR and the forecasted data are validated with set 2 which is benchmark data.

8.2.2 LSSVR Modelling and Approach and Optimization Formulation

LSSVR is based on the principle of structural risk minimization (SRM) in which any undue increase in model complexity (which offers only a slight increase in accuracy) is penalized. Hence, unlike the class of AI techniques based on empirical risk minimization, a trade-off parameter balancing the model complexity and slack variables is present. These trade-off parameters are represented by γ and λ in Fig. 8.8. If γ and λ are set to very large values, then the emphasis would be heavily on restricting the model complexity, i.e., model parameters while a very low values would indicate very less emphasis on the number of parameters. To capture the transient behaviour of the wind speed and direction, some number of previous information (or equivalently data points) needs to be considered as input to the LSSVR model. This helps in evolving the system state from the past states to the future state values. This is represented by the hyper-parameter M. A large value of M results in information overload and low values implies that the LSSVR model is not exposed all the required information to mimic the wind characteristics. Since it is possible to ascribe different kernel parameters to each of the inputs, the kernel parameter σ is also uniquely used for each of the wind characteristics. Finally, there are multiple kernels which can be explored for modelling (as shown in Fig. 8.8). This is because all data sets might not be modelled using Gaussian kernel to acceptable levels of accuracy. Together, M, σ, λ, γ, and kernel type constitute the hyper-parameters of LSSVR which needs to be tuned. The benefit of LSSVR is that it converts a primal formulation into its equivalent dual formulation which is a linear programming problem. Many off-the-shelf solvers can be used to solve this dual formulation. The formulation for the design of

LSSVR is taken from Inapakurthi and Mitra (2022b). It has been demonstrated that this algorithm is successful in determining the optimal hyper-parameters of LSSVR by considering multiple trade-offs in the model development stage. The authors in Inapakurthi and Mitra (2022b) have used non-dominated sorting genetic algorithm (NSGA-II). For the sake of demonstration, the same algorithm has been employed in the current study as well. For the sake of completeness, the algorithm is presented briefly here. In the domain of machine learning (ML), the algorithms need both extensive data as well as parameters to model the system under consideration with reasonably high accuracy. While the discussions around the need for large number of data points is beyond the scope of this book chapter, it can be demonstrated that an increase in parameters of the model would result in increase in the accuracy of the model as well. However, these models might not generalize well, i.e., in the face of unseen data these models may par below their training scores. Since this is against the very idea of model development (most of the time models are developed to forecast the system behaviour or to perform parameter estimation), it is essential that any increase in the parameters of the model does not result in a deterioration in the performance of the model on the unseen data. Combining these requirements together, the authors in Inapakurthi and Mitra (2022b) have presented a mathematical formulation considering the conflicting nature of the parameters and accuracy in an optimization framework. These objectives, while being optimized, are subjected to the tunable parameters of LSSVR such that the need for separate tuning of the parameters is eliminated.

Multi-Objective Optimization:

$$\min_{\gamma,\lambda,\boldsymbol{M},\boldsymbol{\sigma},kt} \; RMSE$$

$$\min_{\gamma,\lambda,\boldsymbol{M},\boldsymbol{\sigma},kt} \; Number\ of\ parameters$$

subject to,

$$\gamma^{LB} \le \gamma \le \gamma^{UB}$$

$$\lambda^{LB} \le \lambda \le \lambda^{UB}$$

$$M_i^{UB} \le M_i \le M_i^{UB}$$

$$\sigma_i^{UB} \le \sigma_i \le \sigma_i^{UB}$$

$$kt = \begin{cases} 1, Gaussian: \exp\left(-\dfrac{\|\boldsymbol{X}_1 - \boldsymbol{X}_2\|^2}{2\boldsymbol{\sigma}^2}\right) \\ 2, Inverse\ quadric: \dfrac{1}{1 + \dfrac{\|\boldsymbol{X}_1 - \boldsymbol{X}_2\|^2}{2\boldsymbol{\sigma}^2}} \\ 3, Inverse\ multi\ quadric: \dfrac{1}{\sqrt{1 + \dfrac{\|\boldsymbol{X}_1 - \boldsymbol{X}_2\|^2}{2\boldsymbol{\sigma}^2}}} \end{cases}$$

Fig. 8.8 Multiobjective optimization formulation used in the current study. The objectives of accuracy and number of parameters are conflicting in nature. These objective functions are subjected to the hyper-parameters of LSSVR

8.2.3 Wind Frequency Map Generation

In any traditional wind farm studies, the wind characteristics are modelled as joint probability distribution using two random variables: wind speed and wind direction and the corresponding probability mass function is called WFM. These WFMs are constructed using the frequentists approach as presented below:

(a) Divide wind speed data into equal intervals called speed bins. In this study, the range of speed is 0–16 m/s. Hence, it is divided into 16 bins, and each bin is in range of 1 m/s.
(b) Similarly divide the direction into 12 equal intervals called sector. Each sector is in the range of 30°.
(c) Count the number of data points in ith speed bin and jth direction sector, N_{ij}.
(d) Since WFM is the probability mass function of speed and direction, the probability value i.e., the frequency is determined as $f_{ij} = \frac{N_{ij}}{N}$, where N is the total number of data points in the time series data.

The annual energy production from a wind farm is a function of WFM and the effective velocity of a turbine. Conventionally, WFMs are modelled using the available past data which is very less. Constructing a WFM with less data may not be efficient to design the wind farm which lasts long for 25–30 years. Hence, in this study, the applicability of LSSVRs in wind energy conversion systems is demonstrated using their efficient use in forecasting the wind characteristics (speed and direction) to construct an accurate WFM while designing the wind farm with maximum power production.

8.3 Results and Discussions

8.3.1 Time Series Analysis LSSVR

The data set consists of wind speed and direction each of length 5566 points where each data point corresponds to a six-hour time resolution. For better model accuracy, a single input single output modelling is performed. Hence, x_t as input would result in x_{t+1} as output where x is wind speed/direction for any time instance t. This would result in a data set of size 5565 points with 2 columns where the first column is the input, and the second column is the output. Of this, 3067 points (55.2%) are used for training LSSVR model, 1315 points (23.6%) are used for validation and the remaining 1183 points (21.2%) are used to forecast the wind speed and direction. The main emphasis here was to have a split in which the validation and testing have more than 20% share and train the model with a smaller number of data points to test the efficacy of the proposed model development algorithm. The parameter values

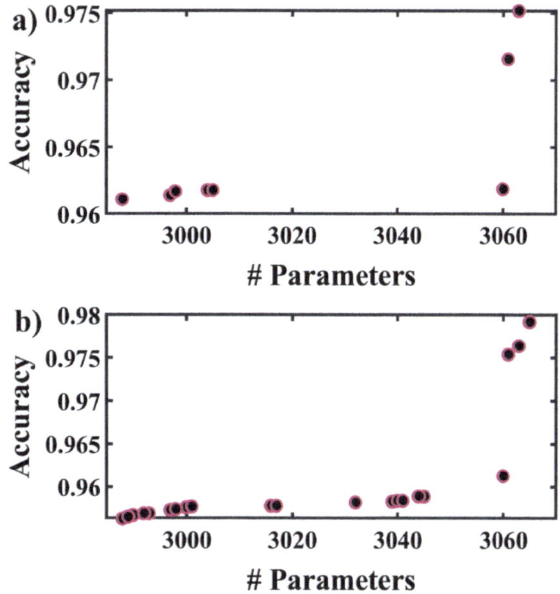

Fig. 8.9 Pareto front for **a** wind speed and **b** wind direction

used for determining LSSVR models for wind speed and direction are as follows: $\gamma^H = 2^{17}, \sigma^H = 2^8, \lambda^H = 2^{12}$, $\gamma^L = 2^{-5}, \sigma^L = 2^{-15}, \lambda^L = 2^{-10}, M^L = 1$ and $M^H = 80$. The simulations were executed till 200 generations of NSGA-II and each generation consisted of 100 populations.

The converged Pareto front at the end of the simulations is reported in Fig. 8.9 for wind speed and direction. From these multiple Pareto optimal LSSVR models for wind speed and direction, one model for each of the wind characteristics is selected using Akaike Information Criterion as reported in Inapakurthi and Mitra (2022a). The selected model is the one with the highest model accuracy. The performance of this model on the validation and test data is shown in Fig. 8.10. One can see that, despite forecasting for long duration (~ 296 days), the error (defined as the difference between the ground truth and the model predictions) is very low. With the performance of the selected models now established, the models are used for the generation of WFM which will be used for calculation of power that can be extracted from the given wind farm layout.

8.3.2 Importance of Forecasting Wind Farm Layout Design

As the utility of the LSSVR model for accurate forecasting of wind speed and wind direction is demonstrated, the significance of forecasting is shown below by determining the energy production from an optimal layout of wind farm using different frequency maps generated using the past and forecasted data. WFM_{avg} is

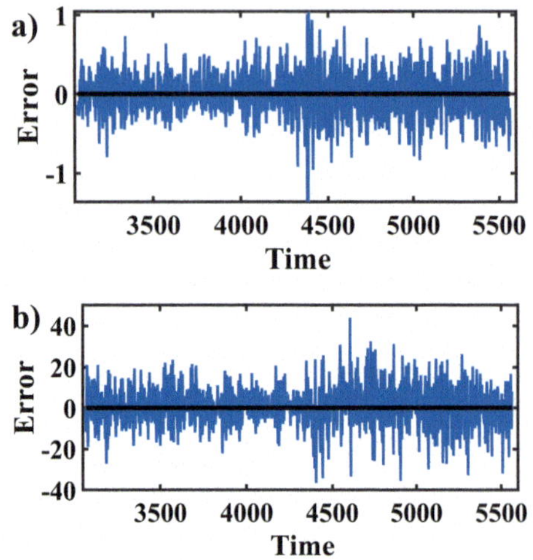

Fig. 8.10 Error plot with the best AIC model for **a** wind speed and **b** wind direction. Plots are shown for validation and forecasting data

the frequency map constructed using the available past data of 3 years, i.e., set 1 as mentioned in Sect. 8.2.1. WFM_{tr} is the frequency map constructed using 3rd year data. WFM_{bm} is the frequency map constructed using the 4 years data, i.e., set 1 and set 2 as mentioned in Sect. 8.2.1. WFM_{pp} is the frequency map constructed using the past data and forecasted data. All these frequency maps are presented in Fig. 8.11. Annual energy production is calculated for all these WFMs using Eq. (8.1), and the results are presented in Table 8.1.

$$\text{Pwr} = \sum_{i=1}^{N_t} \sum_{\text{dir}=1}^{12} \sum_{\text{sp}=1}^{16} p_{\text{curve}}\left(u_{\text{eff}}\left(u_{\text{sp}}, \emptyset_{\text{dir}}, N_t\right)\right) \times \text{WFM}\left(u_{\text{sp}}, \emptyset_{\text{dir}}\right) \quad (8.1)$$

where, Pwr is the expected power produced from a wind farm, N_t is the number of turbines, $p_{\text{curve}}\left(u_{\text{eff}}\left(u_{\text{sp}}, \emptyset_{\text{dir}}, N_t\right)\right)$ is the power curve expressed as a function of wind speed, obtained from turbine manufacturer, u_{eff} is the wind speed obtained for a given turbine after applying a wake model, u_{sp}, $\emptyset_{\text{dir}}$ are the values of speed and direction at sp-th interval and dir-th intervals, respectively. The analysis by determining the power production from several frequency maps has demonstrated that the WFM_{pp} which is constructed with past and forecasted data gave a best approximation with WFM_{bm} which is constructed using the original 4 years data, and this gives the justification of importance of forecasting wind energy conversion systems.

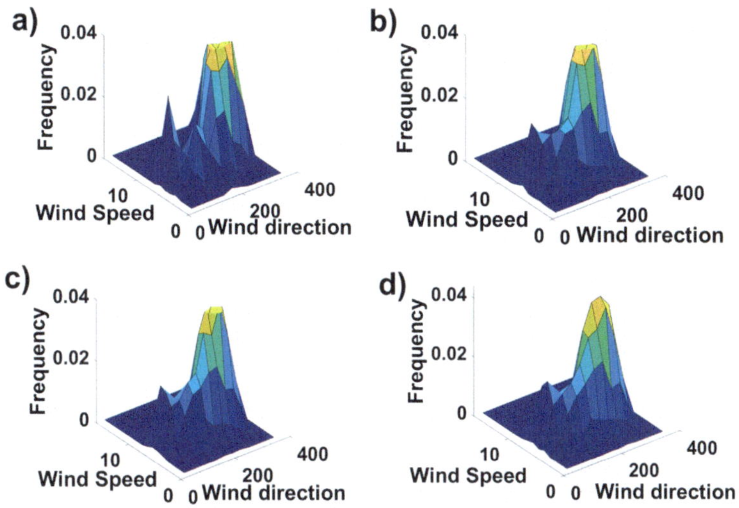

Fig. 8.11 Wind frequency maps for **a** WFM_{pp}, **b** WFM_{bm}, **c** WFM_{avg} and **d** WFM_{tr}

Table 8.1 Comparison of AEP values for considered WFMs

S. No.	Frequency map	AEP (kW)
1	WFM_{avg}	12,273.48
2	WFM_{tr}	14,739.92
3	WFM_{bm}	11,548.02
4	WFM_{pp}	11,535.97

8.4 Conclusion

With growing focus on renewable energy generation and utilization, the focus shifts towards the techno-economical aspects of renewable energy. In the context of capturing wind energy, for example, this includes varying sub-fields of interest like wind farm layout optimization and forecasting of wind speed and direction for a given location. In this study, the focus is on the technical aspect of forecasting wind speed and direction using AI. The potential implications of the study are real and immediate, meaning that the results from this study can be used for understanding wind characteristics in the future with high accuracy. This was accomplished by employing a newly proposed algorithmic development framework. This has resulted in high-accuracy models for both wind speed and direction. As the framework also incorporates the tuning of LSSVR parameters, the heuristic-based approach can be foregone altogether thereby further enhancing the usability of such algorithms. Forecasts of speed and direction from these models are also found to be in good agreement with the ground truth. Since such forecasts are made without the actual data in place, this information can be considered goldmine for evaluating the technical aspects. This has further implications in that the generated power frequency maps paint a more accurate picture thereby providing a better understanding of the wind distribution and enhancing the quality of the assessment of economic viability of the project.

Acknowledgements We would like to acknowledge the support of SPARC project (SPARC/2018-2019/P1084/SL) sponsored by the Ministry of Education (MoE), Government of India and National Supercomputing Mission (NSM) for the project (DST/NSM/R&D_HPC_Applications/2021/23) sponsored by the Department of Science and Technology (DST), Government of India.

References

Acosta SM, Amoroso AL, Sant'Anna ÂMO, Junior OC (2021) Predictive modeling in a steelmaking process using optimized relevance vector regression and support vector regression. Ann Oper Res 2021:1–22. https://doi.org/10.1007/S10479-021-04053-9

Akçay H, Filik T (2017) Short-term wind speed forecasting by spectral analysis from long-term observations with missing values. Appl Energy 191:653–662. https://doi.org/10.1016/J.APENERGY.2017.01.063

Akintunde MO, Oyekunle JO, Olalude GA (2015) Detection of non-linearity in the time series using BDS test. Sci J Appl Math Stat 3:184. https://doi.org/10.11648/J.SJAMS.20150304.13

Arun Jees S, Gomathi V (2019) Load forecasting for smart grid using non-linear model in Hadoop distributed file system. Cluster Comput 22:13533–13545. https://doi.org/10.1007/S10586-018-1997-2/METRICS

Ben-Hur A, Horn D, Siegelmann HT et al (2001) Support vector clustering

Blunden J, Arndt DS (2020) State of the climate in 2019. Bull Am Meteorol Soc 101:S1–S429. https://doi.org/10.1175/2020BAMSSTATEOFTHECLIMATE.1

Dickey DA, Fuller WA (1979) Distribution of the estimators for autoregressive time series with a unit root. J Am Stat Assoc 74:427. https://doi.org/10.2307/2286348

Galicia A, Torres JF, Martínez-Álvarez F, Troncoso A (2018) A novel spark-based multi-step forecasting algorithm for big data time series. Inf Sci (N Y) 467:800–818. https://doi.org/10.1016/J.INS.2018.06.010

Guignard F, Lovallo M, Laib M et al (2019) Investigating the time dynamics of wind speed in complex terrains by using the Fisher–Shannon method. Physica A 523:611–621. https://doi.org/10.1016/J.PHYSA.2019.02.048

Guo ZH, Wu J, Lu HY, Wang JZ (2011) A case study on a hybrid wind speed forecasting method using BP neural network. Knowl Based Syst 24:1048–1056. https://doi.org/10.1016/J.KNOSYS.2011.04.019

Herring SC, Hoerling MP, Peterson TC, Stott PA (2014) Explaining extreme events of 2013 from a climate perspective. Bull Am Meteorol Soc 95:S1–S104. https://doi.org/10.1175/1520-0477-95.9.S1.1

Hong WC, Pai PF (2006) Predicting engine reliability by support vector machines. Int J Adv Manuf Technol 28:154–161. https://doi.org/10.1007/S00170-004-2340-Z

Huang CF (2012) A hybrid stock selection model using genetic algorithms and support vector regression. Appl Soft Comput 12:807–818. https://doi.org/10.1016/J.ASOC.2011.10.009

Huang J-J, Tzeng G-H, Ong C-S. Marketing segmentation using support vector clustering. https://doi.org/10.1016/j.eswa.2005.11.028

Inapakurthi RK, Mitra K (2022a) Machine learning based surrogate assisted multi-objective optimization of continuous casting process, pp 283–288. https://doi.org/10.1109/ICC54714.2021.9703180

Inapakurthi RK, Mitra K (2022b) System identification and process modelling of dynamic systems using machine learning. In: 2022 26th international conference on system theory, control and computing, ICSTCC 2022—proceedings, pp 564–569. https://doi.org/10.1109/ICSTCC55426.2022.9931831

Inapakurthi RK, Naik SS, Mitra K (2022) Toward faster operational optimization of cascaded MSMPR crystallizers using multiobjective support vector regression. Ind Eng Chem Res 61:11518–11533. https://doi.org/10.1021/ACS.IECR.2C00526/SUPPL_FILE/IE2C00526_SI_001.PDF

Islam MR, Mekhilef S, Saidur R (2013) Progress and recent trends of wind energy technology. Renew Sustain Energy Rev 21:456–468. https://doi.org/10.1016/J.RSER.2013.01.007

Kalo L, Kamalanathan P, Pant HJ et al (2019) Mixing and regime transition analysis of liquid-solid conical fluidized bed through RPT technique. Chem Eng Sci 207:702–712. https://doi.org/10.1016/J.CES.2019.07.005

Kumar S, Meena RL (2017) Renewable energy sources-policies in India. Int J Appl Environ Sci 12:293–297

Luo L, Li H, Wang J, Hu J (2021) Design of a combined wind speed forecasting system based on decomposition-ensemble and multi-objective optimization approach. Appl Math Model 89:49–72. https://doi.org/10.1016/J.APM.2020.07.019

Lv SX, Wang L (2022) Deep learning combined wind speed forecasting with hybrid time series decomposition and multi-objective parameter optimization. Appl Energy 311:118674. https://doi.org/10.1016/J.APENERGY.2022.118674

Pujari KN, Miriyala SS, Mittal P, Mitra K (2023) Better wind forecasting using evolutionary neural architecture search driven green deep learning. Expert Syst Appl 214:119063. https://doi.org/10.1016/J.ESWA.2022.119063

Ritchie H, Roser M (2023) Sector by sector: where do global greenhouse gas emissions come from? Our World in Data

Ritchie H, Roser M, Rosado P (2022) Energy. Our World in Data

Sabine CL, Feely RA, Gruber N et al (1979) The oceanic sink for anthropogenic CO_2. Science 305:367–371. https://doi.org/10.1126/SCIENCE.1097403/SUPPL_FILE/SABINE.SOM.PDF

Wang J, Zhang W, Wang J et al (2014) A novel hybrid approach for wind speed prediction. Inf Sci (N Y) 273:304–318. https://doi.org/10.1016/J.INS.2014.02.159

Woldemariam W (2022) A framework for transportation infrastructure cost prediction: a support vector regression approach. Transp Lett. https://doi.org/10.1080/19427867.2021.1985895

Wu Z, Xiao L (2019) A secondary decomposition based hybrid structure with meteorological analysis for deterministic and probabilistic wind speed forecasting. Appl Soft Comput 85:105799. https://doi.org/10.1016/J.ASOC.2019.105799

Xu Y, Liu H, Long Z (2020) A distributed computing framework for wind speed big data forecasting on Apache Spark. Sustain Energy Technol Assess 37:100582. https://doi.org/10.1016/J.SETA.2019.100582

Yin Z, Zhang J (2014) Identification of temporal variations in mental workload using locally-linear-embedding-based EEG feature reduction and support-vector-machine-based clustering and classification techniques. Comput Methods Programs Biomed 115:119–134. https://doi.org/10.1016/J.CMPB.2014.04.011

Zachariah M, Achutarao K, Saeed F et al. Climate change made devastating early heat in India and Pakistan 30 times more likely

(2020) U.S. billion-dollar weather and climate disasters, 1980–present. National Centers for Environmental Information. https://doi.org/10.25921/STKW-7W73

Chapter 9
Handling Wind Stochasticity in Farm Layout Optimization: A Novel Data-Driven Approach

Priyanka Devi Pantula, NagaSree Keerthi Pujari, and Kishalay Mitra

Abstract Wind energy has its own nature of stochasticity which remains a challenge for wind farm designers to extract energy from it as expected. This happens because the sources of uncertainty are not considered during the design calculations for the sake of simplicity. Such adaptation of a deterministic approach for a stochastic problem often gives erroneous as well unrealistic results, which is not implementable as well as useful during a realistic situation. One approach to handle such uncertainty is to consider the layout optimization as a stochastic optimization problem, which has been solved in this case using robust optimization. As several uncertain realizations are necessary to generate such a robust solution, daily average WFMs obtained from the real-life wind data are collected, and Convolutional Auto-Encoders are utilized to extract latent vectors from them. Using a novel clustering algorithm, these latent vectors are clustered into different groups to generate new WFMs, which in turn can be used for wind farm micrositing under uncertain circumstances.

Keywords Wind characteristics modeling · Convolutional auto-encoders · Robust optimization · Optimization under uncertainty · Clustering

---

P. D. Pantula · N. K. Pujari
Department of Chemical Engineering, Indian Institute of Technology, Hyderabad, Kandi, Telangana 502285, India
e-mail: ch20resch11006@iith.ac.in

K. Mitra (✉)
Department of Chemical Engineering, Indian Institute of Technology Hyderabad, Hyderabad, Telangana 502284, India
e-mail: kishalay@che.iith.ac.in

Affiliated Faculty, Department of Climate Change, Indian Institute of Technology Hyderabad, Hyderabad, Telangana 502284, India

Affiliated Faculty, Department of Artificial Intelligence, Indian Institute of Technology Hyderabad, Hyderabad, Telangana 502284, India

K. Mitra et al. (eds.), *Optimization, Uncertainty and Machine Learning in Wind Energy Conversion Systems*, Engineering Optimization: Methods and Applications,
https://doi.org/10.1007/978-981-97-7909-3_9

9.1 Introduction

The power generated by an individual wind turbine in a wind farm depends on stochastic variables, wind speed and direction, though most popularly such stochasticity is ignored and a deterministic formulation is adopted. Adaptation of a deterministic approach, where reality shows the presence of a stochastic variable, is not only erroneous but also unrealistic. Let's discuss this with power calculation formula of a wind farm. Let RPow represent the manufacturer-provided rated power of a turbine. Utilizing the joint Probability Mass Function (PMF), the expected power (EP) is calculated using Eq. (9.1), where *Nt* is the total number of turbines in a wind farm, *i* is the discretized values of wind speed, *j* is the discretized sectors of wind direction and, wind speed and direction are denoted as u_i and θ_j, respectively (Mittal and Mitra 2020)

$$\mathrm{EP} = \sum_{Nt}\sum_{i}\sum_{j}\left(\mathrm{RPow}(u_i)\ \mathrm{WFM}\left(u_i, \theta_j\right)\right) \tag{9.1}$$

As RPow is provided through advanced experimentation by manufacturers, it is presumed to be highly precise. In contrast, wind-frequency maps (WFMs) are created from a limited dataset of speed and direction of wind, potentially leading to inaccuracies (Tony Burton et al. 2011). Consequently, this can introduce errors into the calculation of EP. As EP can be used as an optimization objective in many wind farm layout optimization and control problems (Mosetti et al. 1994; Mittal and Mitra 2018; Pujari et al. 2021), the precision of EP heavily relies on the accuracy of WFMs. Thus, this chapter explores how to model WFMs using Convolutional Auto-Encoders (CAEs) (Guo et al. 2017), followed by a novel generative modeling approach for creating new WFMs over the considered layout, which in turn helps in performing the wind farm micrositing under uncertain circumstances (Aharon Ben-Tal et al. 2009).

In essence, WFMs play a pivotal role in designing and managing wind farms that are expected to operate for over five decades. However, in practice, acquiring a comprehensive dataset spanning such a long duration of wind characteristics remains a challenge. This is in contrast to the vast set of other image-related applications in other domains of research, such as biomedical image analysis (Moen et al. 2019), industrial automation (Mistry et al. 2020), remote sensing (Zhu et al. 2017), and 3D imaging (Hagita et al. 2018). The scarcity of data could arise from various practical limitations, such as economic challenges in recording and storing data, the high expenses associated with sensors for capturing wind characteristics, as well as issues with data accuracy or completeness. These limitations result in the creation of PMFs with a reduced amount of available data. However, to achieve a reliable approximation of the joint PMF, larger the number of data points, better it is (Etemadi 1981). Consequently, due to the insufficient number of data points, a less precise PMF is likely to be constructed leading to inaccurate energy estimates for the wind farms.

Typically, optimization problems exhibit a non-deterministic nature, necessitating decision-making under uncertain conditions (Fang and Sansavini 2019). Assuming deterministic nature for ease of handling may render optimization studies unrealistic and less reliable in practical scenarios. To address such challenges, optimization under uncertainty (OUU) has gained attraction in both industrial and academic domains in recent years (De Jonge et al. 2015; Fang and Sansavini 2019). Various approaches that are probabilistic in nature such as Stochastic Programming (SP) (Shapiro 2008), Chance Constrained Programming (CCP) (Liu et al. 2016), Robust Optimization (RO) (Aharon Ben-Tal et al. 2009), and Expected Value Model (EVM) (Tian et al. 2011) have been well documented for handling OUU. Corresponding fuzzy counterparts such as FCCP (Virivinti and Mitra 2015), FRO (Virivinti and Mitra 2017), and FEVM (Liu and Liu 2002) are equally prevalent. When uncertain data are quantifiable, the probabilistic techniques are typically the preferred choice over their fuzzy counterparts.

Among the techniques for optimization those need to handle uncertainty inherent to the system, RO stands out as a popular alternative due to its efficiency in handling uncertainties, coupled with its computational feasibility (Aharon Ben-Tal et al. 2009; Bertsimas et al. 2018). Coming to how RO addresses an uncertain optimization problem, it formulates an equivalent deterministic model that ensures constraint satisfaction across all possible realizations of the uncertain variables. Such solutions are known as robust as well as feasible for the obvious reasons. Amidst such solutions, one that provides the best objective value under the setting of the worst case of uncertain parameters is referred to as the worst-case value. The conservatism inherent in RO solutions may be attributed to the methodology used to characterize and sample the uncertain parameter hyper-space representing instantiations of uncertain scenarios. This arises from the necessity to generate a finite number of instances of uncertain scenarios to simulate situations effectively. Here, accurate characterization of the uncertain parameter space is extremely important as the idea is to sample such parameter space only from the relevant regions leaving the irrelevant regions and attacking the target of accurate sampling in a very focused manner. One of the popular as well as easier ways of sampling this uncertainty set is the box approach (Bertsimas et al. 2011), where sampling is spread across randomly exploiting the entire bounds of the uncertain parameters. In case the data present in uncertain parameter space are sparse, conservatism may result from such an approach as it samples in regions where data are not present. Addressing this issue may necessitate adaptive sampling techniques to sample data more accurately. However, true identification of uncertain regions, accurate sampling from them, and scaling these methods to higher dimensional problems pose significant challenges in realizing efficient RO formulations.

In this chapter, to overcome the aforementioned issues, a novel deep learning algorithm is amalgamated with RO such that the solutions are not only realistic but also less conservative. As a lot of empty regions are visible in a WFM, the regions of interest from such images are identified using CAE such that the limited amount of wind characteristics data can be used for capturing the important features from them. The primary aim is to cluster the WFMs efficiently, followed by generating more such maps within the clusters as RO needs such scenarios to provide solutions robust with respect to all such scenarios. To achieve this objective, the CAEs are used for identifying the WFMs. During training, the CAE takes these WFMs as image inputs and learns to capture important features from them resulting in the latent vectors. The obtained latent vectors are next passed through a novel clustering algorithm such that the latent vectors of similar nature are grouped together and dissimilar latent vectors are grouped differently. Using another feature of CAE, which uses a decoder network, generally a mirror replica of the encoder network, the clustered latent vectors can be converted into WFM images. Providing several image inputs, CAEs are trained efficiently so that they can replicate the same image with very high accuracy, with the latent vectors as the intermediate product. Thus, a generative modeling framework has been devised for usage in the modeling of wind farm micrositing under uncertainty using robust optimization.

Motivated by the potential of machine/deep learning algorithms and data analytics, uncertainties associated with model parameters are analysed in this chapter using a recently published approach called data-driven robust optimization through intelligent fuzzy transcription (DRIFT), which has been recently developed by the authors (Pantula and Mitra 2020). However, image data, which are more common nowadays, were not considered for analysis in (Pantula and Mitra 2020). The DRIFT algorithm employs the soft clustering technique to identify regions with uncertain data within the provided space, enabling targeted and efficient sampling in those areas. The presence of uncertain data renders it probabilistic, distinguishing it conceptually from Fuzzy RO. Presented below is a concise overview of the method used in this chapter, highlighting its differentiation from other works documented in the literature.

- Assume restricted number of samples are provided in terms of characterization of the uncertain space, and the samples are dispersed throughout the entire uncertain parameter range, not necessarily following any well-known statistical distribution.
- Use novel fuzzy clustering algorithm to cluster the uncertain parameter space efficiently with determination of optimal cluster number.
- Identify cluster boundary using the density of neighbouring points.
- Use Delaunay Triangulations to establish boundaries for each cluster.
- Generate Sobol sampling data points (Dige and Diwekar 2018) inside the hyper-box created using bounds of each cluster.
- Consider only data that lie inside the cluster boundaries.
- Perform RO using data points within boundaries, thereby focusing only on the regions of interest.

This framework presents promising prospects for enhancing the solution quality of RO by mitigating conservatism. As a result, the outcomes are likely to be more realistic and dependable, independent of the dimensionality of the uncertain parameter space. Also, the proposed idea in this work can be extendable to any other images generated from different domains of research.

The rest of the paper is organized as follows: Sect. 9.2 presents the data description and the procedure to construct the WFMs, followed by description of DRIFT. The results and discussions of DRIFT along with the significance of WFMs in wind energy conversion systems are presented in Sect. 9.3. Section 9.4 summarizes the conclusions of current work.

9.2 Formulation

9.2.1 *Creation of WFM Images*

Over the span of four years, four turbines of equal height are positioned at the corners of a 9 sq. km rectangular field in La Haute Borne. ENGIE, a French electric utility company, assists in gathering wind speed and direction data by installing sensors on these turbines (ENGIE 2020). Due to the considerable distance between the turbines, potential wake effects are disregarded in the collected data. After aggregating the data from the four turbines, a time series dataset compromising 210,240 data points for wind direction and speed is generated, with a resolution of 10 min. over the four-year period. To address potential noise in the sensor measurements, a commonly used technique known as moving average is employed to refine the data, enhancing its preprocessing.

Before employing CAEs, how the dataset is compiled is detailed in this section. Utilizing the dataset mentioned above, with a time interval of $T = 1$ day and measurements every 10 min. ($N_T = 24$ hours $\times$ 6 data in an hour), daily WFMs are constructed. Similarly, daily WFMs can be constructed over 4 years, i.e., 1460 of them (365 days in a year $\times$ 4 years). To be more precise, due to the fourth-year data incompleteness, only 1400 WFMs could be prepared. Next, the WFMs are converted into grayscale heatmaps (as shown in Fig. 9.1). With 0 representing black and 1 representing white and the intermediate values representing other gray shades, probability values in the joint PMF matrix (12 direction sector $\times$ 16 different wind speeds) are given different shades of gray. In a grayscale image, pixel values denote brightness or intensity, forming a matrix whose size corresponds to the image resolution. Consequently, the heatmaps of the joint PMF are constructed as grayscale images with probability values P_{ij} signifying pixel intensities on a normalized scale of 0–1. This constitutes the dataset of 1400 WFM images utilized for CAE training, validation and testing. Plotting the frequency values over the z-axis, a 3 dimensional surface plot is generated, as illustrated in the final plot of Fig. 9.1.

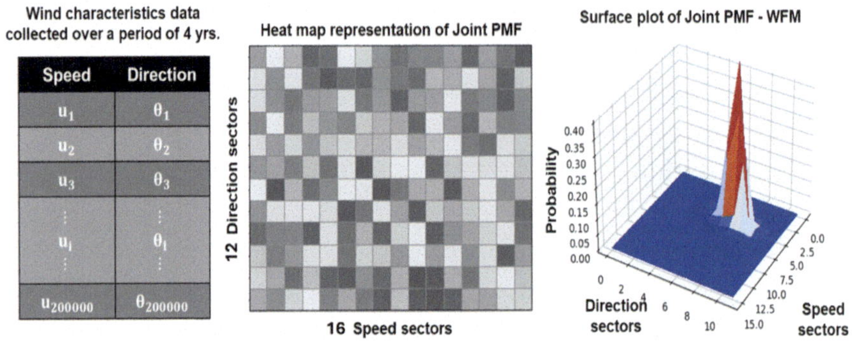

Fig. 9.1 Generation of wind frequency maps using available data

9.2.2 Modeling and Clustering of WFMs Using Convolutional Auto-encoders (CAE) and Image-Neuro-Fuzzy C-means Clustering (I-NFCM) Algorithm

CAEs, as outlined by (Guo et al. 2017), are employed to study these WFMs by converting their pixel values into a feature vector of lower dimensions. Subsequently, the obtained compressed feature vectors are clustered to identify WFMs with similar / dissimilar nature. Initially, CAEs are used to study the WFMs, where the CAE network replicates the input images through encoding and decoding processes. These operations are presented here. First, the convolution operation is applied on the images to extract important features. Typically, in convolutional neural networks (CNNs), following operations occur, as detailed by (Albawi et al. 2017):

(i) The provided image undergoes convolution with a filter (kernel) with bias incorporated into it,
(ii) Transfer function introduces non-linearity,
(iii) Pooling facilitates additional extraction of features.

A single convolutional layer consists of numerous nodes and a series of such operations result in the dimension reduction of the image and the feature extraction happens in parallel. Multilayer perceptron (MLP), a variant of artificial neural network (ANN), takes these features after flattening them into a vector to produce the corresponding output. The extracted features may not hold any specific physical interpretations. The true label is essential along with the network output, enabling the CNN weights to be learned throughout the training process. Figure 9.2 depicts a simplistic representation of CNN. However, in this case, the original output being not known, the challenge of feature extraction from images is intensified.

Handling the aforementioned challenge, an auto-encoder (AE) (Guo et al. 2017), ANN presented in unsupervised way, is employed. This AE learns to decipher the features present in the input image and takes the same image as its target output.

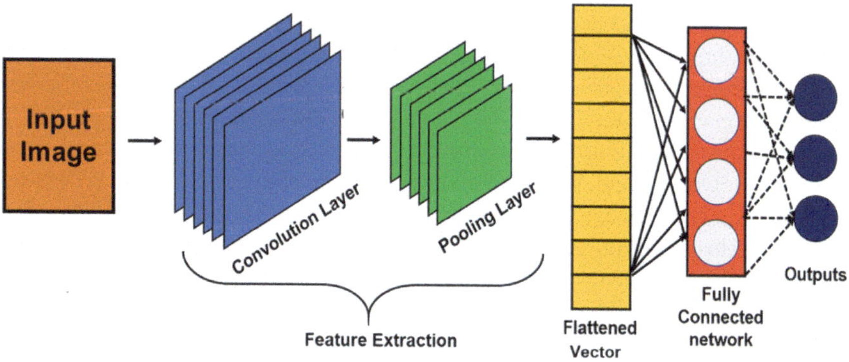

Fig. 9.2 Working of a basic CNN given a random image of size 14×14

Combination of convolutional layers in an AE framework leads to formation of CAEs. AEs consist of two primary components: encoder, responsible for mapping input data to a compressed latent vector, and decoder, tasked with the act of mirror imaging the encoding, i.e. reconstructing the original input. In this context, given that the input data is WFMs, CAEs are utilized to extract features from them via convolution operation (encoding). The resultant latent vectors are then employed to reconstruct the input images through deconvolutional operations (decoding). Putting in other words, first the convolutional layer facilitates feature to be extracted from the provided images followed by the deconvolution operation recreating the input images. Deconvolution is operated in the reverse order compared to the convolution. This involves reshaping the latent vector followed by up-sampling, reverse of the pooling operation. Subsequently, padding is applied to ensure dimensional compatibility, resulting in the desired image size for subsequent operations. After padding, deconvolution is performed to obtain the image.

Like convolution layer, deconvolution layer uses weights, which are to be learnt during training of CAE. In the supervised learning framework, when presented with an image, several operations are carried out sequentially: (i) convolution, (ii) pooling, (iii) flattening [these three constitute convolution layer], (iv) reshaping, (v) up-sampling, (vi) padding, and (vii) de-convolution [these four constitute deconvolution layer]. The total parameters are now the weights associated with both convolution and deconvolution kernels, which are to be evaluated during the training process. Loss function is defined using the disparity between the reconstructed image and the original image, which was given in the beginning as an input and this is minimized during the training process.

Typically, in practical applications, the dimensions of both kernels are maintained to be identical. Additionally, for the sake of versatility, a network compromising multiple layers is employed in CAEs instead of a single operation for each. Furthermore, a symmetric MLP network is used along with the flattened feature vector. Finally, the complete network, comprising convolutional layers, MLPs, and deconvolutional layers, is symmetric with respect to the latent vector. Therefore, given an image, once it is encoded into a latent vector, it can be subsequently decoded to generate an output image due to symmetry around the latent vector.

Once extracted, the latent vectors serve as representations of the images. Subsequently, these vectors are used in the Neuro-Fuzzy C-means (NFCM) algorithm, a novel soft clustering algorithm, for clustering. The resulting clustered vectors are then deconvolved to find WFMs. This process efficiently clusters the WFMs by combining CAE and Image-NFCM (I-NFCM) algorithm. The clustered WFMs play a crucial role in identifying various wind scenarios based on wind speed and direction. Additionally, the frequency of these clusters aids in generating new latent vectors; particularly generating more vectors within dense clusters due to the prevalence of wind characteristics in high-frequency clusters. According to the law of large numbers, as previously mentioned, the original and generated WFMs, collectively, can be utilized for various research applications related to WFMs, such as accurate modeling of WFMs, wind farm layout optimization under uncertainty, and stochastic control of wind farms. Thus, in this chapter, WFMs are considered as images with CAEs employed to extract optimal features from them, followed by the application of the I-NFCM algorithm for clustering the feature vectors and consequently clustering the WFMs. Finally, wind farm layout optimization is conducted using the generated WFMs as the uncertain set.

9.2.3 *Robust Optimization of Farm Layout*

Specifically, in wind farm layout optimization problem, the turbine locations are determined to maximize the energy production (Mosetti et al. 1994). Since wind is highly uncertain in nature, wind farm might obtain a very unrealistic layout if we solve it with deterministic approaches. The deterministic approaches use nominal values of uncertain parameters, thereby resulting in either overestimated or undermined results compared to the actual realization of uncertain parameters. Therefore, it is necessary to consider sources of uncertainty to solve such problems. Robust optimization is one such approach that considers the hard constraints to define the robustness of solutions (Aharon Ben-Tal et al. 2009). Therefore, the constraints have to be satisfied for any uncertain realizations by the robust feasible set. Since there is a possibility of many such robust feasible solutions, two such measures are considered in this study; one is the worst-case scenario from all uncertain realizations and the other is the best-case scenario from all uncertain realizations while solving the optimization problem.

In this work, we aim to maximize the energy production from a wind farm by estimating the optimal locations of a fixed number of turbines under wind state uncertainty. Though the energy production is dependent on a number of turbines, it is fixed in this study to reduce the complexity of the problem. The focus is fixed on maximizing the energy production with these fixed number of turbines and determining the optimal locations of these turbines. Since it is necessary to avoid wake effects between turbines, inter-turbine distance (ITD) is added as a constraint while solving the optimization problem. Hence, the optimization problem is formulated with Annual Energy Production as the objective function, the x and y coordinates of turbines as decision variables and ITD as the constraint to minimize the wake effects. As described earlier, the uncertainty of wind lies with the WFMs. A set of WFMs is considered as uncertain realizations, and these uncertain realizations are generated through the proposed DRIFT approach, which is helpful in generating WFMs covering all regions of different wind speeds and directions.

In this study, 400 such WFMs are generated through DRIFT and considered as uncertain realizations while solving robust optimization formulation. Hence, 400 scenarios generated through DRIFT are used to determine the optimal locations of turbines corresponding to the worst-case scenario and best-case scenario accounting for all the uncertain realizations in the set. Here, the worst-case RO formulation is to maximize the minimum EP value across all the possible scenarios of the uncertainty set, i.e. in one iteration, EP is calculated for all the uncertain realizations, and the aim is to maximize the minimum value obtained through the uncertain realizations; the corresponding formulation is presented in Eq. (9.2).

$$\begin{aligned} &\arg\max_{(X,Y)}\left(\inf_{\varphi_s} \mathrm{EP}(X, Y, \varphi_s)\right) \\ &\text{Subject to}: \\ &\mathrm{ITD}(X_i, Y_i) = n_{\mathrm{sp}} * D - \sqrt{\left(X_i - X_j\right)^2 + \left(Y_i - Y_j\right)^2} < 0 \\ &\forall i, j = 1, 2, \ldots N_T s.t i \neq j \end{aligned} \tag{9.2}$$

where EP (X, Y, φ_s) is the energy production, which is calculated using Eq. (9.1), φ is the WFM in the set of uncertain realizations, φ_s is the set of WFMs generated through DRIFT, (X,Y) are locations of turbines, the minimum of the EPs attained through the uncertain set, φ_s, is denoted as inf (infimum), the diameter of rotor is D, n_{sp} is the allowable distance between turbines. Similarly, the best-case RO formulation is to maximize the maximum EP value across all the possible scenarios of the uncertain set, i.e., in one iteration, EP is calculated for all the uncertain realizations, and the aim is to maximize the maximum value obtained through the uncertain realizations; the corresponding formulation is presented in Eq. (9.3).

$$\begin{aligned} &\arg\max_{(X,Y)}\left(\sup_{\varphi_s} \mathrm{EP}(X, Y, \varphi_s)\right) \\ &\text{Subject to}: \\ &\mathrm{ITD}(X_i, Y_i) = n_{\mathrm{sp}} * D - \sqrt{\left(X_i - X_j\right)^2 + \left(Y_i - Y_j\right)^2} < 0 \\ &\forall i, j = 1, 2, \ldots N_T s.t i \neq j \end{aligned} \tag{9.3}$$

where sup is the supremum of the EPs obtained with φ_s uncertain parameters. The ease of solving an unconstrained problem leads to convert the presented formulation in Eq. (9.2) and Eq. (9.3) into an unconstrained optimization problem as follows:

For worst-case robust optimization:

$$\begin{aligned}
&\delta(X_i, Y_i) = \frac{\sqrt{(X_i - X_j)^2 + (Y_i - Y_j)^2}}{n_{\text{sp}} * D} \forall i, j = 1, 2, \ldots N_T s.t i \neq j \\
&\Lambda_{ij}(X_i, Y_i) = \begin{cases} \Lambda(X_i, Y_i), \ \text{if } \delta(X_i, Y_i) < 1 \\ 1, \ \text{otherwise} \end{cases} \\
&\nu = \prod \Lambda_{ij}(x_i, y_i) \\
&\text{Modified objective}: \max_{X,Y}\left(\inf_{\varphi_s}(\text{EP}(X, Y, \varphi_s) * \nu)\right)
\end{aligned} \tag{9.4}$$

For best-case robust optimization:

$$\begin{aligned}
&\delta(X_i, Y_i) = \frac{\sqrt{(X_i - X_j)^2 + (Y_i - Y_j)^2}}{n_{sp} * D} \forall i, j = 1, 2, \ldots N_T s.t i \neq j \\
&\Lambda_{ij}(X_i, Y_i) = \begin{cases} \Lambda(X_i, Y_i), \ \text{if } \delta(X_i, Y_i) < 1 \\ 1, \ \text{otherwise} \end{cases} \\
&\nu = \prod \Lambda_{ij}(x_i, y_i) \\
&\text{Modified objective}: \max_{X,Y}\left(\sup_{\varphi_s}(\text{EP}(X, Y, \varphi_s) * \nu)\right)
\end{aligned} \tag{9.5}$$

where the normalized distance between turbines, normalized ITD is denoted by δ, where (X_i, Y_i) and (X_j, Y_j) are the coordinates of ith and jth turbines. If δ values are less than 1, they are stored in variable Λ to observe the ITD violation. Hence, ν value will be 1 if ITD is maintained; otherwise, it is less than 1 signifying the ITD violation. Since the P is multiplied with ν in the modified objective, EP value is penalized whenever the ITD is violated which reflects the constraint in the objective function. The Robust formulations for wind farm layout optimization presented in Eqs. (9.4 and 9.5) are addressed utilizing the framework of Bayesian optimization (BO) to identify the optimal locations of turbines for the best-case and the worst-case scenarios.

9.2.4 Bayesian Optimization

The real-world optimization problems suffer from uncertainty and computationally expensive objective functions with multiple local optima. To handle uncertainty, a robust optimization framework is used, whereas the computational costs and multiple local optima can be handled with BO in which the optimal values were found with a smaller number of evaluations. BO employs a surrogate-assisted optimization framework to build a probabilistic model while finding the next probable decision variables for further evaluations and thus minimizes the computational costs. In a broader perspective, two steps are involved in BO: (i) the learning phase, updating the model and characterizing the best guess of possible locations of decision variables with some uncertainty (ii) the choice of acquisition function, which helps to determine the next possible best guess of decision variables (Frazier 2018). A limited number of initial samples that are uniformly distributed are utilized to build a probabilistic surrogate with Gaussian process regressor. The advantage of the Gaussian surrogate is that it provides the measure of uncertainty in the regions that are not sampled; thus, identifying the higher modeling uncertainties becomes easier. The covariance function in the Gaussian process is used to determine the modeling uncertainties and to explore the decision variable space (Rasmussen 2004). The sample points are interpolated using the mean function, thus providing a scope for large confidence bounds on the expensive model. Thus, the location that has the maximum chance of being a local minimum is evaluated by minimizing the mean function exploiting the decision variable space (Rasmussen 2004). Therefore, BO uses an Acquisition function that uses mean and covariance for balancing exploitation and exploration, respectively.

For a single objective optimization, upper confidence bound, expected improvement, and probability of improvement are some of the well-known acquisition functions. The aim of BO is to determine the next possible location of a new sample based on past observations and maximise the acquisition function. The best guess of the new sample is used to determine the expensive objective function and update the probabilistic surrogates, i.e., Gaussian process regressor and the entire process is repeated until the end of the budget for function evaluations. In this work, BOTORCH framework is used to implement BO with expected improvement (Mockus et al. 1978) as an acquisition function to determine the optimal layout of the wind farm for worst-case and best-case by maximizing the objective function presented in Eq. (9.4) and Eq. (9.5).

9.3 Results

9.3.1 WFM Generation Using DRIFT

In this study, out of a total of 1400 WFMs, 70%, 15% and 15% were chosen for training, validation and testing the CAE network respectively. Proposed algorithm obtained an optimal architecture of three encoding and decoding layers each as shown in Table 9.1. Testimony of performance is shown through the original and predicted images (from the test set) in Fig. 9.3. CAE network is observed to successfully model the given images (very low RMSE), by extracting 11 features. Subsequently, these feature vectors were clustered using the I-NFCM algorithm, where four optimal clusters were identified with two hidden layers, 4 and 3 nodes in each hidden layer with a tan-sigmoidal activation function.

Table 9.1 Result of modeling WFMs through convolutional auto-encoders

Parameter	Value
Encoding layers	3
Decoding layers	3
Nodes in encoding layers	[22,19,14]
Feature extracted	11
Transfer function	ReLU
Test set RMSE	0.02

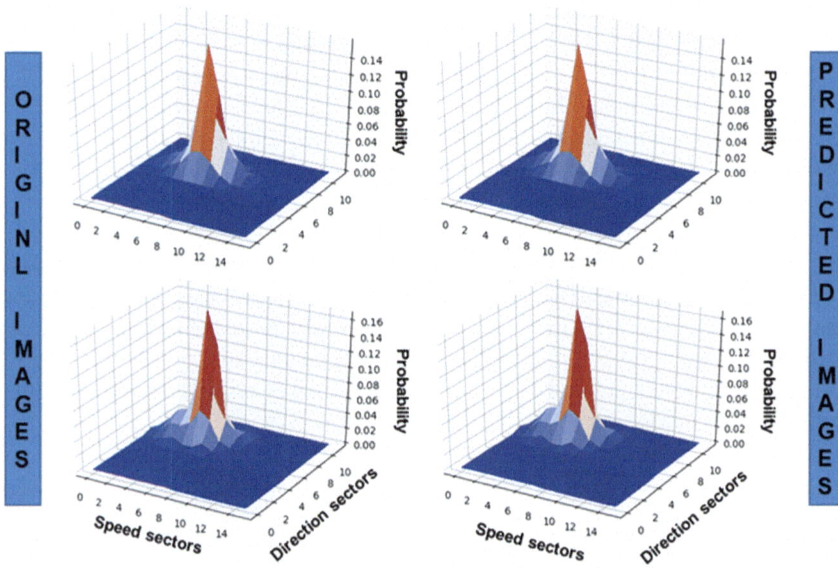

Fig. 9.3 Original and test set WFM comparison

With the help of deconvolution operation, WFMs are generated and the clustered WFMs are presented in Fig. 9.4. Three images are shown as representatives from each cluster. The nature of the WFMs shows similar behaving WFMs are clustered together and dissimilar ones are clustered differently and the number of WFMs in 4 clusters are 223, 392, 338, and 447. The set of extracted 11 feature vectors is presented in one-dimensional format in Fig. 9.5. As it appears, two of the clusters are relatively denser compared to the remaining clusters. This signifies that wind conditions captured by these two clusters are more frequently occurring wind states as per the given data. This way, wind characteristics of the wind farm are captured. As we need to create several WFM scenarios while performing layout optimization using RO, they can be generated from the four clusters maintaining the same ratio as the number of points in them. This also tells that the scenarios are going to come more often from the two clusters where samples were denser compared to the other two instead of sampling in the entire region uniformly/randomly.

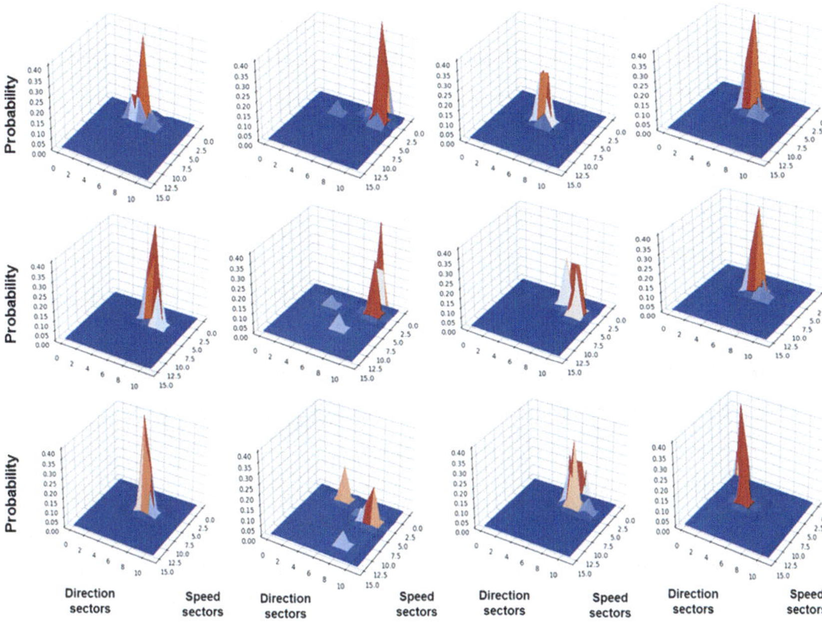

Fig. 9.4 WFM clustering using I-NFCM algorithm (3 images shown for each cluster)

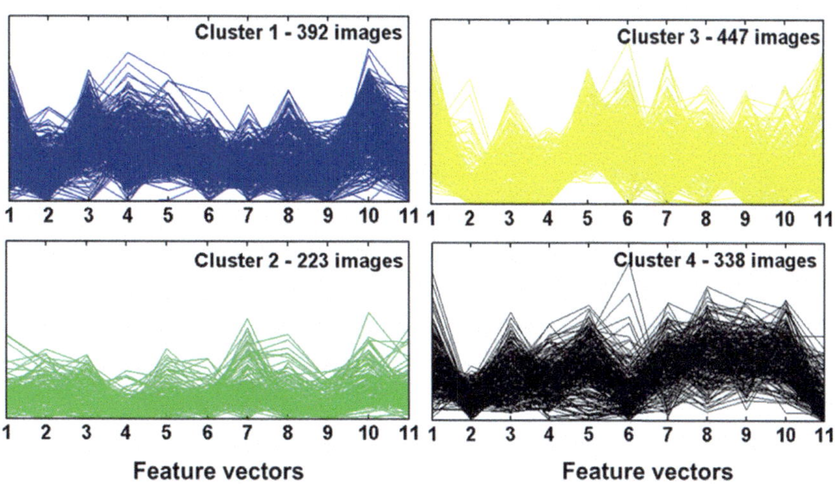

Fig. 9.5 Feature vector (11 in number) representation for each of four clusters

Moreover, this study enables generating more latent vectors in the regions of interest, which can be further decoded into new WFMs. Such WFMs are useful in wind farm optimization under uncertainty using DRIFT. Some of the generated WFMs are presented in Fig. 9.6, where a total of 400 latent vectors are generated within the four clustered regions. Each of these clusters follows a particular distribution, not known, due to which the large number of samples generated assists in performing uncertainty studies.

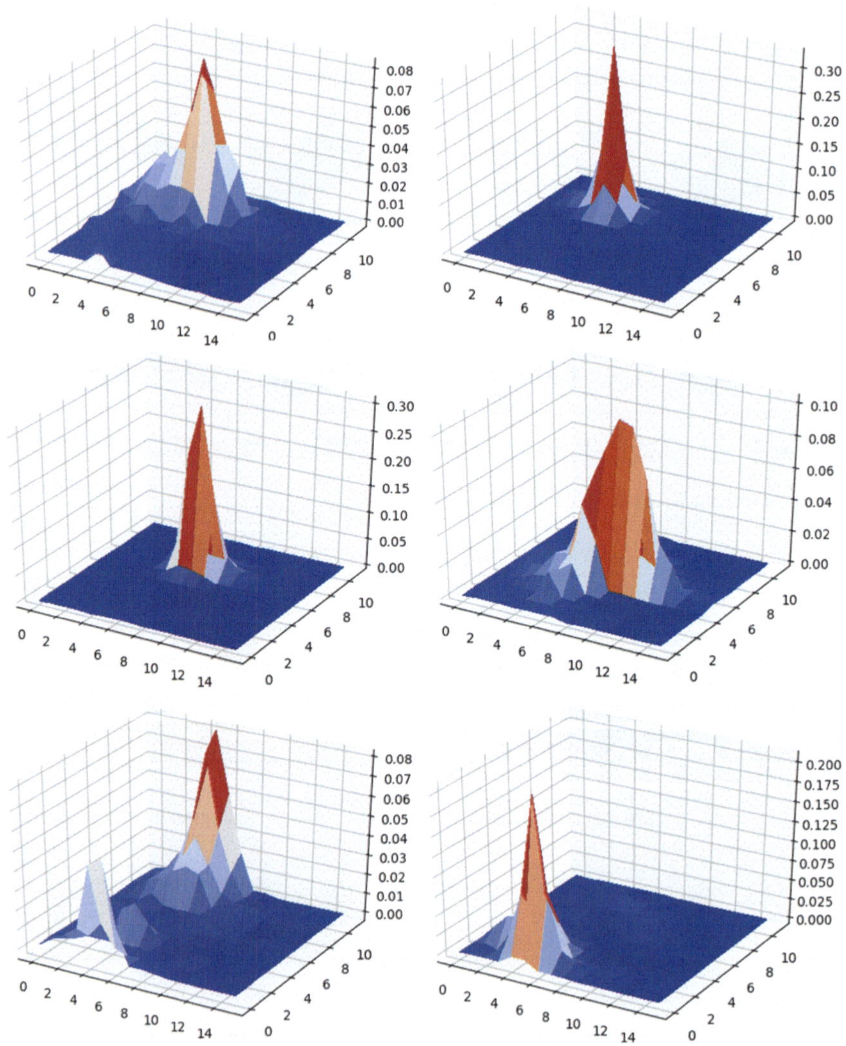

Fig. 9.6 Some generated wind-frequency maps using DRIFT algorithm, where x-axis denotes speed bins, y-axis denotes direction sectors and z-axis denotes the frequency value of the joint PMF

9.3.2 Significance of WFMs in Wind Farm Studies

In the next step, the generated 400 WFMs are considered as uncertain realizations to perform robust wind farm layout optimization with two cases: (i) worst case and (ii) best case, as discussed earlier. For the two cases, a wind farm of 3 km × 3 km is considered with the following assumptions:

- Same wind flow distribution at every location in the wind farm.
- Turbines are identical.
- Wind variations are ignored along the vertical direction.
- Assuming that the ITD of 5 times diameter of rotor is sufficient to handle wake effects (Jensen 1983).
- Wake effects are good enough to be modeled using the Jensen wake model.
- The turbine specifications for the present study are provided in Table 9.2.

BO framework is used for solving RO formulations here. Since BO needs initial samples, the number of turbines is fixed to certain value and the number of decision variables are determined. The initial points, two times the decision variables, are generated through Sobol sampling, and the objective values are calculated using Eq. (9.1). The objective values are evaluated for all the uncertain realizations, and the maximum or minimum EP of all the scenarios is maximized based on the case study, i.e., for the best-case or the worst-case scenario, respectively. With the generated initial samples, Gaussian surrogate is fitted and to sample the next best locations of turbines, expected improvement is used as acquisition function. Then, the Gaussian surrogate is updated with the new best locations and objectives, and the process is repeated till the completion of budget evaluations. To ensure optimality, the process is repeated n times as the Gaussian process regressor is a probabilistic surrogate. Since the initial points are only two times the decision variable space, the advantage is in reducing the sample size determination and the new candidates are sampled along the process by updating the model, ensuring the optimal solutions are reached with

Table 9.2 Turbine and wind farm details

Wind farm and turbine specifications	
Wind farm area	3 km × 3 km
Surface roughness	0.3
Turbine hub height	80 m
Turbine rotor diameter (D)	82 m
Thrust coefficient	0.88
Allowable distance between turbines	5D
Cut-in velocity	3.5 m/s
Rated velocity	15 m/s
Cut-out velocity	22 m/s
Rated power	2000 kw

lower computational costs with online optimization using the Bayesian framework. In this work, the number of turbines is fixed to 15, thus leading to 30 decision variables. Therefore, 60 initial points were determined using Sobol sampling for decision variables, and the objectives were calculated using Eq. (9.1), and the objectives were calculated for a set of 400 WFMs, i.e. the uncertain realizations. The budget of function evaluations for BO is set to 1000, and three repetitions are performed to ensure optimality. From Eqs. (9.4) and (9.5), it is necessary to ensure the σ value as one while maximizing the EP with proper distance between turbines to minimize wake effects. The results of robust wind farm layout optimization with Bayesian framework for best-case and worst-case scenarios are presented below.

9.3.3 Worst-Case Robust Wind Farm Layout Optimization

In the worst-case robust wind farm layout optimization formulation, EP is calculated for all the WFMs in the uncertain set and the minimum of EP obtained from all the scenarios is maximized. This shows the potential of wind farms in the worst situation of lower wind frequencies, being the lower bound of wind farm potential for 15 turbines in an area of 3 km $\times$ 3 km wind farm with the turbine specifications presented in Table 9.2. For 15 turbines, a layout with σ value 1 is obtained within 25 iterations, but the optimal value for the objective is obtained within 180 iterations and the maximum EP for the worst case is obtained as 8801.42 kW. The optimal layout for 15 turbines for worst case is presented in Fig. 9.7.

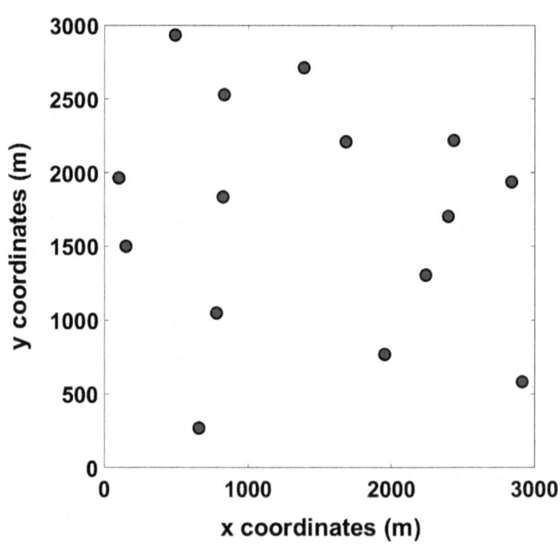

Fig. 9.7 Robust worst case bayesian optimization optimal layout for 15 turbines

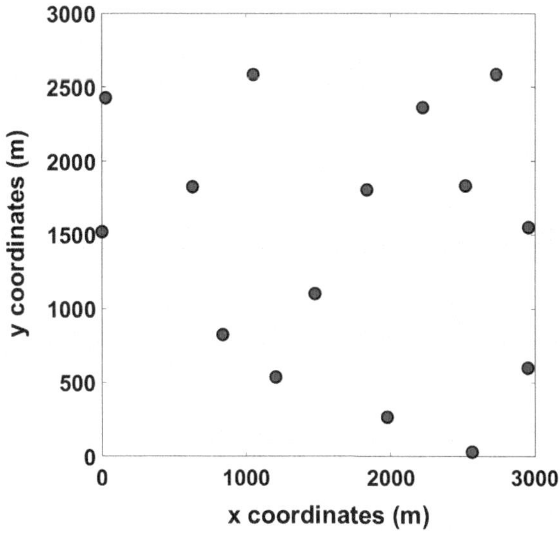

Fig. 9.8 Robust best-case bayesian optimization optimal layout for 15 turbines

9.3.4 Best-Case Robust Wind Farm Layout Optimization

In the best-case robust wind farm layout optimization formulation, EP is calculated for all the WFMs in the uncertain set and the maximum of EP obtained from all the scenarios is maximized. This shows the potential of wind farms in the best situation of higher wind frequencies, being the upper bound of wind farm potential for 15 turbines in an area of 3 km $\times$ 3 km wind farm with the turbine specifications presented in Table 9.2. For 15 turbines, a layout with σ value 1 is obtained within 25 iterations, but the optimal value for the objective is obtained within 400 iterations and the maximum EP for the worst case is obtained as 16,886.73 kW. The optimal layout for 15 turbines for worst case is presented in Fig. 9.8.

In both cases, the best EP is obtained within 400 iterations, thus showing the efficiency of BO in reducing the computational costs and ensuring optimality. The results that are presented for worst-case situation help the wind farm owners to determine the wind farm potential for low wind conditions, and the result of the best-case scenario helps to know the wind farm potential in high wind conditions, thus acting as the boundaries of solutions and the importance of considering the WFMs as uncertain realizations while designing the wind farm.

9.4 Conclusion

In this chapter, an efficient generative modeling framework has been designed for usage in optimization under uncertainty studies. The overall algorithm, including image clustering using efficient soft clustering technique (I-NFCM), is generic enough to be applicable to cluster any kind of image with any resolution. Since the uncertain wind behaviour poses a problem in maintaining the rated power from a wind farm on a sustainable basis, it is necessary to consider the uncertainty while designing the wind farm rather than designing with a fixed WFM obtained from a shorter span of wind behaviour. In this study, the uncertainty of wind is considered while designing the wind farm and ensuring sustainability during the life span of the wind farm. On application of the DRIFT algorithm, 400 new WFMs were generated during stochastic wind farm optimization studies. Further, the generated wind scenarios are considered uncertain realizations and robust optimization using the Bayesian framework is solved for both best-case and worst-case scenarios, which act as boundaries for the energy production for a wind farm with 15 turbines. The computational costs were reduced with the Bayesian framework, resulting in optimal solutions within a few function evaluations. The proposed robust formulation of maximizing the EP from the worst-case scenario of wind allows the wind farm owners to determine the least possibility of power from the wind farm, thus minimizing the risk of overestimation of power. Though the worst-case scenario robust optimization seems to be a conservative approach, it is immune to all possible realizations and thus provides the realistic solution of bare minimum power production of wind farms. Similarly, the best-case scenario determines the best possible power production from the wind farm, minimizing the risk of underestimation of power and providing the solution of maximum power production from the wind farm. Hence, it is advised to consider the uncertainties in wind as an industry practice while designing the wind farms instead of a nominal fixed probability distribution to avoid the risk of overestimations or underestimations of powers and thus reducing the losses for wind farm owners.

Acknowledgements We would like to acknowledge the support of SPARC project (SPARC/2018-2019/P1084/SL) sponsored by the Ministry of Education (MoE), Government of India and National Supercomputing Mission (NSM) for the project (DST/NSM/R&D_HPC_Applications/2021/23) sponsored by the Department of Science and Technology (DST), Government of India.

References

Albawi S, Mohammed TA, Al-Zawi S (2017) Understanding of a convolutional neural network. Proceedings of 2017 international conference on engineering and technology, ICET 2017 2018 January, pp 1–6. https://doi.org/10.1109/ICENGTECHNOL.2017.8308186

Ben-Tal A, El Ghaoui L, Nemirovski A (2009) Robust optimization. Princeton University Press

Bertsimas D, Brown DB, Caramanis C (2011) Theory and applications of robust. Optimization 53:464–501. https://doi.org/10.1137/080734510

Bertsimas D, Gupta V, Kallus N (2018) Data-driven robust optimization. Math Program 167:235–292. https://doi.org/10.1007/S10107-017-1125-8/TABLES/4

Burton T, Jenkins N, Sharpe D, Bossanyi E (2011) Wind energy handbook. John Wiley & Sons

De Jonge B, Klingenberg W, Teunter R, Tinga T (2015) Optimum maintenance strategy under uncertainty in the lifetime distribution. Reliab Eng Syst Saf 133:59–67. https://doi.org/10.1016/J.RESS.2014.09.013

Dige N, Diwekar U (2018) Efficient sampling algorithm for large-scale optimization under uncertainty problems. Comput Chem Eng 115:431–454. https://doi.org/10.1016/J.COMPCHEMENG.2018.05.007

ENGIE (2020) La haute borne-vaudeville-le-haut (France)—wind farms—online access—the wind power. https://www.thewindpower.net/windfarm_en_3354_la-haute-borne.php. Accessed 1 Mar 2024

Etemadi N (1981) An elementary proof of the strong law of large numbers. Zeitschrift Für Wahrscheinlichkeitstheorie und Verwandte Gebiete 55:119–122. https://doi.org/10.1007/BF01013465/METRICS

Fang YP, Sansavini G (2019) Optimum post-disruption restoration under uncertainty for enhancing critical infrastructure resilience. Reliab Eng Syst Saf 185:1–11. https://doi.org/10.1016/J.RESS.2018.12.002

Frazier PI (2018) Bayesian optimization. Inform Tutorials Oper Res. https://doi.org/10.1287/EDUC.2018.0188

Guo X, Liu X, Zhu E, Yin J (2017) Deep clustering with convolutional autoencoders. Lect Notes Comput Sci including Subseries Lect Notes Artif Intell Lect Notes Bioinform 10635:373–382. https://doi.org/10.1007/978-3-319-70096-0_39/FIGURES/3

Hagita K, Higuchi T, Jinnai H (2018) Super-resolution for asymmetric resolution of FIB-SEM 3D imaging using AI with deep learning. Sci Rep 8:1–8. https://doi.org/10.1038/s41598-018-24330-1

Jensen NO (1983) A note on wind generator interaction

Liu B, Liu YK (2002) Expected value of fuzzy variable and fuzzy expected value models. IEEE Trans Fuzzy Syst 10:445–450. https://doi.org/10.1109/TFUZZ.2002.800692

Liu S, Farid SS, Papageorgiou LG (2016) Integrated optimization of upstream and downstream processing in biopharmaceutical manufacturing under uncertainty: a chance constrained programming approach. Ind Eng Chem Res 55:4599–4612. https://doi.org/10.1021/ACS.IECR.5B04403/ASSET/IMAGES/IE-2015-04403D_M049.GIF

Mistry I, Tanwar S, Tyagi S, Kumar N (2020) Blockchain for 5G-enabled IoT for industrial automation: a systematic review, solutions, and challenges. Mech Syst Signal Process 135:106382. https://doi.org/10.1016/J.YMSSP.2019.106382

Mittal P, Mitra K (2018) Determining layout of a wind farm with optimal number of turbines: a decomposition based approach. J Clean Prod 202:342–359. https://doi.org/10.1016/J.JCLEPRO.2018.08.093

Mittal P, Mitra K (2020) In search of flexible and robust wind farm layouts considering wind state uncertainty. J Clean Prod 248:119195. https://doi.org/10.1016/J.JCLEPRO.2019.119195

Mockus J, Tiesis V, Zilinskas A (1978) The application of Bayesian methods for seeking the extremum. L Dixon G Szego toward Global Optim 2:2

Moen E, Bannon D, Kudo T et al (2019) Deep learning for cellular image analysis. Nat Methods 16:1233–1246. https://doi.org/10.1038/s41592-019-0403-1

Mosetti G, Poloni C, Diviacco B (1994) Optimization of wind turbine positioning in large windfarms by means of a genetic algorithm. J Wind Eng Ind Aerodyn 51:105–116. https://doi.org/10.1016/0167-6105(94)90080-9

Pantula PD, Mitra K (2020) Towards efficient robust optimization using data based optimal segmentation of uncertain space. Reliab Eng Syst Saf 197:106821. https://doi.org/10.1016/J.RESS.2020.106821

Pujari KN, Srivastava V, Miriyala SS, Mitra K (2021) Comparison of deep reinforcement learning techniques with gradient based approach in cooperative control of wind farm. 2021 7th Indian control conference, ICC 2021—proceedings, pp 400–405. https://doi.org/10.1109/ICC54714.2021.9703186

Rasmussen CE (2004) Gaussian processes in machine learning. Lect Notes Comput Sci including Subseries Lect Notes Artif Intell Lect Notes Bioinform 3176:63–71. https://doi.org/10.1007/978-3-540-28650-9_4/COVER

Shapiro A (2008) Stochastic programming approach to optimization under uncertainty. Math Program 112:183–220. https://doi.org/10.1007/S10107-006-0090-4/METRICS
Tian G, Chu J, Liu Y et al (2011) Expected energy analysis for industrial process planning problem with fuzzy time parameters. Comput Chem Eng 35:2905–2912. https://doi.org/10.1016/J.COMPCHEMENG.2011.05.012
Virivinti N, Mitra K (2015) Intuitionistic fuzzy chance constrained programming for handling parametric uncertainty: an industrial grinding case study. Ind Eng Chem Res 54:6291–6304. https://doi.org/10.1021/IE504109V/ASSET/IMAGES/IE-2014-04109V_M056.GIF
Virivinti N, Mitra K (2017) Fuzzy robust optimization for handling feed stream and model parameter uncertainties during comminution process. J Taiwan Inst Chem Eng 70:411–425. https://doi.org/10.1016/J.JTICE.2016.10.046
Zhu XX, Tuia D, Mou L et al (2017) Deep learning in remote sensing: a comprehensive review and list of resources. IEEE Geosci Remote Sens Mag 5:8–36. https://doi.org/10.1109/MGRS.2017.2762307

Chapter 10
Generative Adversarial Networks for Modelling Uncertainties in Wind Farm Design

NagaSree Keerthi Pujari, Srinivas Soumitri Miriyala, and Kishalay Mitra

Abstract Energy Crisis and Global warming highlight the exploitation of renewable energy sources such as wind energy. The biggest challenge lies in the stochastic nature of wind, which makes the wind energy conversion systems inefficient. The important aspect in the design and control of wind farm is the probability mass function which is constructed using the wind speed and direction time series, called Wind Frequency Map. Due to the limited amount of available wind time series data, the construction of Wind Frequency Map is inaccurate and results in unrealistic estimation of power production from a wind farm. Though wind speed and direction forecasts can be used for construction of Wind Frequency Map, it needs long-term forecasts for accurate wind frequency maps which is not reliable. Therefore, the authors, in this paper, have explored data-driven probabilistic technique, called Generative Adversarial Networks (GANs) to capture the distribution of wind scenarios from the limited wind time series data. The ability of GANs to capture the hidden distributions from the available data can be utilized to generate new scenarios from the data which leads to efficient design of Wind Energy Conversion Systems under uncertainty by performing Robust Optimization under Bayesian framework.

Keywords Wind characteristics modeling · Generative adversarial networks · Effective wind farm design · Optimization under uncertainty

---

N. K. Pujari · S. S. Miriyala
Department of Chemical Engineering, Indian Institute of Technology, Hyderabad, Kandi, Telangana 502285, India
e-mail: ch20resch11006@iith.ac.in

K. Mitra (✉)
Department of Chemical Engineering, Indian Institute of Technology Hyderabad, Hyderabad, Telangana 502284, India
e-mail: kishalay@che.iith.ac.in

Affiliated Faculty, Department of Climate Change, Indian Institute of Technology Hyderabad, Hyderabad, Telangana 502284, India

Affiliated Faculty, Department of Artificial Intelligence, Indian Institute of Technology Hyderabad, Hyderabad, Telangana 502284, India

K. Mitra et al. (eds.), *Optimization, Uncertainty and Machine Learning in Wind Energy Conversion Systems*, Engineering Optimization: Methods and Applications,
https://doi.org/10.1007/978-981-97-7909-3_10

10.1 Introduction

Depletion of fossil reserves and an upsurge of environmental issues associated with climate change driven national governments to introduce policies towards the usage of renewable energy sources. As part of transition to renewable energy, India has become a global champion in adopting renewable energy sources. The installed capacity of renewable energy sources rose from a few MW to 163 GW in this decade and is expected to grow rapidly with annual additions of 35–40 GW by next decade (IEEFA 2023). Among all the renewable energy sources, wind energy is the fastest growing renewable source due to its abundance, clean energy production and wide application range, thus becoming the important player in global energy market. A significant increase of wind capacity in India has been seen with a total installed capacity of 40 GW, fourth largest installed capacity in the world (MNRE 2023). Despite the growth of installed wind capacity, the biggest challenge is to utilize the wind energy to its full potential with the stochastic and volatile nature of wind. To capture the wind energy, wind turbines are aggregated in a planning area called wind farm which is designed in the development stage and have long lasting effects as the operation stage spans over 20–25 years. Therefore, to ensure a reliable and stable operation of wind power systems, there is a need to develop a robust and efficient wind energy conversion systems with consistent wind energy generation. The important aspect in modelling a wind energy conversion system is the construction of probability mass function with two random variables, wind speed and wind direction time series, known as Wind Frequency Map (WFM). Since the WFMs are reliant on wind time series data, the inadequate wind data, and the poor maintenance of available data leads to inaccurate construction of WFM. The inaccurate WFM might create a mismatch with the real power production from a wind farm while designing a wind farm, thereby resulting in unrealistic power production and incur losses to wind farmers (Mittal and Mitra 2020). Therefore, it is necessary to create an accurate WFMs with the available wind time series data to model the efficient wind energy systems for realistic power production. Extensive research on forecasting approaches has been performed in the wind energy community to develop efficient wind energy conversion systems. Here it is important to say that research conducted by Hu et al. indicated that 1% increase in the accuracy of wind energy forecasting might lead to savings of 1.38 billion USD for one year (Hu et al. 2018).

Since the construction of WFM relies on wind speed and wind direction, industrialists and researchers have shown interest in developing models to forecast the wind characteristics efficiently which can be divided into two categories: short-term forecasting and long-term forecasting; short-term forecasts can minimize the scheduling errors and highly used in wind farm control studies and long-term forecasts can be helpful in planning the location and designing of wind farm. Based on the review of previous studies, it is noted that several models have been exploited by the researchers to forecast the wind characteristics accurately. The developed models can be categorized into three broad categories: physical model, statistical models, and data-driven models. To model the wind characteristics using the physical models involves real-time Numerical Weather Prediction (NWP) (Zhao et al. 2018) and atmospheric data (Wang et al. 2019). NWP techniques play a major role for certain wind resource assessments with good accuracies (Sile et al. 2014). A boundary layer scaling technique has been used to predict long-term near surface wind speeds and power densities. The technique incorporates the scaling of reference climatological data obtained from NWP and thus includes the detailed parameterization of surface dynamics (Allen et al. 2017). The physics-based modelling of wind characteristics involves complexities and computational costs and led the research community to utilize statistical tools such as Gaussian regression processes and system identification techniques.

Models for wind speed prediction from 10 min interval to 1 h interval have been built based on linear and nonlinear autoregressive models. Four different models were developed for wind speed forecasting with and without exogeneous variables. The parameters of the linear and nonlinear autoregressive models are estimated using the Gauss–Newton algorithm and data mining algorithms (Lydia et al. 2016). A hybrid autoregressive moving average nonparametric model has been proposed to forecast the wind speed in which the residuals of auto regressive model are used by the nonparametric models to study the nonlinearity in the data (Han et al. 2017). The stochastic nature of wind encouraged the researchers to utilize Gaussian process regression for capturing the randomness in wind characteristics data. But the computational complexity and the inability of Gaussian processes to adapt the time varying systems became a challenging task. Therefore, a time-variant Gaussian process regression was proposed in (Yan et al. 2016) to enhance the accuracy of prediction and minimize the computational costs. In order to train the Gaussian processes, a teaching learning-based optimization is used which accelerates the learning rate (Yan et al. 2016). A probabilistic forecasting technique has been proposed which uses selective ensemble learning with finite mixture Gaussian process regression models with high accuracies by handling time varying changes in the wind data (Jin et al. 2021). In the recent years, data-driven techniques such as Artificial Neural Networks (ANNs) and Support Vector Machines (SVMs) became popular due to their ability to predict the data accurately. ANNs have been used for short-term forecasting of wind speed using the historical time series data (Kaur et al. 2016). A novel hybrid method which combines empirical mode decomposition (EMD) and feature selection with ANNs and SVMs are proposed in (Zhang et al. 2016). EMD is used to decompose the data set into sub-series and the models are constructed between input variables and targets with the sub-series and original series. Then the informative features have been selected using the feature selection process. Later, ANN or SVM can be modelled using the important features and the models are used to forecast the wind speed data (Zhang et al. 2016). A hybrid method was proposed based on support vector regression and autoregressive models in (Santamaría-Bonfil et al. 2016). Feature selection and time delayed coordinates are estimated using the autoregressive models with phase space reconstruction procedure. Then support vector regression model is trained and the parameters of support vector regression model are tuned using the genetic algorithm (Santamaría-Bonfil et al. 2016). Though these techniques have great training accuracies, they fail to capture the extreme cases of transient, high nonlinearities, and long-term dependencies in the data. Therefore, the application of deep learning techniques such as Long Short-Term Memory networks (LSTMs), Autoencoders, Gated Recurrent Units (GRUs), and Generative Adversarial networks (GANs) has observed an upswing in the recent times to improve the accuracy of the wind time series forecasting.

A hybrid model using Autoencoder and LSTMs is used to forecast the real-time wind speed data (Liu et al. 2020). A combination of several Autoencoders and a regression layer is stacked up, called as Stacked Denoising Autoencoders, and used to extract the intrinsic features in the data and LSTMs are used to forecast the data. The effectiveness of the proposed model was validated by comparing with the conventional stacked Autoencoder and LSTMs (Liu et al. 2020). A hybrid decomposition-based wind speed forecasting model has been developed using the Bidirectional LSTM networks (Jaseena and Kovoor 2021). The decomposition techniques have been employed to denoise the wind speed data and to divide the time series into several low- and high-frequency signals. Bidirectonal LSTMs are utilized to forecast the low and high-frequency signals separately and then aggregated for final outcomes (Jaseena and Kovoor 2021). A novel data-driven model was proposed in (Kisvari et al. 2021) which integrates data pre-processing, anomaly detection, feature engineering, and hyper-parameter tuning of GRUs and is compared with LSTMs. It is reported that the proposed algorithm using GRUs outperformed the LSTMs with better accuracy on test data sets (Kisvari et al. 2021). A comparison of statistical tools, such as Autoregressive Moving Average technique (ARMA) and ANNs in the form of Feed forward Neural Networks (FNN), Recurrent Neural Network (RNN), LSTM, and GRUs is performed for forecasting of wind speed data. The GRU model has shown better accuracy when compared to other techniques (Zaman et al. 2021). A hybrid model is proposed with a combination of Convolutional Autoencoders (CAE) and LSTMs, where CAE is used to extract the important features of data and then the data is modelled with LSTMs. The results have shown that the Convolutional Autoencoders (CAE)-LSTM model predicted the data more accurately compared to other techniques (Zhao et al. 2019). In Jiang et al. (2018) power scenarios are generated from a wind farm using the GANs showing that GANs surpassing the other models. A conventional high fidelity numerical solver is used to generate a very coarse scale wind field and the resolution is refined with GANs for accurate aerodynamic simulations (Tran et al. 2020).

To summarize, most of the studies in the literature used statistical or data-driven techniques to model the wind characteristics such as wind speed and direction as time series modelling and then used to model the wind energy conversion systems. In certain studies, the modelling tools are used to forecast the wind power scenarios to model the uncertainty in wind power systems. In any wind energy conversion systems, the wind forecasts are used to design the wind farms which are heavily dependent of Wind Frequency Maps, a joint probability distribution of wind speed and direction. To construct the WFMs accurately, large number of wind characteristic predictions are required which is not reliable as many of the forecasting techniques are known to be accurate only for short-term predictions. Since the WFMs play an important role in designing a wind farm with reliable operation, the accurate construction of WFMs is explored with Generative Modelling approach with the available data. The idea is to construct several WFMs with the available data and these WFMs are used to generate several scenarios using the probabilistic data-driven techniques. In this study, the authors used GANs, a data-driven generative modelling approach which models the probability distributions hidden in the historical data. First, WFMs are generated with the available wind time series data and are considered as images in this study. The ability of GANs in generating the images is explored by considering the WFMs as images and trained using an improved version of GANs known as, Deep Convolutional GANs (DCGANs) (Radford et al. 2016). In the current work, a real wind characteristics data of four years was collected and several WFMs are constructed with 6-hr resolution. The constructed WFMs are utilized for training DCGANs, and new wind frequency maps are generated. To demonstrate the importance of forecasting, the generated scenarios are used for calculating annual power production from a wind farm and compared with the WFM constructed through the original data. Conventionally, the wind farm owners design the layout by constructing an average WFM from the available and forecasted data which is not reliable for long-term due to its uncertain nature. Thus, robust formulations are employed under Bayesian framework in this study to estimate the worst and best potential of wind farm by considering the WFMs generated through GANs as uncertain realizations. To the best of our knowledge, no work has been reported which models the wind frequency maps with GANs for accurate construction of WFMs and designed the layout of wind farm by considering the WFMs generated through GANs as uncertain set. The main features of the proposed work are presented in Fig. 10.1 and are summarized as follows:

- Designing accurate wind farm layout with Wind Frequency Maps estimated using data-driven approach, DCGANs.

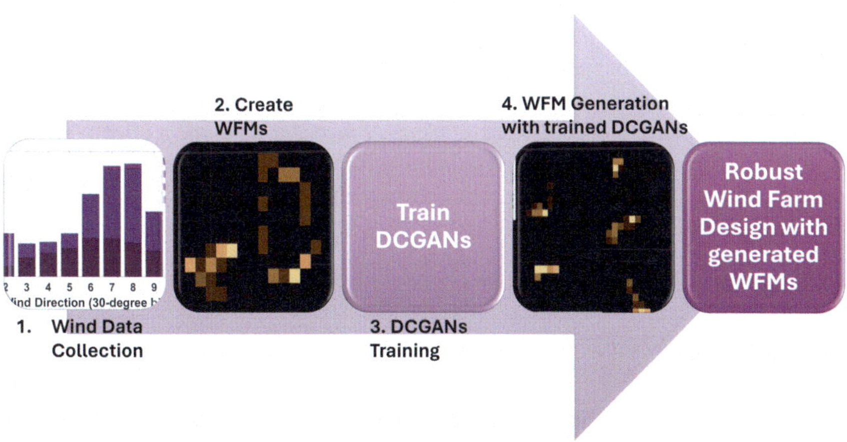

Fig. 10.1 Summary of the proposed work

- A joint probability mass function of two random variables, wind direction and wind speed were estimated by a model free generative modelling approach in contrast to Weibull distribution.
- Efficient wind farm design under uncertainty using the Bayesian framework and considering the WFMs generated through GANs as uncertain realizations.

The rest of the paper is organized as follows: Sect. 10.2 presents the data description and the procedure to construct the WFMs, followed by description of GANs. The results and discussions along with the significance of WFMs in wind energy conversion systems are presented in Sect. 10.3. Section 10.4 summarizes the conclusions of current work.

10.2 Formulation

The data-driven techniques are growing rapidly in the contemporary times to enhance the accuracy of the modelling techniques. GANs are powerful tools to learn the probabilistic distribution in the data and generate new samples with the learnt distribution. Therefore, the authors in this work used GANs to build the new scenarios of wind while enhancing the efficiency of wind energy conversion systems. The data description is presented followed by a brief description on GANs. The robust formulations for the best and the worst-case are presented, and the description of Bayesian optimization is presented in Chapter 9 (Sect. 9.2).

10.2.1 Data Description

Wind time series data of wind speed and direction of four years was collected from an open-source repository called ENGIE, a French Electric Utility company (ENGIE 2020). The data was collected from a wind farm located in La Haute Borne, France in which four turbines are located at the corners of wind farm. The data was measured using the anemometers on the wind turbines. Since the data was collected using sensors, the data was pre-processed to remove the noise present in the data using moving average technique. After pre-processing the wind time series data of wind speed and direction, it is divided into several disjoint intervals to construct the WFM.

10.2.2 Procedure for Wind Frequency Map Construction

The collected wind speed and direction data are pre-processed to remove the noise. Then the data is divided into several disjoint intervals. The wind speed data is divided into several intervals with equal range and these intervals are called speed bins. Then

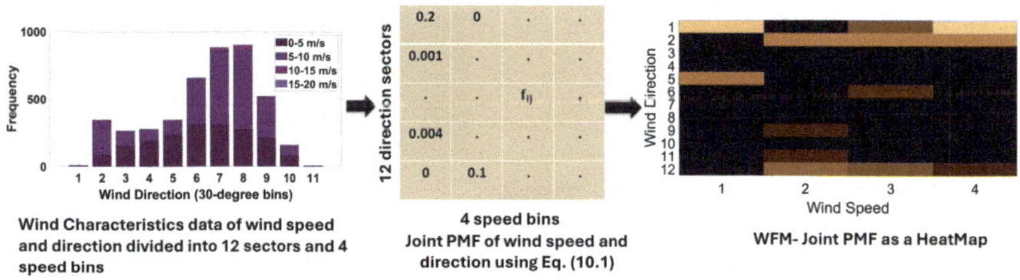

Fig. 10.2 Procedure of WFM construction with the wind characteristics data

the wind direction data is divided into several intervals called direction sectors. In this work, the wind speed ranges from 0 to 16 m/s, therefore, the wind speed is divided into 16 speed bins with a range of 1 m/s in each bin. Similarly, the wind direction is divided into 12 sectors with 30 °C in each sector. After dividing the time series data into several intervals, the WFM is constructed using the frequentist's approach described as follows: Let the total number of data points in the collected time series be $\mathbb{N}$. First count the number of points in sth speed bin and dth direction sector and is denoted by $\mathfrak{n}_{sd}$. The probability value of the joint PMF of speed and direction is given by Eq. (10.1).

$$\mathfrak{f}_{sd} = \frac{\mathfrak{n}_{sd}}{\mathbb{N}} \tag{10.1}$$

Now, the WFM, joint PMF of wind speed and direction, is constructed with frequency values, $\mathfrak{f}_{sd}$ calculated through Eq. (10.1) for all the speed bins and direction sectors. Then a grey scale image with a size of 12 × 16 is constructed by considering each probability value, $\mathfrak{f}_{sd}$, as a pixel value. With the available data of four years, several such WFMs can be constructed and the procedure of WFM construction is presented pictorially in Fig. 10.2. In this study, a data set comprising 1400 is generated at a resolution of 1 day, with 1000 images allocated for training DCGANs, and 400 images designated for testing the model. Figure 10.3 presents some of the WFMs constructed with the available data.

10.2.3 Deep Convolutional Generative Adversarial Networks

Generative Adversarial Networks, GANs, proposed by Ian Goodfellow belongs to the category of generative modelling, which uses deep learning techniques such as convolutional neural networks (Ian Goodfellow 2016). The aim of generative modelling is to produce samples that appear to be part of original data set by identifying the consistent patterns within the input data. The working of GANs can be categorized into two sub models with a game theoretic approach: Generator and Discriminator. The samples are generated through the generator and the generated samples are distinguished from the original input data set using the discriminator. Hence, the training of

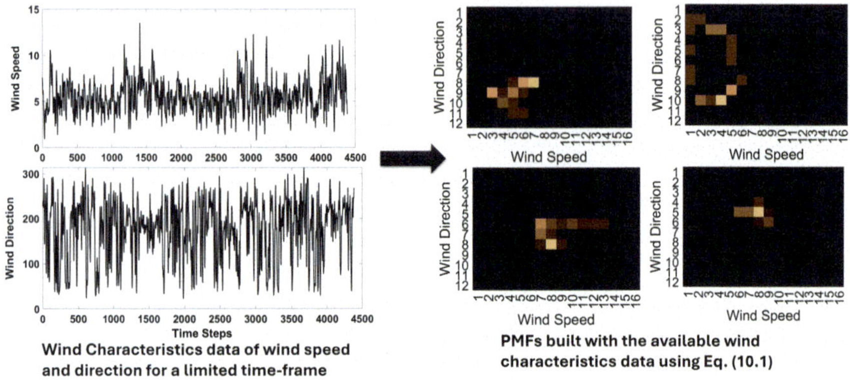

Fig. 10.3 Sample of PMFs constructed with the available data

GANs focuses on strengthening the ability of generator and discriminator to produce samples resembling the real ones and to distinguish between real and fake samples, respectively, until reaching an equilibrium point. During this stage of equilibrium, the discriminator cannot differentiate between real samples and fake samples which are generated using generator, representing the state of Nash equilibrium.

The mathematics behind the GANs is discussed as follows: Let R represent the available data and P_{R} be the true distribution of the original samples. Suppose we sample random noise, X, from a readily understood distribution, such as the Normal distribution, P_X. Now, the objective is to transform a sample X drawn from P_X into desired sample that adheres to the distribution P_{R}. To train the model for sampling from a desired distribution, two networks, Generator and Discriminator, are used. In the training process, the generator network tries to generate plausible samples using a combination of deconvolutional layers and fully connected layers and which can fool the discriminator as real samples. The generator loss function, G_{L} is structured in a way to prevent that the discriminator from distinguishing between real data samples and generated fake samples and thus updating the parameters of generator network. The process of updating the parameters of generator network by generating fake samples as close to real samples is called training.

During training process, a random variable, X, which is sampled using a distribution P_X, is given to the generator as input. The resulting distribution of outputs from the generator (G(X)) is denoted as $\mathrm{P_G}$. If the samples generated by P_{G} closely resemble the real samples according to the discriminator, G_{L} tends to be very low and the loss function, G_{L} is represented in Eq. (10.2)

$$G_{\mathrm{L}} = -E_{X \sim P_X}\left\{D\left[G\left(X, \emptyset_{\mathrm{g}}\right), \emptyset_{\mathrm{d}}\right]\right\} \tag{10.2}$$

where $\emptyset_{\mathrm{g}}$ describes the generator network parameters, $\emptyset_{\mathrm{d}}$ describes the discriminator network parameters, $G\left(X, \emptyset_{\mathrm{g}}\right)$ is the generator network with input, X,

$D\left[G\left(X, \emptyset_{\mathrm{g}}\right), \emptyset_{\mathrm{d}}\right]$ is the Discriminator output with samples generated through generator, $G\left(X, \emptyset_{g}\right)$ as input and the noise distribution expectation is represented as E_X. The real samples and fake samples which are obtained from generator provided as input to discriminator and trained to maximize the difference between P_X and P_{R} distributions. Hence, the discriminator loss function is formulated to ensure the loss value D_{L} and increases significantly if the discriminator fails to differentiate between original and fake samples. Hence, the loss function of discriminator, D_{L}, is described as Eq. (10.3).

$$D_{\mathrm{L}} = -E_{\mathrm{R}\sim \mathrm{P_R}}\{D[R, \emptyset_{\mathrm{d}}]\} + E_{X\sim \mathrm{P}_X}\left\{D\left[G\left(X, \emptyset_{\mathrm{g}}\right), \emptyset_{\mathrm{d}}\right]\right\} \quad (10.3)$$

where E_{R} is the expectation of true distribution of real samples, $D[R, \emptyset_{\mathrm{d}}]$ represents the discriminator output with real samples as input. Thereby, it can be inferred that the expected output of D_{L} approaches 1 when the sample originates from real data, whereas if D_{L} tends to 0, the sample is from generator. Hence, the objective function of GANs can be formulated as a minimax game problem and is described in Eq. (10.4)

$$\min_{G}\max_{D} V(G,\ D) = E_{\mathrm{R}\sim P_{\mathrm{R}}}\{D[R,\ \emptyset_{\mathrm{d}}]\} - E_{X\sim P_X}\left\{D\left[G\left(X, \emptyset_{\mathrm{g}}\right), \emptyset_{\mathrm{d}}\right]\right\} \quad (10.4)$$

Therefore, the aim of GANS while generating samples is to minimize the objective function by generating fake samples as close as to real samples, whereas it tries to maximize the objective function while training the discriminator to distinguish the fake samples and real samples, by learning the distribution of generated data and real data. Thus, the two networks progress during the training, and the generator succeeds in generating the samples that convincingly adheres to the true distribution of the real samples.

The advantage of GANs is that it can learn good feature representation without any specific cost function but it becomes very unstable while training and can generate nonsensical outputs. Therefore, DCGANs are developed to achieve stable training which can allow for deeper generative modes with high resolution of training (Radford et al. 2016). DCGANs employ convolutional neural networks in both the generator and discriminator models, leveraging their capacity for robust feature extraction to enhance image quality. The Convolutional Neural Networks are adopted in GANs with three modifications: replacing the spatial pooling functions such as maxpooling with strided convolutions, eliminating fully connected layers, introducing the batch normalization to stabilize the training. Therefore, the authors have explored DCGANs due to their added advantage of stabilized training for the current work while generating new WFMs. Figure 10.4 presents the pictorial representation of DCGANs.

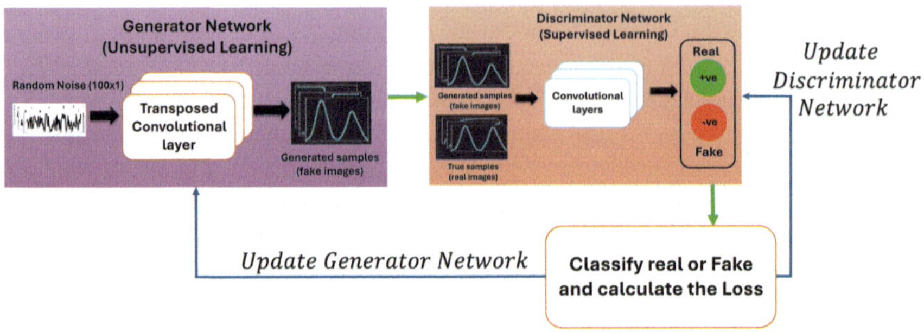

Fig. 10.4 Pictorial representation of DCGANs

10.2.4 Wind Farm Design Under Uncertainty

In wind farm design, the locations of turbines are estimated optimally such that the energy harnessed is maximized from the wind farm. Conventionally, the design of wind farm layout is determined using deterministic approaches, where an average WFM is constructed and is used in wind farm design which lasts for 20–25 years, thus making it unreliable due to either overestimation or undermined results depending on the variations of uncertain parameters, i.e. WFMs, when compared to the actual nominal values. Therefore, in this study, a robust optimization approach is employed considering the WFMs as uncertain realizations defining the robustness of solutions. In this work, two cases are considered, suggesting the boundaries of wind farm potential, i.e. the worst-case scenario and the best-case scenario, while aiming to maximize the energy production from a wind farm. In this study, the number of turbines is fixed to reduce complexity and aims only at estimating the optimal locations of turbines while maximizing the power from a wind farm. To minimize wake effects, the distance between turbines (Interturbine Distance, ITD) should be maintained and this is a constraint in this formulation. However, this constraint is embedded in the objective function by modifying this as a multiplication factor, factor value of 1 signifying the constraint satisfaction and a fractional value of the factor (<1) signifying constraint dissatisfaction. Since the uncertainty of wind is expressed through the WFMs, the WFMs generated through GANs are considered as the set of uncertain realizations while solving the robust optimization problem of maximizing the energy production of a wind farm. The worst-case and the best-case robust optimization formulations are presented below:

10.2.4.1 Worst-Case Robust Formulation

In the worst-case robust optimization formulation, the WFM which gives the minimum energy is identified among various realizations and the optimal locations are estimated in such a way that it returns the maximum power from the wind farm. Here, the worst-case robust optimization formulation is to maximize the minimum of power value across all the possible scenarios of uncertain set. Thus, this case determines the worst-potential of a wind farm, acting as the lower boundary of the wind farm potential. Since ITD is also incorporated in the objective, the formulation for worst-case scenario is presented in Eq. (10.5).

$$\delta(X_i, Y_i) = \frac{\sqrt{(X_i - X_j)^2 + (Y_i - Y_j)^2}}{n_{sp} * D} \forall i, j = 1, 2, \ldots N_T s.t i \neq j$$

$$\Lambda_{ij}(X_i, Y_i) = \begin{cases} \Lambda(X_i, Y_i), & \text{if } \delta(X_i, Y_i) < 1 \\ 1, & \text{otherwise} \end{cases} \tag{10.5}$$

$$\nu = \prod \Lambda_{ij}(x_i, y_i)$$

$$\text{Final objective : } \max_{X,Y}\left(\inf_{\varphi_s}(P(X, Y, \varphi_s) * \nu)\right)$$

where $\mathrm{P}(X, Y, \varphi_s) = \sum_{\mathrm{Nt}} \sum_{\mathrm{s}} \sum_{\mathrm{d}} (\mathrm{RP}(u_s)\varphi)$, and RP, the rated power, is presumed to be precise as it is supplied by the manufacturers of turbine, φ is the WFM in the set of uncertain realizations, φ_s, i.e. the set of WFMs generated through GANs, (X, Y) are locations of turbines, the minimum of the Ps attained through the uncertain set, φ_s, is denoted as inf, the diameter of rotor is D, n_{sp} is the allowable distance between turbines, the normalized distance between turbines, normalized ITD is denoted by δ, where (X_i, Y_i) and (X_j, Y_j) are the coordinates of ith and jth turbines. If δ values are less than 1, they are stored in variable Λ to observe the ITD violation. Hence, ν value will be 1 if ITD is maintained otherwise it is less than 1 signifying the ITD violation. Since the P is multiplied with ν in the modified objective, P value is penalized whenever the ITD is violated which reflects the constraint in the objective function.

10.2.4.2 Best-Case Robust Formulation

In best-case robust formulation, the WFM which gives the maximum energy is considered and the optimal locations are estimated such that it gives the maximum power from a wind farm. Here, the best-case robust optimization formulation is to maximize the maximum of power value across all the possible scenarios of uncertain set. Thus, this case determines the best potential of a wind farm, acting as the upper boundary of the wind farm potential. In the similar lines in Eq. (10.5), the formulation for best-case scenario is presented in Eq. (10.6).

Table 10.1 Specifications of wind farm and turbine

Wind farm and turbine specifications	
Wind farm area	3 km × 3 km
Surface roughness	0.3
Hub height of turbine	80 m
Rotor diameter of turbine (D)	82 m
Thrust coefficient	0.88
n_{sp}, allowable distance between turbines	5D
Cut in velocity	3.5 m/s
Rated velocity	15 m/s
Cut out velocity	22 m/s
Rated power	2000 kw

$$\begin{aligned}
&\delta(X_i, Y_i) = \frac{\sqrt{\left(X_i - X_j\right)^2 + \left(Y_i - Y_j\right)^2}}{n_{\text{sp}} * D} \forall i, j = 1, 2, \ldots N_{\text{T}} s.t i \neq j \\
&\Lambda_{ij}(X_i, Y_i) = \begin{cases} \Lambda(X_i, Y_i), & \text{if } \delta(X_i, Y_i) < 1 \\ 1, & \text{otherwise} \end{cases} \\
&\nu = \prod \Lambda_{ij}(x_i, y_i) \\
&\text{Final objective} : \max_{X,Y}\left(\sup_{\varphi_s}(P(X, Y, \varphi_s) * \nu)\right)
\end{aligned} \tag{10.6}$$

where sup is the supremum of the P_s obtained with φ_s uncertain parameters. The robust formulations for wind farm layout optimization presented in Eqs. (10.5) and (10.6) are addressed utilizing the framework of Bayesian Optimization to identify the optimal locations of turbines for the best-case and the worst-case scenarios. The description of Bayesian Optimization is presented in Chapter 9, Sect. 9.2 and restricted here for the sake of brevity. For the two cases mentioned above, a wind farm of 3 km × 3 km is considered with the following assumptions:

- Same wind flow distribution at every location in the wind farm.
- Turbines are identical.
- Wind variations are ignored along the vertical direction.
- Assuming that the ITD of 5 times the diameter of rotor is adequate to minimize wake effects.
- Wake effects are modelled using Jensen wake model, under the assumption that it provides a satisfactory estimation of wake effects as referred in the literature.
- The turbine specifications for the present study are provided in Table 10.1

10.3 Results

10.3.1 WFM Generation Using DCGANs

In the present work, wind characteristics data, i.e. wind speed time series and wind direction time series of four years was collected and is used to construct 1400 wind scenarios (WFMs) with 1 day resolution. Each WFM is represented as a matrix of size 12×16 representing the 16 speed bins and 12 direction sectors and these are transformed into grey scale images with each probability value of matrix as pixel intensities. From the 1400 images, 1000 images, i.e. 70% of the data is used for training the DCGANs and the rest 400 images, i.e. 30% of the data was used for testing the model. The two networks of DCGANs, Generator and Discriminator, are constructed as follows. The generator is constructed with a convolution network of 5 layers, in which the first layer with random noise of size 100×1 is input layer and can be called as fully connected layer. To achieve the desired size, 12×16, it undergoes upsampling through an additional 4 layers. Given that the target image is a WFM, a probability mass function, Softmax layer is appended as the final layer ensuring all the values in the 12×16 matrix range between 0 and 1, with the all the frequencies summing up to 1. The discriminator consists of 2 convolutional layers, and the last layer is flattened and is activated by tanh, which is used to distinguish between original and fake samples. The training of the model ensures that real images are distinguished from fake images with positive and negative values while the network parameters are updated using Binary Cross Entropy Loss and Adam optimizer (Kingma and Ba 2014). The training process is represented pictorially in Fig. 10.5, which shows the decrease in loss function discriminator depicting that the discriminator is not able to distinguish between real samples and fake samples. It also represents that the models are trained in such a way that the generator has generated images which are close to the real images. Figure 10.6 presents some of the new samples of WFMs generated by the trained DCGANs.

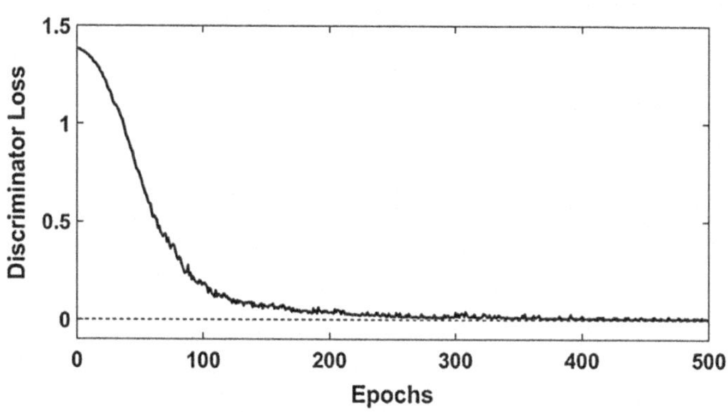

Fig. 10.5 Training loss vs epochs of discriminator network of DCGANs

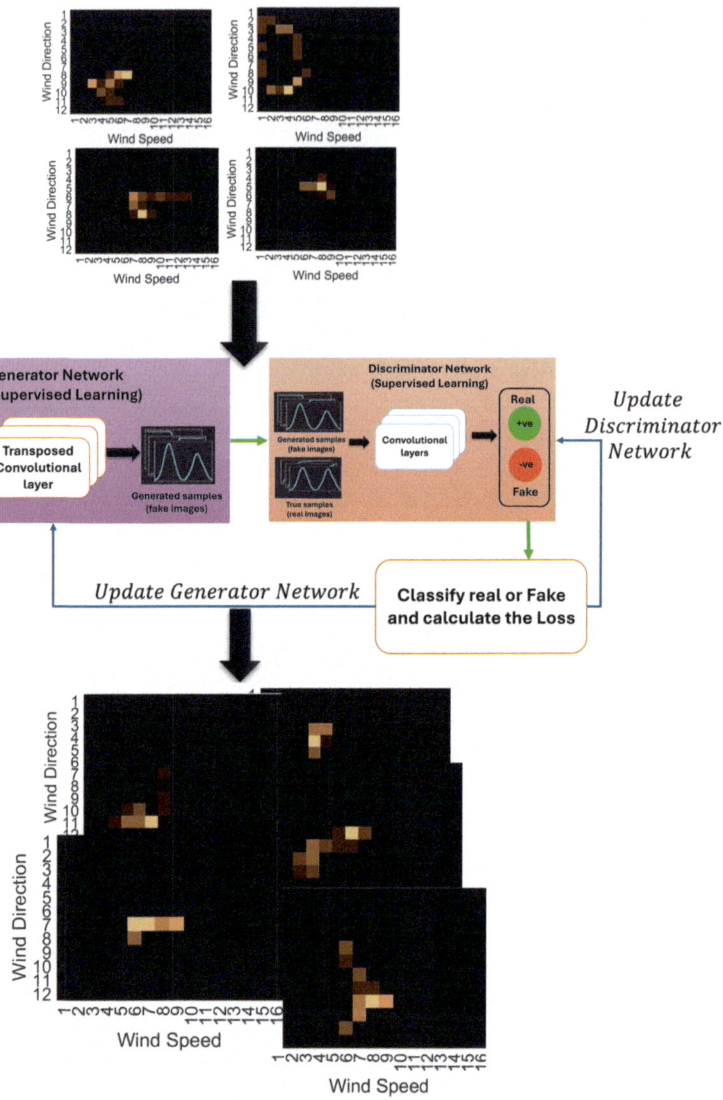

Fig. 10.6 Samples generated through DCGANs

A total of 400 new scenarios have been generated by GANS and an average WFM was constructed from the generated samples id described from Eqs. (10.7–10.8).

$$\mathfrak{n}_{sd} = \mathfrak{f}_{sd} * \mathbb{N} \tag{10.7}$$

$\mathfrak{n}_{sd}$ is calculated through $\mathfrak{f}_{sd}$ obtained from generated scenarios (see Eq. 10.1). Then an average WFM constructed with N number of WFMs is presented in Eq. (10.8):

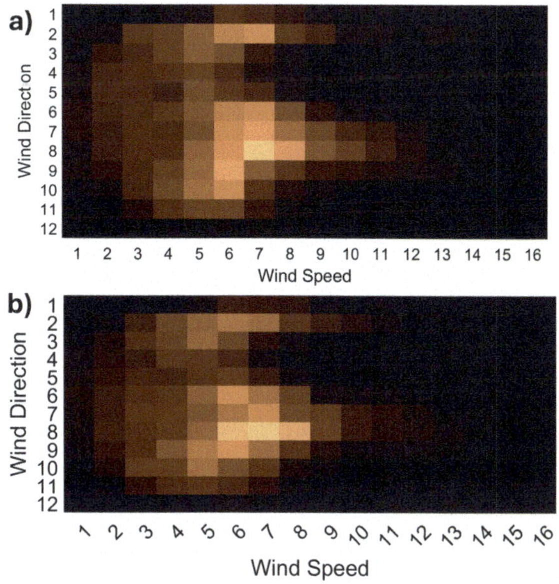

Fig. 10.7 Comparison between WFMs constructed from test data and those generated by DCGANs. **a** Average WFM constructed from test data **b** Average WFM constructed from scenarios generated through DCGANs

$$\mathbb{F}_{\mathrm{sd}} = \frac{\sum_{k=1}^{N} \mathrm{n}_{\mathrm{sd}}^{k}}{\mathbb{N} * N} \tag{10.8}$$

Similarly, an average WFM was also created from the test data set of 400 original samples, and the comparison of both the averaged WFMs of original data and generated data is presented in Fig. 10.7. The outcome indicates a comparable average between both the WFMs, suggesting that the DCGANs trained model successfully captures the true distribution of training data set. Hence, it is concluded that GANs effectively learn the true WFMs distribution, enabling the generation of scenarios while maintaining the average characteristics of the distribution.

10.3.2 Significance of WFMs in Wind Farm Studies

DCGANs are used in this study to generate 400 images of WFMs by utilizing the 1000 images of training. The generated WFMs are validated with the existing test data set as shown in Fig. 10.7. Further an analysis has done by calculating the power from a simple layout of wind farm with the use of generated scenarios and the existing scenarios to show the importance of WFMs in the wind farm studies. Equation (10.9) presents the power calculated from the given layout and the WFMs.

$$P = \sum_{i=1}^{N_t} \sum_{\mathrm{d}=1}^{12} \sum_{\mathrm{s}=1}^{16} p_{\mathrm{c}}(u_{\mathrm{d}}(u_{\mathrm{s}}, \theta_{\mathrm{d}}, N_{\mathrm{t}})) \times F(u_{\mathrm{s}}, \theta_{\mathrm{d}}) \tag{10.9}$$

Table 10.2 Power values calculated with average WFMs obtained through test data and DCGANs

WFMs obtained	Power (kW)
DCGANs	12,328.55
Test data	13,305.47

u_s, θ_d are the values of wind speed and wind direction at sth and dth intervals, u_d is effective speed of wind from a turbine after wake characterization, $p_c(u_d(u_s, \theta_d, N_t))$ is the power curve obtained from turbine manufacturer and is a function of effective wind speed, u_d. F is the wind frequency which is a joint probability mass function of two random variables: wind speed and wind direction.

Two WFMs are considered for analysis of power calculations to quantify the results obtained through GANs: (1) An average WFM constructed with 400 images generated using GANs (2) an average WFM constructed with 400 images of test data using the procedure shown in Sect. 10.3.1. The analysis was done on an optimal layout of wind farm with 33 turbines. The calculated power values from the layout with the 2 WFMs are presented in Table 10.2.

The results have shown some similarity in the power values obtained through the 2 WFMs which reports that GANs can preserve the average characteristics of the WFMs. From Table 10.2, it can be observed that the power obtained with the WFM constructed through test data set is higher than the WFM generated through GANs. The capability of GANs in generating the new scenarios which are not present in the original data set might be the underlying reason for the decreased power output observed through the WFMs generated with GANs. However, the GANs capacity to capture the underlying distribution within the data can serve as scenario generator during the design of wind farm with robust wind energy conversion systems.

Thus, the generated scenarios of WFM are considered as uncertain realizations while solving the robust worst-case and best-case formulations as described in Sect. 10.2. The wind farm design under uncertainties for best-case and worst-case are solved using the Bayesian framework. Since the study aims at maximizing the potential of a wind farm, the number of turbines is fixed to 15 in this study to reduce the complexity. Since the Bayesian framework requires initial samples to build the surrogates, sobol sampling is used to generate initial samples of decision variables, i.e. the (x, y) coordinates of turbines and the power is calculated. Then the Gaussian probabilistic surrogates are built using the initial samples. Later the expected improvement is used as acquisition function to determine the next best possible guess while updating the surrogate model and the process is repeated till a fixed number of function evaluations (a give budget) is reached. Since it is a probabilistic surrogate model, the entire procedure is repeated 3 times to ensure optimality. In this work, the number of turbines is fixed to 15 and thus leading to 30 decision variables. Therefore, 60 initial points were determined using sobol sampling for decision variables and the objectives are calculated for a set of 400 WFMs, i.e. the uncertain realizations. The budget of function evaluations for Bayesian Optimization is set to 1000 and 3 repetitions are performed to ensure optimality. From Eqs. (10.5) and (10.6), it is necessary to ensure σ value as one while maximizing the power with proper distance

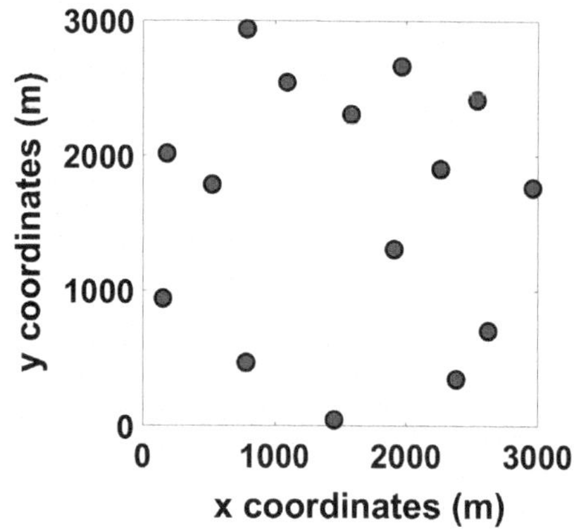

Fig. 10.8 Optimal layout configuration of wind farm for 15 through robust worst-case scenario

between turbines to minimize wake effects. The results of robust wind farm layout optimization with Bayesian framework for the best-case and the worst-case scenarios are presented below:

10.3.2.1 Results for the Worst-Case Robust Design of Wind Farm

In the worst-case robust wind farm layout optimization formulation, the power calculated for all the WFMs in the uncertain set and the minimum of the obtained power from all the scenarios is maximized. Thus, this case shows the potential of wind farm in the worst situation of lower wind frequencies and can act as the lower bound of wind farm potential for 15 turbines in an area of 3 km $\times$ 3 km wind farm with the turbine specifications presented in Table 10.1. For 15 turbines, a layout with σ value 1 is obtained within 35 iterations but the optimal value for objective is obtained within 37 iterations and the maximum power for the worst-case is obtained as 18.75 kW. The optimal layout for 15 turbines for worst-case is presented in Fig. 10.8.

10.3.2.2 Results for the Best-Case Robust Design of Wind Farm

In the best-case robust wind farm layout optimization formulation, the power calculated for all the WFMs in the uncertain set and the maximum of the obtained power from all the scenarios is maximized. Thus, this case shows the potential of wind farm in the best situation of higher wind frequencies and can act as the upper bound of wind farm potential for 15 turbines in an area of 3 km $\times$ 3 km with the turbine specifications presented in Table 10.1. For 15 turbines, a layout with σ value 1 is obtained within 290 iterations but the optimal value for objective is obtained within 440 iterations and the maximum power for the best-case is obtained as 25,537.63 kW. The optimal layout for 15 turbines for best-case is presented in Fig. 10.9.

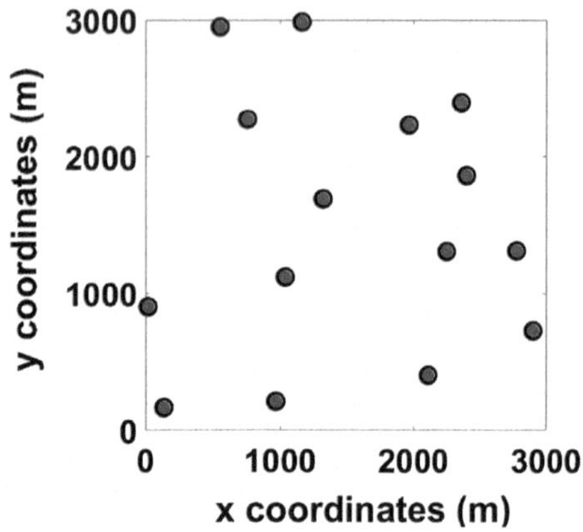

Fig. 10.9 Optimal layout configuration of wind farm for 15 through robust best-case scenario

In both the cases, the best power is obtained within 450 iterations thus showing the efficiency of Bayesian optimization in reducing the computational costs and ensuring the optimality. The results that are presented for the worst-case situation helps the wind farm owners to determine the wind farm potential for low wind conditions and the results of the best-case scenario helps to know the wind farm potential in high wind conditions thus acting as the bounds for power generation potential for the given wind farm.

10.4 Conclusion

One of the important aspects in wind farm designing is the accurate construction of WFM, a joint distribution of two random variables, i.e. wind speed and wind direction. Since the lifetime of wind farms is almost 35–40 years, it is necessary to consider the wind characteristics over the past as well as future years while designing a wind farm at a particular location. Since the wind characteristics are stochastic and volatile in nature, it is important to consider the past and future data as uncertain realizations while designing the wind farm. The lack of accurate techniques for modelling the wind sequential data for long-term accurate forecasts leads to inaccurate construction of WFMs and thereby unrealistic estimates of power production from wind farms. This leads to explore a method capable of capturing the distribution inherent in a limited data set, which can be subsequently utilized to generate various potential scenarios. In the present work, the authors demonstrated the significance of generative modelling approach for modelling the wind characteristics efficiently using DCGANs. The significance of accurate estimation of WFM is presented using a study on power production from an optimal layout. Later, the importance of optimization under uncertainty with WFMs as uncertain realizations is presented showing the significance of robust optimization to determine the true potential of a wind farm. The work is summarized below:

- The ability of generative modelling approach for accurate construction WFMs is explored using DCGANs on real-time data collected from ENGIE.
- DCGANs are shown to be successful in generating images by learning the distribution of the training data set.
- The projected power output of a wind farm, computed with scenarios generated using DCGANs alongside original data, indicated that the GANs have produced scenarios capable of maintaining the average characteristics of true distribution.
- DCGANs prove highly beneficial for generating numerous new scenarios during robust wind farm design.
- The generated scenarios are helpful to be considered as uncertain realizations while determining the best- and the worst-cases of a wind farm thus revealing the true potential of it to its owners.

Acknowledgements We would like to acknowledge the support of SPARC project (SPARC/2018-2019/P1084/SL) sponsored by the Ministry of Education (MoE), Government of India and National Supercomputing Mission (NSM) for the project (DST/NSM/R&D_HPC_Applications/2021/23) sponsored by the Department of Science and Technology (DST), Government of India.

References

Allen DJ, Tomlin AS, Bale CSE et al (2017) A boundary layer scaling technique for estimating near-surface wind energy using numerical weather prediction and wind map data. Appl Energy 208:1246–1257. https://doi.org/10.1016/J.APENERGY.2017.09.029

ENGIE (2020) La haute borne-vaudeville-le-haut (France)—wind farms—online access—the wind power. https://www.thewindpower.net/windfarm_en_3354_Frankreich.php. Accessed 10 Mar 2024

Ian Goodfellow (2016) NIPS 2016 tutorial: generative adversarial networks. NIPS 2016 tutorial

Han Q, Meng F, Hu T, Chu F (2017) Non-parametric hybrid models for wind speed forecasting. Energy Convers Manag 148:554–568. https://doi.org/10.1016/J.ENCONMAN.2017.06.021

Hu J, Heng J, Tang J, Guo M (2018) Research and application of a hybrid model based on Meta learning strategy for wind power deterministic and probabilistic forecasting. Energy Convers Manag 173:197–209. https://doi.org/10.1016/J.ENCONMAN.2018.07.052

IEEFA (2023) IEEFA | Institute for energy economics and financial analysis. In: IEEFA. https://ieefa.org/. Accessed 10 Mar 2024

Jaseena KU, Kovoor BC (2021) Decomposition-based hybrid wind speed forecasting model using deep bidirectional LSTM networks. Energy Convers Manag 234:113944. https://doi.org/10.1016/J.ENCONMAN.2021.113944

Jiang C, Mao Y, Chai Y et al (2018) Scenario generation for wind power using improved generative adversarial networks. IEEE Access 6:62193–62203. https://doi.org/10.1109/ACCESS.2018.2875936

Jin H, Shi L, Chen X et al (2021) Probabilistic wind power forecasting using selective ensemble of finite mixture Gaussian process regression models. Renew Energy 174:1–18. https://doi.org/10.1016/J.RENENE.2021.04.028

Kaur T, Kumar S, Segal R (2016) Application of artificial neural network for short term wind speed forecasting. In: 2016—Biennial International conference on power and energy systems: towards sustainable energy, PESTSE 2016. https://doi.org/10.1109/PESTSE.2016.7516458

Kingma DP, Ba JL (2014) Adam: a method for stochastic optimization. 3rd International conference on learning representations, ICLR 2015—conference track proceedings

Kisvari A, Lin Z, Liu X (2021) Wind power forecasting—a data-driven method along with gated recurrent neural network. Renew Energy 163:1895–1909. https://doi.org/10.1016/J.RENENE.2020.10.119

Liu X, Zhang H, Kong X, Lee KY (2020) Wind speed forecasting using deep neural network with feature selection. Neurocomputing 397:393–403. https://doi.org/10.1016/J.NEUCOM.2019.08.108

Lydia M, Suresh Kumar S, Immanuel Selvakumar A, Edwin Prem Kumar G (2016) Linear and non-linear autoregressive models for short-term wind speed forecasting. Energy Convers Manag 112:115–124. https://doi.org/10.1016/J.ENCONMAN.2016.01.007

Mittal P, Mitra K (2020) In search of flexible and robust wind farm layouts considering wind state uncertainty. J Clean Prod 248:119195. https://doi.org/10.1016/J.JCLEPRO.2019.119195

MNRE (2023) Wind overview | ministry of new and renewable energy | India. https://mnre.gov.in/wind-overview/. Accessed 1 Mar 2024

Radford A, Metz L, Chintala S (2016) Unsupervised representation learning with deep convolutional generative adversarial networks. 4th International conference on learning representations, ICLR 2016—conference track proceedings

Santamaría-Bonfil G, Reyes-Ballesteros A, Gershenson C (2016) Wind speed forecasting for wind farms: a method based on support vector regression. Renew Energy 85:790–809. https://doi.org/10.1016/J.RENENE.2015.07.004

Sile T, Bekere L, Cepite-Frisfelde D et al (2014) Verification of numerical weather prediction model results for energy applications in Latvia. Energy Procedia 59:213–220. https://doi.org/10.1016/J.EGYPRO.2014.10.369

Tran DT, Robinson H, Rasheed A et al (2020) GANs enabled super-resolution reconstruction of wind field. J Phys Conf Ser 1669:012029. https://doi.org/10.1088/1742-6596/1669/1/012029

Wang H, Han S, Liu Y et al (2019) Sequence transfer correction algorithm for numerical weather prediction wind speed and its application in a wind power forecasting system. Appl Energy 237:1–10. https://doi.org/10.1016/J.APENERGY.2018.12.076

Yan J, Li K, Bai E et al (2016) Time series wind power forecasting based on variant Gaussian Process and TLBO. Neurocomputing 189:135–144. https://doi.org/10.1016/J.NEUCOM.2015.12.081

Zaman U, Teimourzadeh H, Sangani EH et al (2021) Wind speed forecasting using ARMA and neural network models. 2021 IEEE electrical power and energy conference. EPEC 2021:243–248. https://doi.org/10.1109/EPEC52095.2021.9621650

Zhang C, Wei H, Zhao J et al (2016) Short-term wind speed forecasting using empirical mode decomposition and feature selection. Renew Energy 96:727–737. https://doi.org/10.1016/J.RENENE.2016.05.023

Zhao X, Liu J, Yu D, Chang J (2018) One-day-ahead probabilistic wind speed forecast based on optimized numerical weather prediction data. Energy Convers Manag 164:560–569. https://doi.org/10.1016/J.ENCONMAN.2018.03.030

Zhao X, Han X, Su W, Yan Z (2019) Time series prediction method based on convolutional autoencoder and LSTM. proceedings—2019 Chinese automation congress. CAC 2019:5790–5793. https://doi.org/10.1109/CAC48633.2019.8996842

Chapter 11
Exploring Reinforcement Learning for Efficient Wind Farm Control

NagaSree Keerthi Pujari and Kishalay Mitra

Abstract The energy crisis and climate change led the global community towards renewable resources and wind energy, being clean and abundant, has a significant potential to generate power on a large scale. Though designing the layout of wind farm is crucial part to maximize energy production, the control settings of turbine give an advantage to optimize the control parameters based on the variations of wind while maximizing the energy production from a wind farm. But it is a challenging task to optimize the control settings of a turbine through model-based methods due to the nonlinear interactions of wake between the turbines. In contemporary times, reinforcement learning (RL) has been emerging as a promising area of research in the field of wind farm control. In this work, the authors aimed to use model-free deep deterministic policy gradient (DDPG) method-based reinforcement learning techniques to optimize the control settings of wind turbines to deflect the wakes and increase the power production. As a demonstration case study, yaw misalignment is considered as control parameter while maximizing the power production from a wind farm layout, which was obtained through robust layout optimization studies presented earlier.

Keywords Reinforcement learning · Wind farm control · Nonlinear cooperative control · Deep deterministic policy gradient

---

N. K. Pujari · K. Mitra (✉)
Department of Chemical Engineering, Indian Institute of Technology Hyderabad, Hyderabad, Telangana 502284, India
e-mail: kishalay@che.iith.ac.in

N. K. Pujari
e-mail: ch20resch11006@iith.ac.in

K. Mitra
Department of Climate Change, Indian Institute of Technology Hyderabad, Hyderabad, Telangana 502284, India

Department of Artificial Intelligence, Indian Institute of Technology Hyderabad, Hyderabad, Telangana 502284, India

K. Mitra et al. (eds.), *Optimization, Uncertainty and Machine Learning in Wind Energy Conversion Systems*, Engineering Optimization: Methods and Applications,
https://doi.org/10.1007/978-981-97-7909-3_11

11.1 Introduction

Energy crisis has a significant impact on economic growth of a nation and highlights the urgent need to address the issue of dependence on fossil fuels, their depletion, and the effects on climate change. To mitigate these challenges, several nations have introduced policies toward the usage of renewable energy sources, promoting environmental sustainability, energy security, and economic growth (Subodh and Meena 2017). Wind energy stands out as the promising renewable energy source due to its ability to produce clean energy, accessibility, and economic viability. Further advancements in wind turbine technology led to increased efficiency and reduced costs, making it flexible to deploy quickly at various scales, from small wind turbines at roof top of homes to large-scale wind farms (Ovgor et al. 2012). Therefore, wind energy offers a clean, reliable solution to meet our growing energy needs while mitigating the challenges of climate change and energy crisis. Wind being unpredictable and intermittent nature, stable and maximum energy production is the biggest challenge when the wind turbines are installed in large scales. Hence, the wind farm owners aim for wind farm energy maximization under uncertainties associated with wind. One way to increase the energy production from a wind farm is to increase the number of turbines. But increasing the number of turbines in a limited space leads to extra loads on turbines downstream by the upstream turbines and thereby decreasing the energy production. To maintain the proper interturbine distance between turbines, a vast area is required which is not possible due to the geographical constraints. Therefore, many studies are available in literature which optimize the locations of turbines while maximizing energy production (Mittal and Mitra 2018; Mosetti et al. 1994). Though these studies are effective to optimize the locations of turbines, these formulations have many theoretical assumptions on control parameters of turbine such as wind strikes the turbine rotor perpendicularly (yaw misalignment) and turbine operates at optimal axial induction factor which are not valid and might result in inaccuracies. Hence, several studies aimed at optimizing the control parameters of turbine to maximize the energy production from a wind farm leading to wind farm control studies (Desalegn et al. 2023; Knudsen et al. 2015). In the previous chapters, we have seen the methods of wind farm layout optimization and the importance of forecasting before the design of the layout. Since the layout is designed with many assumptions on control parameters, this chapter focusses on optimizing the control parameters of turbines while aiming at energy maximization.

The control formulations can be either adapted in wind farm design studies by considering the control parameters (e.g., yaw misalignment, etc.) as decision variables along with the turbine locations or we can solve the control formulations by considering the control parameters as decision variables after the optimal design of layout. The second approach is adopted widely as it gives the advantage of changing the control parameters based on the uncertainties associated with the wind speed and direction even after designing the layout. The conventional control studies first focused on building the turbine models and then optimized the control parameters to maximize the energy production from a wind farm (Orlando et al. 2013). But these models do not consider the aerodynamic effects of wake on the downstream turbines from the upwind turbines which leads to inaccuracies. To decrease the wake effects on downwind turbines, a cooperative control strategy has been developed which aims at yaw misalignment of upwind turbines which deflect the wake on downwind turbines. Though the cooperative control strategy focusses on deflecting the wakes of upstream turbines by yaw misalignment, the upstream turbines face lower wind speeds leading to lower energy extraction. The deflection of wakes led to more energy extraction from downwind turbines, thus leading to overall increased energy produced (Gebraad and Van Wingerden 2015). In reported studies, the control laws are formulated using a dynamic model of a wind farm (Miao et al. 2015). But the stochastic fluctuations and nonlinearities of wind make the model to be complex. Thus, several techniques free of models have been adapted to reduce the complexities associated with model-based methods. One of such approaches is the game-theory approach, which is decentralized and maximizes the energy produced without modeling the turbine aerodynamics explicitly (Marden et al. 2013). In Gebraad and Van Wingerden (2015), a data-driven adaptive approach which accounts for wake interactions with a gradient-based optimization approach has been developed, maximum power point tracking method (MPPT), utilizing the cooperative control strategy. The results have shown that the proposed gradient-based approach is more computationally efficient when compared to the game theoretic approach while maximizing the production of wind farm energy (Gebraad and Van Wingerden 2015).

In modern times, the researchers in wind farm domain are shifting toward the machine learning techniques realizing their abilities in solving the complex systems accurately. The research community of control accelerated the use of RL in several real-world applications in control studies such as robotics, traffic control, and healthcare systems (Al-Masrur Khan et al. 2020; Coronato et al. 2020; Han et al. 2023). Compared to other machine learning techniques which rely on repetitive examples, RL focuses on the experience obtained through previous experiments while learning the tasks. RL finds the best control policy by leveraging on the interaction of environment with the agent, proving its efficiency in several process control industries while tuning the PI controllers (Shipman and Coetzee 2019). The self-learning ability of the controllers based on RL makes them to be easily adaptable for online optimizations or online control studies. In a broader perspective, there are two types of RL techniques: model-based algorithms and model-free algorithms. In model-based algorithms, the dynamics of the system is used while learning phase of the agent, i.e., the agent and reward function are connected through the system dynamics and the policy is learnt through a learning strategy. Example of such model-based RL method is iterative linear quadratic regulator (iLQR) method which can be used in applications such as self-driving (Pan et al. 2020). The model-free RL algorithms focus on value-estimation methods, which are learnt through the experience obtained from the interactions with the environment. Deep deterministic policy gradients (DDPG) is an emerging model-free RL algorithm, which estimates Q values through the agent and is proven to be efficient with legged locomotion and car driving (Lillicrap et al. 2015). With the emerging trends of RL in real-world applications, the industrialists and wind farm owners are paying attention towards using RL in wind farm control (Greaves et al. 2022). An actor-critic method is employed in Fernandez-Gauna et al. (2022) while solving a joint-control problem using blade pitch and generator torque as control parameters. Model-free and model-based RL algorithms, DDPG and iLQR, are compared with the MPPT technique in Pujari et al. (2021) to optimize the control parameter and yaw misalignment while maximizing the power production, and it is reported that DDPG gave better results when compared to other techniques. Therefore, the current work aims at optimizing the wind farm control parameters using the model-free RL algorithm, DDPG, while maximizing the wind farm energy production.

In this work, four optimal layouts are considered, designed through robust optimization by considering the wind frequency maps, obtained through GANs and DRIFT, as uncertain realizations (see Chaps. 9 and 10). Four optimal layouts are designed in an area of 3 km $\times$ 3 km with 15 turbines. Yaw misalignment is used as control parameter, which is to be optimized through the control strategy DDPG for variable wind speed and direction. Since wake characterization has significant impact on power production, flow reduction and induction in steady state (FLORIS) wake model developed by National Renewable Energy Laboratory's (NREL) is used to model the wake dynamics. The unique contributions for this work are summarized below:

- The model-free RL algorithm, DDPG, is implemented to determine the optimal yaw angles of the turbines to maximize the energy production.
- Four layouts obtained through robust optimization are considered to demonstrate the control studies.

In the rest of the chapter, a brief description on FLORIS wake model and DDPG is presented in Sect. 11.2. Section 11.3 presents data considered for this work followed by the results of the control strategy. Section 11.4 outlines the summary of the proposed work.

11.2 Formulation

Data-driven techniques are growing rapidly in contemporary times to enhance the accuracy of the wind farm control techniques. RL is the emerging control strategy which has proven its efficiency in various control problems. Therefore, in this work, the authors used the model-free RL algorithm for control strategy in wind farms to optimize the yaw angles for maximum energy production. The depiction of yaw angle is presented in Fig. 11.1. The wake model description followed by a brief description on RL algorithm, DDPG, is presented.

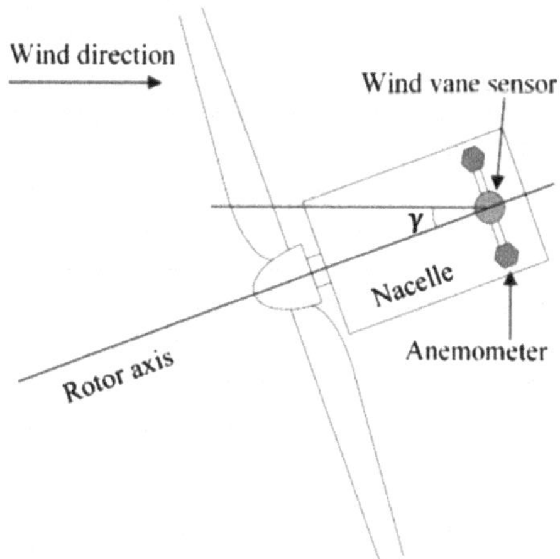

Fig. 11.1 Yaw angle (γ) depiction for a wind turbine

11.2.1 FLORIS Wake Model

One way to increase energy production is to increase the number of turbines. If the number of turbines is increased in a fixed area, the upstream turbines pass the turbulence on the downwind turbines leading to lower wind velocities for down wind turbines. This reduces the energy extracted from the downwind turbines, and this phenomenon is called wake. To characterize these wake effects efficiently, several formulations are developed in the literature from simple analytical models to the complex large eddy simulation models (Annoni et al. 2014). Here, FLORIS wake model is used developed by NREL, which balances the simple analytical model and large eddy simulations while characterizing the wake effects (Gebraad et al. 2016), and it is presented in Eq. (11.1).

$$u_n = u_\infty \left(1 - u_{\text{eff}} e^{\frac{(\beta-\delta)^2}{2\sigma_y^2}} e^{\frac{(\zeta - z_h)^2}{2\sigma_z^2}} \right) \tag{11.1}$$

$$P_n = \frac{\rho \pi D_n^2}{8} (u_n \cos(\gamma_n))^3 C_{p_n}(\alpha_n, \gamma_n) \quad n = 1, 2, \ldots N$$

$$C_{p_n} = 4\alpha_n (\cos(\theta \gamma_n) - \alpha_n)^2 \text{ and } \alpha_n = \frac{u_\infty - u_r}{u_\infty}$$

where u_{eff} is effective velocity after characterizing wake, δ is wake deflection, σ_y, σ_z represent the wake growth rate in radial and vertical directions, respectively, u_∞ is the free stream velocity, z_h is the height of turbine hub, P_n represents the power of nth turbine, D_n is the diameter of the rotor of nth turbine, γ_n is the yaw angle of the nth turbine, ρ is the air density, C_{p_n} is the thrust coefficient, and α_n is the axial induction factor and θ is the model coefficient.

11.2.2 Deep Deterministic Gradient Method

RL is based on a trial-and-error approach with an agent reacting with an environment repeatedly. Once an observation is received by the agent, it will take an action resulting in a feedback signal called reward. Thus, the agent learns the action to be taken through the rewards achieved by different experiments such that the future reward is maximized. The sum of all the reward achieved by the agent for a specific time horizon is called return (Lillicrap et al. 2015) which is presented in Eq. (11.2).

$$R_t = r_t + \omega r_{t+1} + \omega^2 r_{t+2} + \cdots \tag{11.2}$$

where r_t is the reward attained by the agent at t^{th} time step, ω is the discounting factor, lies between 0 and 1, and R_t is the return. The behavior of the agent can be assessed by the following attributes: State action value function, state value function, and policy. The policy of an agent is used to map states with actions using either deterministic methods or stochastic methods. The policy in deterministic methods is defined as Eq. (11.3) where s refers to the states and a as action. Similarly, the policy in stochastic methods relates the states to actions as probability distributions which is defined as Eq. (11.4).

$$\pi : S \rightarrow a \,\text{where}\, a = \pi(S) \,\text{where}\, \pi \,\text{is the policy} \tag{11.3}$$

$$\pi : S \rightarrow P(a) \,\text{where}\, \pi(a|s) = P(A_t = a|S_t = s) \tag{11.4}$$

The quality of states is determined by the state value function represented in Eq. (11.5), and the Q value related to state action pair is represented by state action value function described in Eq. (11.6). The agent tries to optimize the Q value by learning a policy which maximizes the reward, and therefore, the policy which gives optimal Q value is called optimal policy. The optimal value function and its corresponding optimal policy are presented in Eq. (11.7).

$$v_\pi(s) = E_\pi[R_t|S_t = s] \tag{11.5}$$

$$Q_\pi(s, a) = E_\pi[R_t|S_t = s, A_t = a] \tag{11.6}$$

$$Q^*(s, a) = \max_\pi Q_\pi(s, a) \tag{11.7}$$

$$\pi^* = \arg\max_\pi Q_\pi(s, a)$$

Therefore, the aim of RL is to maximize the expected return by learning the optimal policy through several experiments by interacting with the environment. Broadly, RL is divided into two different techniques: model-based methods which include system dynamics, model-free methods which learn the Q value. The inefficiency of value-based methods for continuous state action space gave a scope for deep reinforcement learning where neural networks are utilized to estimate the Q value, also known as deep Q networks (DQN). The two features of DQNs, target networks and replay buffer, add advantages of memorizing past actions and stable training. Target networks are cloned networks to damp out the short-term oscillations and helpful for stable training. Replay buffer memorizes the agent's past actions and correlated observations, and therefore, avoids repetitions of actions. One of such model-free methods are actor-critic networks which are developed for continuous action spaces. In actor-critic methods, the policy generating action is represented by actor network and the maximum possible return achieved from the current state of action is determined by critic network. These actor and critic networks are considered as deep neural networks while solving complex environments. DDPG is one of the popular actor-critic networks which is used in this study while tuning the control parameters of turbine to maximize energy production. In DDPG, the actor and critic networks act as target networks to add stability in the training. The action space is represented as the change in states, and the past action space is stored in experience replay which acts as a memory buffer. In the buffer, the states, change in states, actions, and rewards are stored, and any of the stored values are selected during the training to determine the correlations between inputs and outputs having similar action.

In the current work, yaw angle is considered as the control parameter which should be tuned while maximizing the energy production. In RL terminology, wind farm is the environment, and the controller is the agent, yaw angle is the manipulated variable, yaw angle of a turbine and power from turbines are states, and change in yaw angle is action space. In this work, we have considered a stochastic policy, i.e., the policy is determined through probabilistic distribution of states. The DDPG is presented in Eq. (11.8)

$$\begin{aligned}&\text{Observation vector: } \left[\gamma_i, P_i, P\right]^{\mathrm{T}} \; i = 1, 2, \ldots N\\&\text{and } \gamma_i \in \left[-45^\circ, 45^\circ\right]; \; P = \sum_{i=1}^{N} P_i\end{aligned} \tag{11.8}$$

Action space: Change in yaw angle; $\Delta\gamma_i \in [-1^\circ, 1^\circ]$
Memory buffer: $\{a, s, s', r\}$

$$a_i = \max_{a_i}(\hat{r}(s_i, a_i))$$

where γ_i is the yaw angle of ith turbine, P_i is the power of ith turbine, P is the total power of turbines, and change in yaw angle is in between -1 and 1 to ensure the stability. $\hat{r}$ is the reward function described in Eq. (11.9), s_i, a_i represent the states and actions. The reward function is defined in such a way that the yaw angles have minimum fluctuations during training to reduce the turbine loads and power is maximized.

$$\hat{r} = -\sum_{i=1}^{N} (\gamma_i)^2 + P(t) \tag{11.9}$$

To account for the previous controller's performance and the previous rewards, the reward function is modified to Eq. (11.10).

$$r(t) = r + \sum_{k=1}^{\mathbb{N}} \gamma^k r_{t+k} \tag{11.10}$$

Therefore, in the current work, the DDPG focuses on maximizing the reward function in Eq. (11.10) while changing the yaw angles by interacting with the wind farm and the results are presented in Sect. 11.3. The working principle of DDPG is presented in Fig. 11.2.

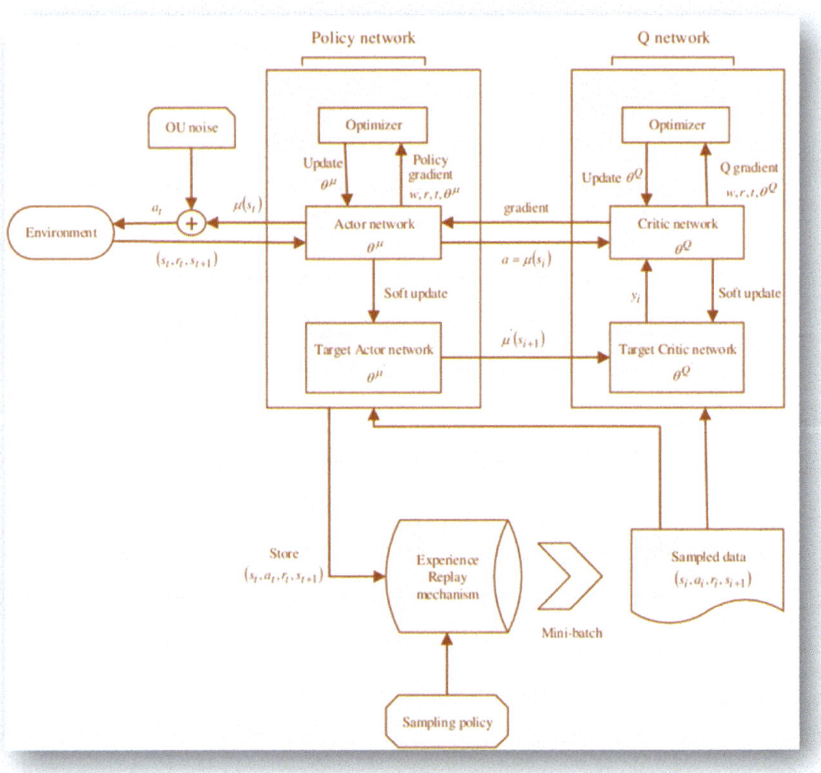

Fig. 11.2 Schematic of DDPG algorithm

11.3 Results

11.3.1 Data Description

Since wind varies with time, control settings of parameters can be changed to maximize wind energy production. Yaw angle is considered as manipulated variable in this study, and FLORIS model is used to characterize the wake effects which has significant impact on power production. As mentioned earlier, four optimized layouts are considered to demonstrate the wind farm control studies obtained through robust optimization. In the four layouts, two layout represents the best-case scenarios and other two present the worst-case scenarios (see Chaps. 9 and 10 for more details). The considered layouts are presented in Fig. 11.3. In all these layouts, the number of turbines is 15, and hence, the yaw angles for all these 15 turbines should be optimized for attaining maximum wind energy.

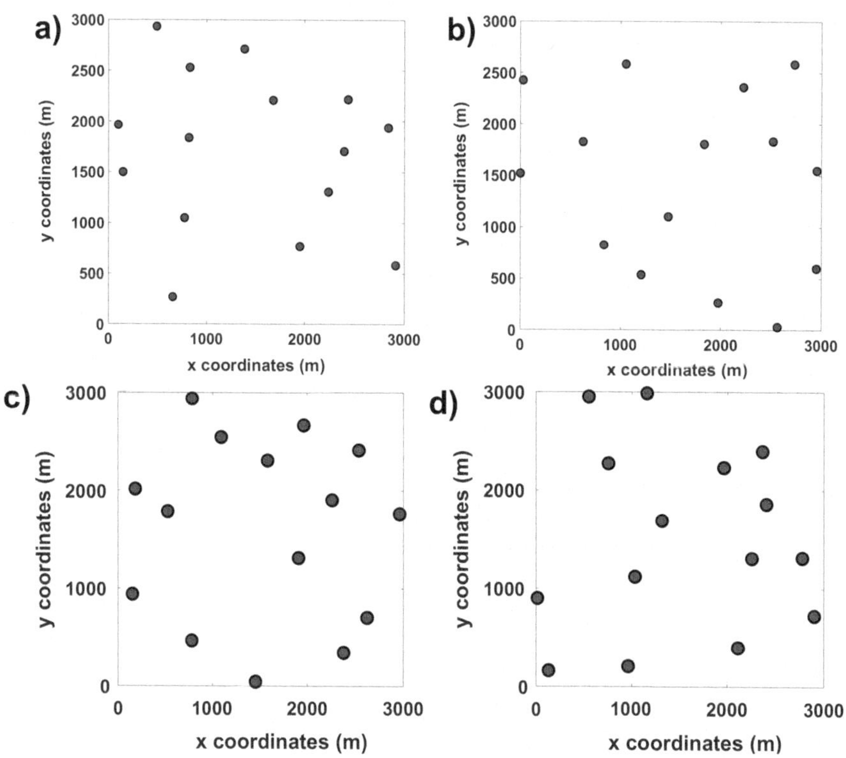

Fig. 11.3 Schematic of layouts considered. **a** Presents the layout obtained from worst-case scenario of DRIFT RO. **b** Presents the layout obtained from best-case scenario of DRIFT RO. **c** Presents the layout obtained from worst-case scenario of GANs RO. **d** Presents the layout obtained from best-case scenario of GANs RO

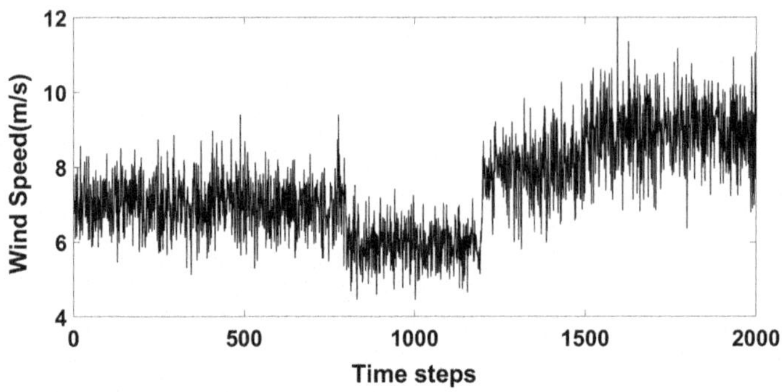

Fig. 11.4 Pictorial representation of wind speed

A variable wind signal with different speed and directions presented in Figs. 11.4 and 11.5 are considered for this study. The turbine and wind farm specifications are presented in Table 11.1.

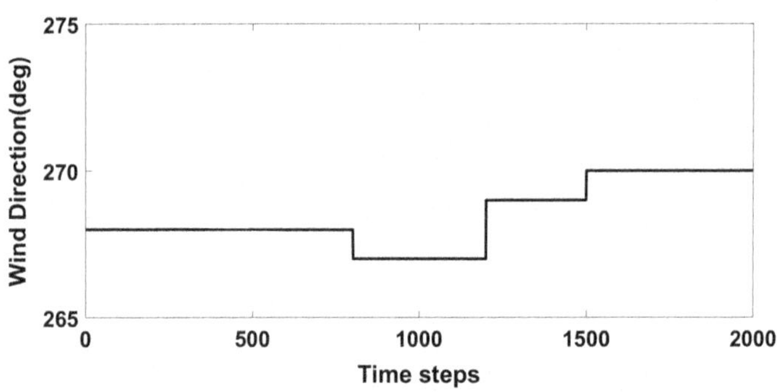

Fig. 11.5 Pictorial representation of wind direction

Table 11.1 Specifications of wind farm and turbine

Wind farm and turbine specifications	
Wind farm area	3 km × 3 km
Surface roughness	0.3
Hub height of turbine	80 m
Rotor diameter of turbine (*D*)	82 m
Thrust coefficient	0.88
n_{sp}, allowable distance between turbines	5*D*
Cutin velocity	3.5 m/s
Rated velocity	15 m/s
Cutout velocity	22 m/s
Rated power	2000 kW

Specifications of DDPG

In this work, actor-critic networks in DDPG are considered as deep neural networks with an architecture of [2–500–400–1]. The hidden layers have ReLu activation, and the output layer has linear activation. However, the policy network has tanh activation in the output layer. The learning rates for actor and critic networks are 10^{-4} and 10^{-3}, respectively. The DDPG is ran for 200 episodes with batch size of 128 and horizon as 10.

With the specifications described above, DDPG is trained to optimize the yaw angles by maximizing the reward function described in Eq. (11.10). The results for the considered four layouts are presented as follows:

11.3.2 Results for Case 1: Layouts Obtained Through DRIFT RO

In this case, the layouts are obtained through the robust optimization performed with the WFMs obtained through DRIFT as uncertain set. The wind farm design problem is solved using Bayesian framework. Two robust optimization formulations are solved considering the worst- and the best-case scenarios. For detailed explanation about these formulations, please refer to Chap. 9. The resulting layouts are presented in Fig. 11.3a, b. For these layouts, the obtained powers are 8801.42 kW and 16,886.73 kW for one year, respectively, presenting lower and upper bounds on potentials of wind farm. For these two layouts, the yaw angles are optimized to maximize power production. The total power obtained for the layout presented in Fig. 11.3a, i.e., for the worst-case DRIFT RO, is presented in Fig. 11.6a, and the total power obtained for the considered wind signals (see Figs. 11.4 and 11.5) is obtained as 9.91 GW. The total power obtained for the layout presented in Fig. 11.3b, i.e., for the best-case DRIFT RO, is presented in Fig. 11.6b, and the total power obtained for the considered wind signals (see Figs. 11.4 and 11.5) is obtained as 11.06 GW. The yaw angle profile for one of the turbines for the case of DRIFT RO is presented in Fig. 11.7.

11.3.3 Results for Case 2: Layouts Obtained Through GANs RO

In this case, the layouts are obtained through the robust optimization performed with the WFMs obtained through GANs as uncertain set. The wind farm design problem is solved using Bayesian framework. Two robust optimization formulations are solved considering the worst- and the best-case scenarios. For detailed explanation about these formulations, please refer to Chap. 10. The resulting layouts are presented in Fig. 11.3c, d. For these layouts the obtained powers are 18.75 kW

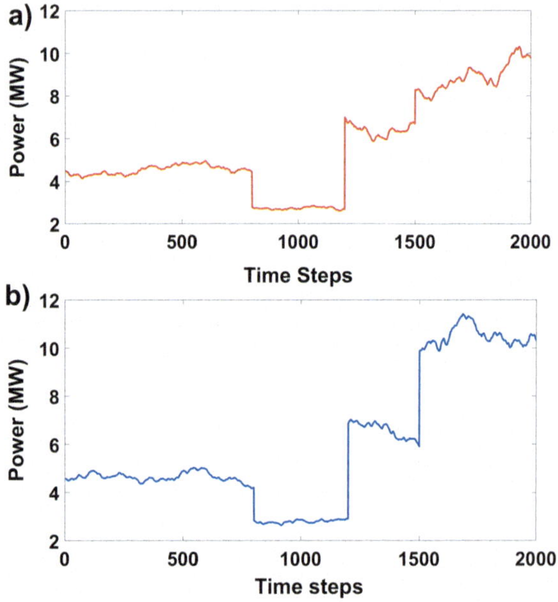

Fig. 11.6 Power obtained from DDPG. **a** Presents the power obtained from worst-case scenario of DRIFT RO layout. **b** Presents the power obtained from best-case scenario of DRIFT RO layout

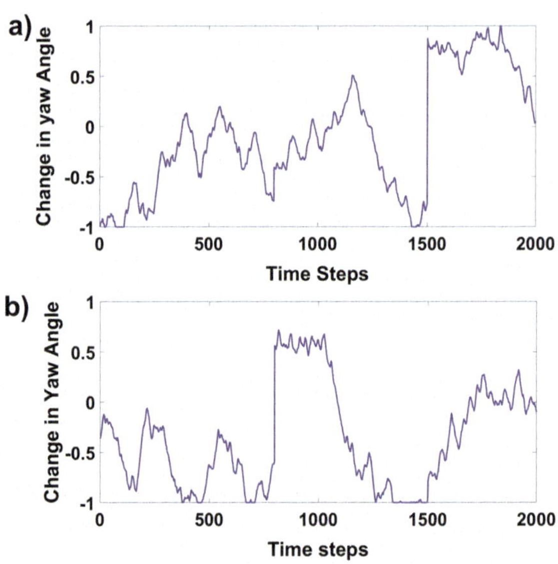

Fig. 11.7 Profiles for change in yaw angle (normalized scale). **a** Presents the yaw angle profile worst-case scenario of DRIFT RO layout. **b** Presents the yaw angle profile from best-case scenario of DRIFT RO layout

and 25,537.63 kW for one year, respectively, presenting lower and upper bounds on potentials of wind farm. For these two layouts, the yaw angles are optimized to maximize power production. The total power obtained for the layout presented in Fig. 11.3c, i.e., for the worst-case GANs RO, is presented in Fig. 11.8a, and the total power obtained for the considered wind signals (see Figs. 11.4 and 11.5) is obtained as 9.2 GW. The total power obtained for the layout presented in Fig. 11.3d, i.e., for the best-case GANs RO, is presented in Fig. 11.8b, and the total power obtained for the considered wind signals (see Figs. 11.4 and 11.5) is obtained as 10.65 GW. The yaw angle profile for one of the turbines for the case of GANs RO is presented in Fig. 11.9.

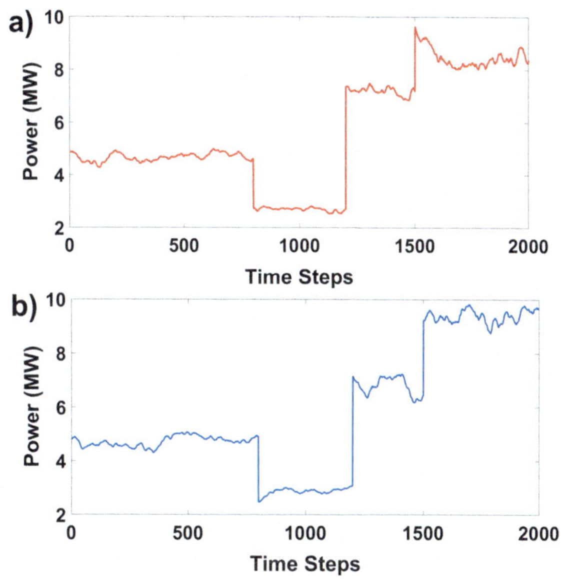

Fig. 11.8 Power obtained from DDPG. **a** Presents the power obtained from worst-case scenario of GANs RO layout. **b** Presents the power obtained from best-case scenario of GANs RO layout

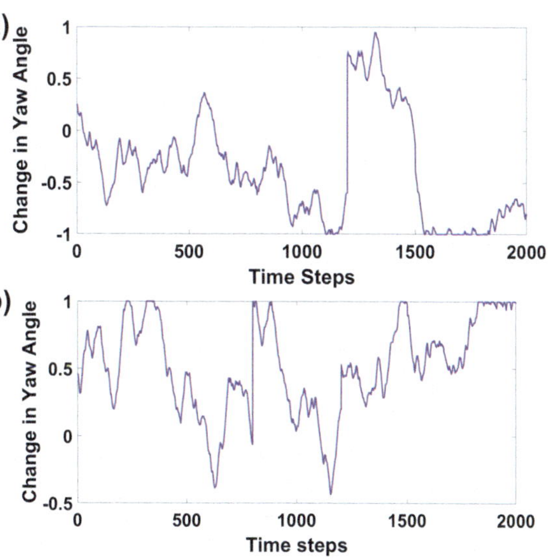

Fig. 11.9 Profiles for change in yaw angle (normalized scale). **a** Presents the yaw angle profile worst-case scenario of GANs RO layout. **b** Presents the yaw angle profile from best-case scenario of GANs RO layout

In this study, the results have shown that the power obtained through the DDPG algorithm in all the cases is reported as around 10 GW, and it is proven that the DDPG has the capability to optimize the yaw angles to obtain the maximum power production from a wind farm.

11.4 Conclusion

One of the important aspects in wind modeling is the wind farm control that attributes to the maximization of energy production. Though the wind farm design plays a crucial role in energy maximization, there is a scope to increase energy production through optimizing the control parameters such as yaw angle, axial induction factor, generator torque, and blade pitch. In this work, yaw angle is considered as the manipulated variable to optimize the power production. As RL has been coming up as an emerging technique in wind farm control in recent years, DDPG is used as control strategy in this work. The results have shown that the DDPG worked effectively in optimizing the yaw angles resulting in maximum power production. Thus, the wind farm control studies can be efficiently incorporated with short-term forecasting of wind characteristics to reduce the loads and fatigue of wind turbines.

Acknowledgements We would like to acknowledge the support of SPARC project (SPARC/2018-2019/P1084/SL) sponsored by the Ministry of Education (MoE), Government of India and National Supercomputing Mission (NSM) for the project (DST/NSM/R&D_HPC_Applications/2021/23) sponsored by the Department of Science and Technology (DST), Government of India.

References

Al-Masrur Khan MD, Khan MRJ, Tooshil A, Sikder N, Parvez Mahmud MA, Kouzani AZ, Nahid AA (2020) A systematic review on reinforcement learning-based robotics within the last decade. IEEE Access 8:176598–176623. https://doi.org/10.1109/ACCESS.2020.3027152

Annoni J, Seiler P, Johnson K, Fleming P, Gebraad P (2014) Evaluating wake models for wind farm control. In: Proceedings of the American control conference, pp 2517–2523. https://doi.org/10.1109/ACC.2014.6858970

Coronato A, Naeem M, De Pietro G, Paragliola G (2020) Reinforcement learning for intelligent healthcare applications: a survey. Artif Intell Med 109:101964. https://doi.org/10.1016/J.ARTMED.2020.101964

Desalegn B, Gebeyehu D, Tamrat B, Tadiwose T (2023) Wind energy-harvesting technologies and recent research progresses in wind farm control models. Front Energy Res 11:1124203. https://doi.org/10.3389/FENRG.2023.1124203/BIBTEX

Fernandez-Gauna B, Graña M, Osa-Amilibia JL, Larrucea X (2022) Actor-critic continuous state reinforcement learning for wind-turbine control robust optimization. Inf Sci 591:365–380. https://doi.org/10.1016/J.INS.2022.01.047

Gebraad PMO, Van Wingerden JW (2015) Maximum power-point tracking control for wind farms. Wind Energy 18(3):429–447. https://doi.org/10.1002/WE.1706

Gebraad PMO, Teeuwisse FW, Van Wingerden JW, Fleming PA, Ruben SD, Marden JR, Pao LY (2016) Wind plant power optimization through yaw control using a parametric model for wake effects—a CFD simulation study. Wind Energy 19(1):95–114. https://doi.org/10.1002/WE.1822

Greaves D, Jin S, Wong P, Volker PJ, Hahmann AN, Badger J, Smits C-J, Gonzalez Silva J, Chabaud V, Dong H, Xie J, Zhao X (2022) Wind farm control technologies: from classical control to reinforcement learning. Prog Energy 4(3):032006. https://doi.org/10.1088/2516-1083/AC6CC1

Han Y, Wang M, Leclercq L (2023) Leveraging reinforcement learning for dynamic traffic control: a survey and challenges for field implementation. Commun Transp Res 3:100104. https://doi.org/10.1016/J.COMMTR.2023.100104

Knudsen T, Bak T, Svenstrup M (2015) Survey of wind farm control—power and fatigue optimization. Wind Energy 18(8):1333–1351. https://doi.org/10.1002/WE.1760

Lillicrap TP, Hunt JJ, Pritzel A, Heess N, Erez T, Tassa Y, Silver D, Wierstra D (2015) Continuous control with deep reinforcement learning. In: 4th international conference on learning representations, ICLR 2016—conference track proceedings. https://arxiv.org/abs/1509.02971v6

Marden JR, Ruben SD, Pao LY (2013) A model-free approach to wind farm control using game theoretic methods. IEEE Trans Control Syst Technol 21(4):1207–1214. https://doi.org/10.1109/TCST.2013.2257780

Miao L, Wen J, Xie H, Yue C, Lee WJ (2015) Coordinated control strategy of wind turbine generator and energy storage equipment for frequency support. IEEE Trans Ind Appl 51(4):2732–2742. https://doi.org/10.1109/TIA.2015.2394435

Mittal P, Mitra K (2018) Determining layout of a wind farm with optimal number of turbines: a decomposition based approach. J Clean Prod 202:342–359. https://doi.org/10.1016/J.JCLEPRO.2018.08.093

Mosetti G, Poloni C, Diviacco B (1994) Optimization of wind turbine positioning in large windfarms by means of a genetic algorithm. J Wind Eng Ind Aerodyn 51(1):105–116. https://doi.org/10.1016/0167-6105(94)90080-9

Orlando NA, Liserre M, Mastromauro RA, Dell'Aquila A (2013) A survey of control issues in PMSG-based small wind-turbine systems. IEEE Trans Ind Inf 9(3):1211–1221. https://doi.org/10.1109/TII.2013.2272888

Ovgor B, Lee SK, Lee S (2012) A method of micrositing of wind turbine on building roof-top by using joint distribution of wind speed and direction, and computational fluid dynamics. J Mech Sci Technol 26(12):3981–3988. https://doi.org/10.1007/S12206-012-0897-2/METRICS

Pan Y, Lin Q, Shah H, Dolan JM (2020) Safe planning for self-driving via adaptive constrained ILQR. In: IEEE international conference on intelligent robots and systems, pp 2377–2383. https://doi.org/10.1109/IROS45743.2020.9340886

Pujari KN, Srivastava V, Miriyala SS, Mitra K (2021) Comparison of deep reinforcement learning techniques with gradient based approach in cooperative control of wind farm. In: 2021 7th Indian control conference, ICC 2021—proceedings, pp 400–405. https://doi.org/10.1109/ICC54714.2021.9703186

Shipman WJ, Coetzee LC (2019) Reinforcement learning and deep neural networks for PI controller tuning. IFAC-PapersOnLine 52(14):111–116. https://doi.org/10.1016/J.IFACOL.2019.09.173

Subodh K, Meena RL (2017) Renewable energy sources-policies in India. Int J Appl Environ Sci 293–297

Chapter 12
Maximizing Energy Production from Wind Farm Layout Using Model Predictive Control

Ravi Kiran Inapakurthi, NagaSree Keerthi Pujari, and Kishalay Mitra

Abstract Wind farms are optimally designed to capture wind energy considering the applicability of certain wind speed and direction patterns over a long duration. With growing deviation in the weather patterns across the globe, this assumption might not be applicable after installation resulting in sub-optimal power production from the wind farm. Hence, control of yaw angles of turbines to maximize the overall power production becomes essential. In this study, Bastankhah wake effect model is used for model predictive control with an aim of maximizing the power production from a given layout. For this, the best-case and worst-case scenarios of generative adversarial networks (GAN) and data-driven robust optimization through intelligent fuzzy transcription (DRIFT) are selected. The maximum power produced for the best-case scenario with GANs and DRIFT are 5.93 GW and 7.08 GW, respectively. For the worst-case scenarios, the corresponding figures are 5.78 GW and 5.64 GW, respectively. While the central focus of this chapter is wind energy, the approach can be followed with other sources of energy like solar energy.

Keywords Artificial intelligence · Evolutionary optimization · Renewable energy · Control · Unsupervised machine learning · Clustering

R. K. Inapakurthi · N. K. Pujari · K. Mitra (✉)
Department of Chemical Engineering, Indian Institute of Technology Hyderabad, Hyderabad, Telangana 502284, India
e-mail: kishalay@che.iith.ac.in

N. K. Pujari
e-mail: ch20resch11006@iith.ac.in

K. Mitra
Department of Climate Change, Indian Institute of Technology Hyderabad, Hyderabad, Telangana 502284, India

Department of Artificial Intelligence, Indian Institute of Technology Hyderabad, Hyderabad, Telangana 502284, India

K. Mitra et al. (eds.), *Optimization, Uncertainty and Machine Learning in Wind Energy Conversion Systems*, Engineering Optimization: Methods and Applications,
https://doi.org/10.1007/978-981-97-7909-3_12

12.1 Introduction

It is understood and widely accepted that human activities are primarily responsible for the abnormal weather changes, increase in extreme weather events, and global level harm to flora and fauna (Cavanaugh et al. 2014; Freeman and Class Freeman 2014). With this acceptance comes the need to act responsibly such that fellow living beings are not made to suffer or worse still, driven to extinction. A report puts several animal species are driven out of their natural habitat due to human energy needs driven activities (Pecl et al. 2017). To preserve the ecological diversity and balance within the ecosystem that humans live, it is essential to identify energy pathways that do not cause harm or minimize the harm to the ecology, and environment. This needs to be done while keeping the economic growth prospectus in mind. It is widely argued that renewable sources of energy are the go-to option for minimizing this damage.

While many of the earlier chapters focus on the yet-to-be-installed wind farms, this chapter focuses on an existing wind farm layout. The yaw angle of the wind turbines can be controlled to maximize the total power output from a layout. Many studies have used control strategies for the optimization of power production from wind layout. In Munters and Meyers (2018), control of dynamic induction and yaw angle was explored on six different wind farm layouts. Control of wind farms was reported to have a positive effect on the maximum power produced from the existing layout. It was noted that the most profitable strategy among the dynamic induction control and yaw angle control is a function of the wind flow direction. In Gebraad et al. (2016), computational fluid dynamics (CFD) approach was followed to factor the wake effect of the upstream turbine on the downstream turbine. For the CFD simulations, flow redirection and induction in steady-state (FLORIS) model was used. FLORIS model is a prominent model used to capture the wake effects, flow velocity, induction, and yaw angle on the power generation from a turbine. The use of the FLORIS model has resulted in maximization of the power production from the plant while, as an added advantage, has resulted in the decrease in load on the turbines.

The broad philosophy of the control of wind turbines for maximization of power production is a measly sacrifice of power production from the upstream turbines by adjusting the yaw angles of the upstream turbines (Pujari et al. 2023). This provides the benefit of reduced wake effects on the downstream turbines. This means the velocity of the free stream wind reaching the downstream turbines is more, thereby facilitating a gain in the power production from the downstream turbines (Miriyala et al. 2022). This act of sacrifice-to-overcompensate-later is widely reported in the literature. For e.g., experimental evaluations in Zong and Porté-Agel (2021) indicate that the power production from full wake effect results in a local minimum (peak power gain of 5.4%) while a misalignment-induced partial wake case has resulted in a peak power gain of 18%.

With the growth of wind data, there is a similar growth in the number of publications on wind farm control using data-based techniques. Within the machine learning (ML) class of techniques, reinforcement learning (RL) has received particular attention due to its similarity in finding the optimal path to attain a stated objective. In RL, an agent interacts with the dynamic environment (in this case the wind farm) such that the state of the agent is changed (the yaw angle of the turbine) to maximize the stated objective of power production (Barto and Richard 2018). In Pujari et al. (2021), a model-free deep deterministic policy gradient, which is a RL method, was used along with linear quadratic regulator (LQR) for the optimal control of the yaw angle. The authors reported that the use of RL results in superior power production than the LQR.

However, with the growing population, the growth in energy needs of humans would only increase. This has a cascading effect on the size of the wind farm. As the complexity of existing wind farms increases, the computational power needed for the RL agent to run down all the many possible combinations of yaw angles to optimize the wind farm will rise exponentially (Li et al. 2024). To address this issue, researchers have developed grouping strategies in which small batches of wind turbines would be grouped. The authors in Li et al. (2024) have grouped the wind turbines based on the strength of the wake interacting with the turbines. This results in a decrease of computational overhead, thereby enhancing the controlling efficiency. In another study (Zhang et al. 2023), along with grouping the turbines, the communication mechanism between neighboring aggregates of wind turbine groups is factored into the maximum power calculation. While the approach shows promising results, its application on only thirteen-turbine wind farm raises questions on the scalability of the proposed approach.

In addition to the use of RL, other data-driven techniques have also been explored in the literature. In Dittmer et al. (2023), the authors used data-driven wind farm control approach for the Koopman model predictive control (MPC) to predict the wake effects of the upstream turbine on the downstream turbine. For simulations, the effect of yaw angle and the thrust control is considered on a two-turbine wind farm. In another set of studies, evolutionary algorithms (EAs)-based control of turbines is explored for optimizing the power production from the wind farms. In Wang et al. (2024), the authors have used particle swarm optimization for the optimization of yaw angle in the Chapman Ranch Wind Farm.

From this analysis of the literature, the use of ML techniques to forecast the wind characteristics like speed and direction to be used in the control exercise seems to be missing. The argument in favor of this approach is that once high accuracy ML models are developed, they can be used to forecast the long-term trends of the wind speed and direction. This information is then used in the MPC framework in which the deviations of the wind characteristics assumed during the design stage of the wind farm can be accounted for. This helps in maximizing the power production from the wind farm by control of the yaw angle. To facilitate the MPC, forecasted values of wind speed and direction from Chap. 8 are taken for reference as they are demonstrated to be highly accurate. The novelties of this chapter are as follows:

(a) The maximization of the power production from the layout is achieved using MPC in which yaw angle is the manipulated variable.
(b) The wind speed and direction forecasted using least square support vector regression is used in this study to perform control of yaw angle.
(c) Best-case and worst-case scenarios of DRIFT and GAN (see Chaps. 9 and 10 for more details) robust optimization wind farm layouts are considered.

The rest of the book chapter is organized as follows. In Sect. 12.2, the model describing the wake effect within a wind farm is described first followed by an explanation of the MPC framework used in this study. In Sect. 12.3, the results of the MPC with the four cases are presented along with an analysis of the obtained results. The conclusions of the chapter are then summarized in the conclusion section.

12.2 Formulation

12.2.1 Bastankhah Model

The wake model used in this study is taken from Bastankhah and Porté-Agel (2016). A wind farm with 15 turbines are considered. While all the turbines are assumed to be identical, the wake influence on a particular turbine can be from more than one turbine. First, axis rotation is applied on the turbines as shown in Eqs. (12.1) to (12.3).

$$\theta^j_{\mathrm{ww}} = \theta^j_{\mathrm{w}} \frac{\pi}{180} \tag{12.1}$$

$$X_i = x_i \cos\left(\theta^j_{\mathrm{ww}}\right) - y_i \sin\left(\theta^j_{\mathrm{ww}}\right) \tag{12.2}$$

$$Y_i = x_i \sin\left(\theta^j_{\mathrm{ww}}\right) + y_i \cos\left(\theta^j_{\mathrm{ww}}\right) \tag{12.3}$$

Here, θ^j_{w} is the wind direction at jth time instance, and x and y represent the streamwise and spanwise directions, respectively. The subscript i represents the turbine number. Next, the influence of the wake effect of the upstream turbine on the downstream turbine is checked using Eq. (12.4).

$$\Delta Y_i \geq \frac{\sigma_{y_i} + d}{2}, \quad \text{no wake effect} \tag{12.4}$$

For a turbine which is not under wake effect, the velocity is taken as free stream velocity, and for a turbine under wake effect of its upstream turbine, velocity is calculated using Eq. (12.5).

$$\frac{\Delta u_i}{u_\infty} = \left[1 - \sqrt{1 - \frac{C_T \cos(\gamma_{ij})}{8\left(\frac{\sigma_{y_i}\sigma_{z_i}}{d^2}\right)}}\right] e^{-0.5\left(\frac{Y_i - \delta_i}{\sigma_{y_i}}\right)^2} \tag{12.5}$$

where σ_y and σ_z are the wake diameters corresponding to the lateral and vertical directions, respectively. The yaw angle of the turbine is represented by γ. C_T is the thrust coefficient of the turbine whose rotor diameter is d and transformed coordinate Y. δ is the wake deflection. u_∞ is the free stream velocity. The wake diameters are calculated using the formulas shown in Eqs. (12.6) and (12.7), respectively.

$$\frac{\sigma_{y_i}}{d} = k_y \frac{X_i - x_0}{d} + \frac{\cos(\gamma_{ij})}{\sqrt{8}} \tag{12.6}$$

$$\frac{\sigma_{z_i}}{d} = k_z \frac{X_i - x_0}{d} + \frac{1}{\sqrt{8}} \tag{12.7}$$

where k_y and k_z are the spanwise and vertical wake growth rates, respectively. x_0 is the length of the potential core which is calculated using Eq. (12.8).

$$\frac{x_0}{d} = \frac{\cos(\gamma_{ij})(1 + \sqrt{1 - C_T})}{\sqrt{2}\beta^*(1 - \sqrt{1 - C_T})} \tag{12.8}$$

The wake deflection coefficient is calculated using Eq. (12.9)

$$\frac{\delta_i}{d} = \begin{cases} \frac{x_i\theta_{co_i}}{d}, & \text{for } x \le x_0 \\ \frac{x_i\theta_{co_i}}{d} + \frac{a_{ij}\theta_{co_i}}{14.7}\sqrt{\frac{\cos(\gamma_{ij})}{k_y k_z C_T}}\left(2.9 - C_T + 1.3\sqrt{1 - C_T}\right), & \text{for } x > x_0 \end{cases} \tag{12.9}$$

with $a_{ij} = \ln\left(\frac{(1.6+\sqrt{C_T})\left(1.6\sqrt{\frac{8\sigma_{y_i}\sigma_{z_i}}{\cos(\gamma_{ij})d^2}} - \sqrt{C_T}\right)}{(1.6-\sqrt{C_T})\left(1.6\sqrt{\frac{8\sigma_{y_i}\sigma_{z_i}}{\cos(\gamma_{ij})d^2}} + \sqrt{C_T}\right)}\right)$. The deflection angle in the potential core, θ_{co_i} is calculated using Eq. (12.10).

$$\theta_{co_i} = u_{ij}\left[1 - \frac{\Delta u_i}{u_\infty}\right] \tag{12.10}$$

For downstream turbine under the influence of the wake effect of more than one upstream turbine, the velocity is calculated using Eq. (12.11).

$$u_{i+1} = u_i \left[1 - \sqrt{\sum_{r=1}^{n} \left(\frac{\Delta u_r}{u_{r\infty}} \right)^2} \right] \tag{12.11}$$

Here, n is the number of upstream turbines whose wake effects are having an impact on the downstream turbine. Based on the calculated velocity of the wind, power generation from ith turbine at the jth time instance can be calculated using Eq. (12.12).

$$P_{ij} = \frac{8}{27} \rho A u_{ij}^3 \cos^2(\gamma_{ij}) \tag{12.12}$$

with ρ being the density of air, A being the rotor area, and u_i being the velocity deficit.

12.2.2 Model Predictive Control

In MPC, the goal is to optimize an objective function subject to certain constraints on the manipulated and controlled variable (Inapakurthi et al. 2020a). Usually, the yaw angles of wind turbines are changed from one value to another value, and this change is also restricted between limits to restrict drastic changes within the yaw angle of the turbine (Inapakurthi et al. 2020b). The objective function considered is the maximization of the power production from the wind farm layout as shown in Eq. (12.13). The considered wind farm layouts are best-case and worst-case scenarios of DRIFT and GAN (for more details on these, interested readers are referred to Chaps. 9 and 10). The change in yaw angle is restricted between limits as shown in Eq. (12.14). Practical constraints on the maximum yaw angle are also applicable as indicated in Eq. (12.15).

$$\max_{\Delta \overline{\gamma}} P_j = \sum_{i=1}^{N} P_{ij} \tag{12.13}$$

subject to,

$$\Delta \gamma_i^{\min} \leq \Delta \gamma_i^k \leq \Delta \gamma_i^{\max} \tag{12.14}$$

$$\gamma_i^{\min} \leq \gamma_i^l \leq \gamma_i^{\max} \tag{12.15}$$

$$\forall i = 1, \ldots N;\ j = 1, \ldots, M;\ k = 1, \ldots L;\ l = 1, \ldots Q \tag{12.16}$$

Here, N represents the number of turbines in the considered wind farm layout, M is the number of time instances for which wind speed and direction data are available for control of yaw angle. L and Q are the control horizon and prediction horizon, respectively. $\Delta\overline{\gamma}$ in the argument of the objective function represents the vector of change in yaw angle values for all the turbines. $\Delta\gamma$ represents the change in yaw angle between successive time instances. The superscript min and max indicate the lower and upper bounds on the variables, respectively.

For each time instance j, the optimizer provides a feasible set of manipulated variables ($\Delta\overline{\gamma}$) for all the instances k to $k+L$, and Bastankhah wake model is evaluated. After the satisfaction of the termination or convergence criteria, the optimizer would return optimal value of the power for the time instance j. This would correspond to optimal yaw angles for the selected wind farm layout such that the power production for all the time instances till $k+L$ are maximum. The optimal yaw angle obtained from the optimizer at time instance $k+L$ is fed to the wake model for the next $L+Q$ time instances to check for the constraint satisfaction. If the constraints are satisfied, the change in yaw angle at time instance k is implemented in the wake model. This helps in identifying constraint violations at the early stage and taking necessary action. After this, the time instance is updated as $k=k+1$, and the entire process is repeated for all the M time instances.

12.3 Results and Discussions

12.3.1 Time-Series Least Square Support Vector Regression Results

For the control of wind turbines installed in a wind farm, at every time instance, the wind speed and direction used in this chapter are taken as the same as those obtained in Chap. 8. The least square support vector regression model predictions obtained in Chap. 8 are shown in Fig. 12.1. The wind characteristic trend is noted to be highly fluctuating and covers a wide spectrum of values since the prediction duration is roughly 296 days.

12.3.2 Best-Case Scenario for GAN and DRIFT

The uncertain variables considered for the control of wind turbine yaw angle are described in Chaps. 9 and 10. For the best-case scenario, GAN and DRIFT simulations are run to obtain the layout of the wind farm. With the obtained layout and the simulated wind characteristics data, control of the yaw angle is performed. The simulation settings for the control of wind farm yaw angle are presented in Table 12.1. The values of parameters used in Bastankhah wake effect model is presented in Table 12.2. In Fig. 12.2, the yaw angles of GAN for some representative turbines

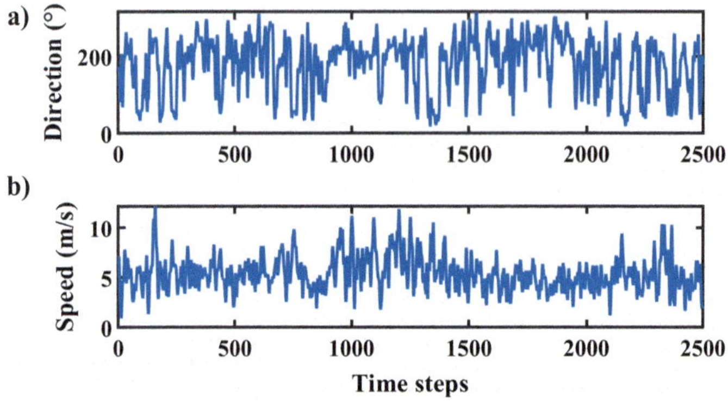

Fig. 12.1 Model predictions obtained for **a** wind speed and **b** wind direction from Chap. 8. Plots are shown for validation and forecasting data

Table 12.1 Comparison of AEP values for considered WFMs

Parameter	Value used in simulation
Lower limit for change in yaw angle	− 5
Upper limit for change in yaw angle	+ 5
Max. generations for optimization	100
Population in each generation	100
Minimum acceptable yaw angle	− 20
Maximum acceptable yaw angle	+ 20
Number of control horizons	1
Number of prediction horizons	5

(2nd, 6th, 10th, and 14th turbine) are shown while the corresponding power generated from these turbines are depicted in Fig. 12.3. For all the turbines and for all the time instances, the set constraints were satisfied, and the maximum power was obtained using GA. Similar plots for DRIFT are shown in Figs. 12.4 and 12.5. The total power produced for the GAN and DRIFT at each time instance is pictorially depicted in Fig. 12.6. The total power from the GAN case is 5.93 GW while that for the DRIFT case is 7.08 GW.

12.3.3 Worst-Case Scenario for GAN and DRIFT

For the worst-case scenario, GAN and DRIFT simulations are run to obtain the layout of the wind farm. In Fig. 12.7, the yaw angles of GAN for some representative turbines (2nd, 6th, 10th, and 14th turbine) are shown while the corresponding power generated from these turbines are depicted in Fig. 12.8. As with the previous case, constraint satisfaction is achieved, and the maximum power was obtained using GA. Similar

Table 12.2 Values of constants used in Bastankhah model

Parameter	Units	Value used in simulation
C_T		0.76
k_y		0.022
k_z		0.022
β^*		0.154
d	m	82
ρ	kg/m^3	1.23
$u_{\text{cut-in}}$	m/s	3.5
u_{rated}	m/s	15
$u_{\text{cut-out}}$	m/s	22
P_{rated}	kW	2000
z_h	m	82
Surface roughness		0.3

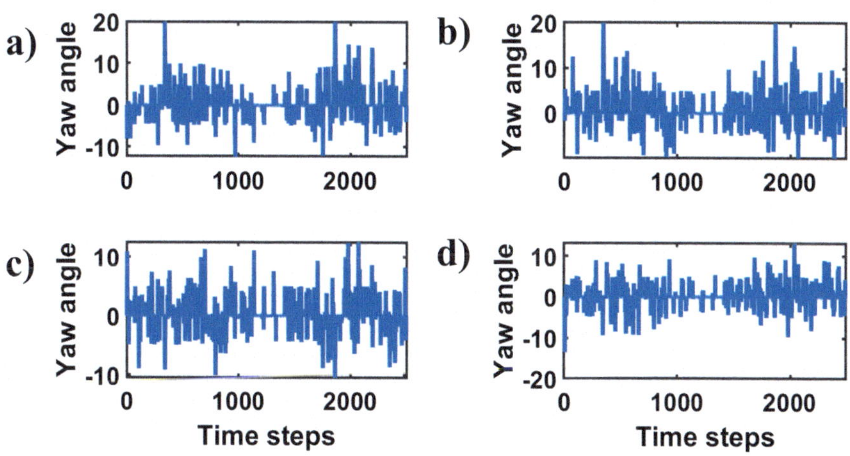

Fig. 12.2 Yaw angles obtained for the best-case scenario of GAN for **a** 2nd, **b** 6th, **c** 10th, and **d** 14th turbine

plots for DRIFT are shown in Figs. 12.9 and 12.10. The total power produced for the GAN and DRIFT at each time instance is pictorially depicted in Fig. 12.11. For the worst-case, the total power production is slightly higher for GAN which managed to generate 5.78 GW while simulations with DRIFT indicate a total power production of 5.64 GW.

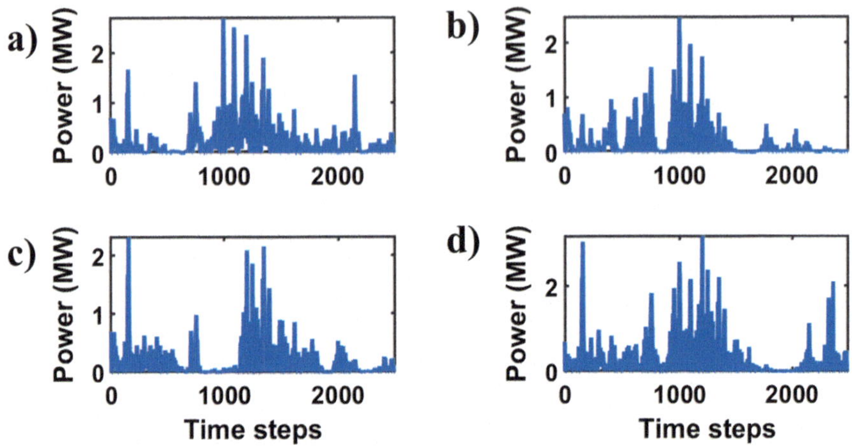

Fig. 12.3 Power generated by individual turbines for the best-case scenario of GAN for **a** 2nd, **b** 6th, **c** 10th, and **d** 14th turbine

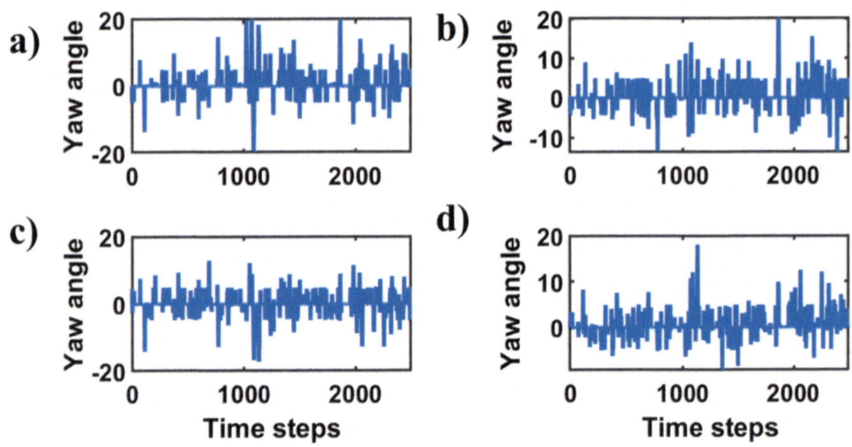

Fig. 12.4 Yaw angles obtained for the best-case scenario of DRIFT for **a** 2nd, **b** 6th, **c** 10th, and **d** 14th turbine

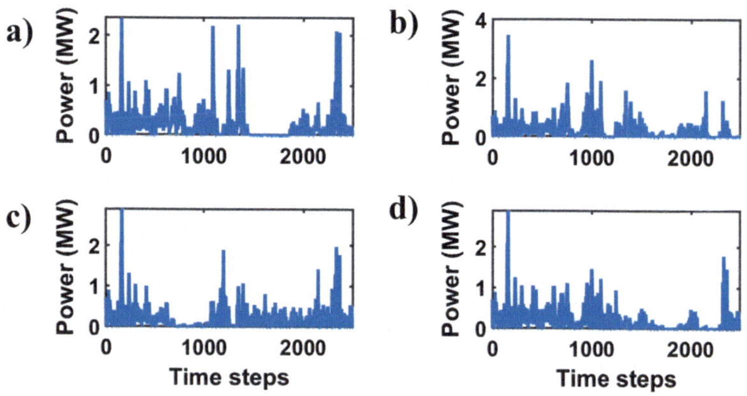

Fig. 12.5 Power generated by individual turbines for the best-case scenario of DRIFT for **a** 2nd, **b** 6th, **c** 10th, and **d** 14th turbine

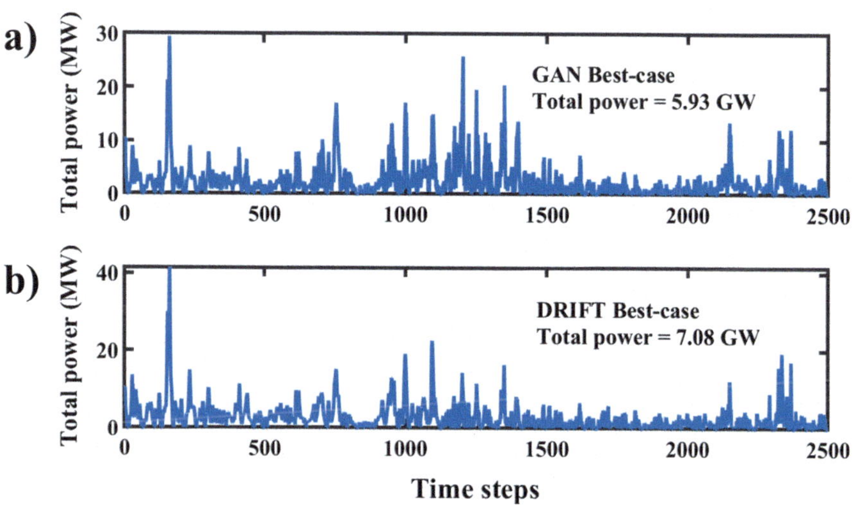

Fig. 12.6 Total power production for best-case scenario for **a** GAN and **b** DRIFT

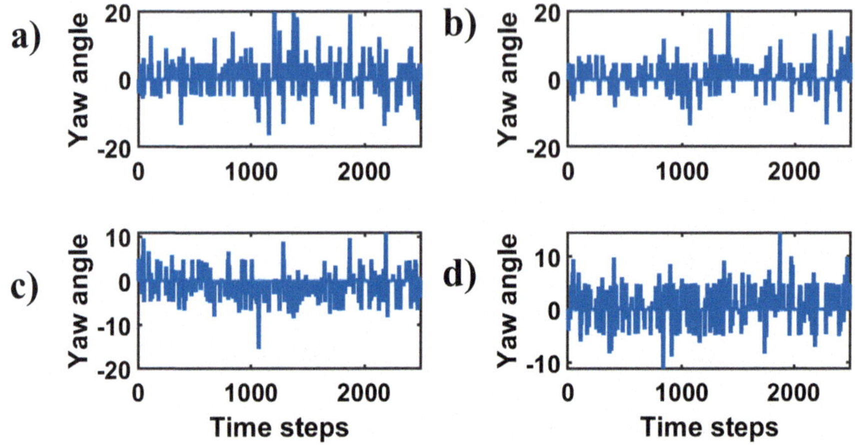

Fig. 12.7 Yaw angles obtained for the worst-case scenario of GAN for **a** 2nd, **b** 6th, **c** 10th, and **d** 14th turbine

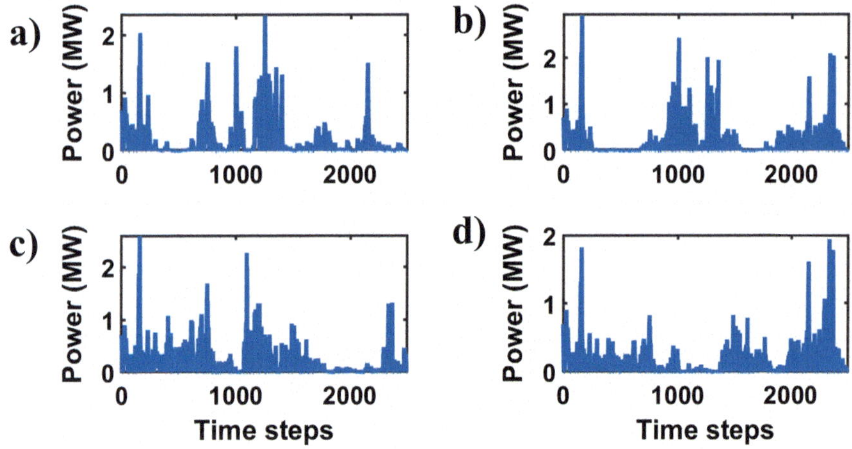

Fig. 12.8 Power generated by individual turbines for the worst-case scenario of GAN for **a** 2nd, **b** 6th, **c** 10th, and **d** 14th turbine

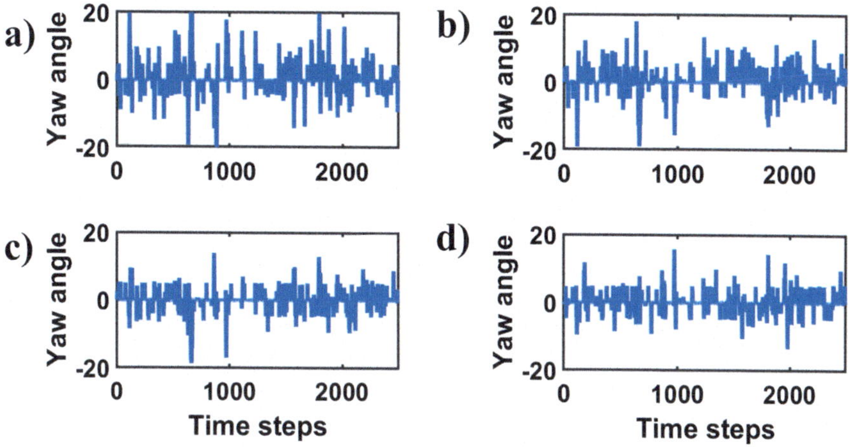

Fig. 12.9 Yaw angles obtained for the worst-case scenario of DRIFT for **a** 2nd, **b** 6th, **c** 10th, and **d** 14th turbine

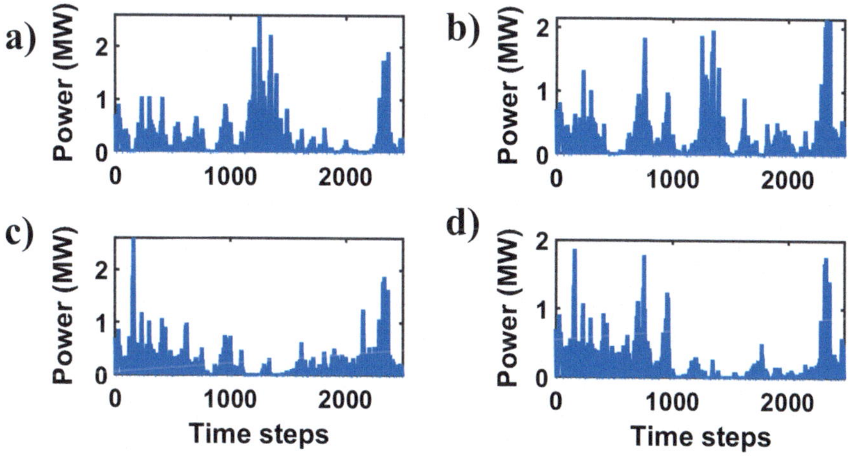

Fig. 12.10 Power generated by individual turbines for the worst-case scenario of DRIFT for **a** 2nd, **b** 6th, **c** 10th, and **d** 14th turbine

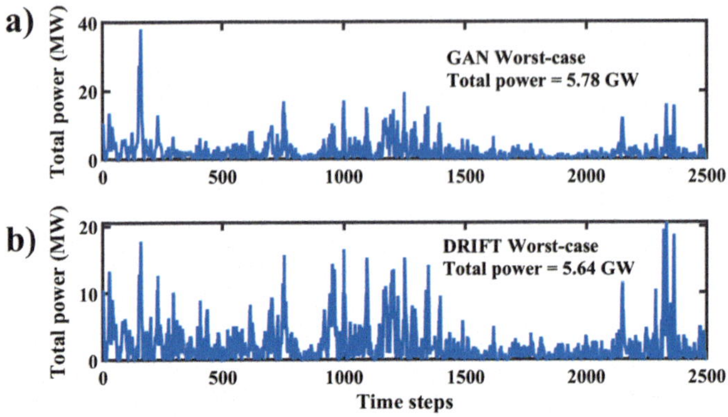

Fig. 12.11 Total power production for worst-case scenario for **a** GAN and **b** DRIFT

12.4 Conclusion

As the global average temperature continues to rise, it is certain that the number and severity of extreme weather-related events would only go upwards. This places the onus on scientists, engineers, and governments to find, design, and implement ways to reduce carbon emissions, so that global warming is contained. While many ways exist, in this study, the control of yaw angles of turbines installed in a wind farm is explored. The tuning of the turbine yaw angle to maximize power production from a given wind farm layout implies less energy generation from fossil fuels. This has further implications on the national import bill while ensuring energy security. In this study, control of turbines in a wind farm for a selected layout is considered. Considering both the best-case and worst-case, MPC is used for the maximization of power generation. This has resulted in a generation of 5.93 GW with GAN in the best-case scenario while the use of DRIFT implies a generation of 7.08 GW. On the other hand, use of GAN indicates a generation of 5.78 GW power for the worst-case scenario while the use of DRIFT implies a power production of 5.64 GW.

Acknowledgements We would like to acknowledge the support of SPARC project (SPARC/2018-2019/P1084/SL) sponsored by the Ministry of Education (MoE), Government of India and National Supercomputing Mission (NSM) for the project (DST/NSM/R&D_HPC_Applications/2021/23) sponsored by the Department of Science and Technology (DST), Government of India.

References

Barto A, Richard SS (2018) Reinforcement learning: an introduction, 2nd edn. The MIT Press, Cambridge, MA

Bastankhah M, Porté-Agel F (2016) Experimental and theoretical study of wind turbine wakes in yawed conditions. J Fluid Mech 806:506–541. https://doi.org/10.1017/JFM.2016.595

Cavanaugh KC, Kellner JR, Forde AJ et al (2014) Poleward expansion of mangroves is a threshold response to decreased frequency of extreme cold events. Proc Natl Acad Sci U S A 111:723–727. https://doi.org/10.1073/PNAS.1315800111/SUPPL_FILE/PNAS.201315800SI.PDF

Dittmer A, Sharan B, Werner H (2023) Koopman model predictive control for wind farm yield optimization with combined thrust and yaw control. IFAC-PapersOnLine 56:8420–8425. https://doi.org/10.1016/J.IFACOL.2023.10.1037

Freeman BG, Class Freeman AM (2014) Rapid upslope shifts in New Guinean birds illustrate strong distributional responses of tropical montane species to global warming. Proc Natl Acad Sci U S A 111:4490–4494. https://doi.org/10.1073/PNAS.1318190111/SUPPL_FILE/PNAS.201318190SI.PDF

Gebraad PMO, Teeuwisse FW, Van Wingerden JW et al (2016) Wind plant power optimization through yaw control using a parametric model for wake effects—a CFD simulation study. Wind Energy 19:95–114. https://doi.org/10.1002/WE.1822

Inapakurthi RK, Miriyala SS, Kolluri S, Mitra K (2020a) Nonlinear model predictive control of industrial grinding circuits using machine learning. In: 2020 IEEE symposium series on computational intelligence, SSCI 2020. Institute of Electrical and Electronics Engineers Inc., pp 1921–1927

Inapakurthi RK, Miriyala SS, Mitra K (2020b) Recurrent neural networks based modelling of industrial grinding operation. Chem Eng Sci 219:115585. https://doi.org/10.1016/j.ces.2020.115585

Li C, Liu L, Lu X (2024) A grouping strategy for reinforcement learning-based collective yaw control of wind farms. Theor Appl Mech Lett 14:100491. https://doi.org/10.1016/J.TAML.2024.100491

Miriyala SS, Pujari KNS, Naik S, Mitra K (2022) Evolutionary neural architecture search for surrogate models to enable optimization of industrial continuous crystallization process. Powder Technol 405:117527. https://doi.org/10.1016/J.POWTEC.2022.117527

Munters W, Meyers J (2018) Optimal dynamic induction and yaw control of wind farms: effects of turbine spacing and layout. J Phys Conf Ser 1037:032015. https://doi.org/10.1088/1742-6596/1037/3/032015

Pecl GT, Araújo MB, Bell JD et al (2017) Biodiversity redistribution under climate change: impacts on ecosystems and human well-being. Science 355. https://doi.org/10.1126/SCIENCE.AAI9214/SUPPL_FILE/PECL.SM.PDF

Pujari KN, Srivastava V, Miriyala SS, Mitra K (2021) Comparison of deep reinforcement learning techniques with gradient based approach in cooperative control of wind farm. In: 2021 7th Indian control conference, ICC 2021—proceedings, pp 400–405. https://doi.org/10.1109/ICC54714.2021.9703186

Pujari KN, Miriyala SS, Mittal P, Mitra K (2023) Better wind forecasting using evolutionary neural architecture search driven green deep learning. Expert Syst Appl 214:119063. https://doi.org/10.1016/J.ESWA.2022.119063

Wang Q, Xu T, von Terzi D et al (2024) Synchronized optimization of wind farm start-stop and yaw control based on 3D wake model. Renew Energy 223:120044. https://doi.org/10.1016/J.RENENE.2024.120044

Zhang Y, Chen X, Gong S, Chen J (2023) Collective large-scale wind farm multivariate power output control based on hierarchical communication multi-agent proximal policy optimization. Renew Energy 219:119479. https://doi.org/10.1016/J.RENENE.2023.119479

Zong H, Porté-Agel F (2021) Experimental investigation and analytical modelling of active yaw control for wind farm power optimization. Renew Energy 170:1228–1244. https://doi.org/10.1016/J.RENENE.2021.02.059

Chapter 13
Data Analysis in Wind Power Prediction: An Essential Step Before Data-Based Modeling

Aswitha Tadepalli, NagaSree Keerthi Pujari, and Kishalay Mitra

Abstract Wind energy, being clean and sustainable, has a significant potential to generate large-scale renewable energy. However, the intermittency and randomness of wind cause a huge barrier to the safety and control of the energy generation grid. Thus, constant forecasting of the wind data is crucial for efficient wind power generation. Lately, the support of machine learning (ML) models has been exploited in wind forecasting because of their quick and reliable predictions. However, reliable data is a primary requirement in the training of ML models. Since raw wind data is obtained from various sensors, the elimination of abnormal data and in-depth data analysis is essential to improve the quality of the data, thereby improving the model's performance. This study, therefore, focuses on performing exploratory data analysis (EDA) on the ENGIE wind data to get deeper insights from the data and eliminate any futile information. Further, this study aims to identify the significant parameters in wind power prediction. From the results, it can be concluded that the performance of random forest regressor (RFR) and extra trees regressor (ETR) in the wind power prediction is excellent and that the wind speed is the only considerable parameter in the wind power prediction.

A. Tadepalli · N. K. Pujari · K. Mitra (✉)
Department of Chemical Engineering, Indian Institute of Technology Hyderabad, Hyderabad, Telangana 502284, India
e-mail: kishalay@che.iith.ac.in

A. Tadepalli
e-mail: ch23resch11002@iith.ac.in

N. K. Pujari
e-mail: ch20resch11006@iith.ac.in

K. Mitra
Department of Climate Change, Indian Institute of Technology Hyderabad, Hyderabad, Telangana 502284, India

Department of Artificial Intelligence, Indian Institute of Technology Hyderabad, Hyderabad, Telangana 502284, India

K. Mitra et al. (eds.), *Optimization, Uncertainty and Machine Learning in Wind Energy Conversion Systems*, Engineering Optimization: Methods and Applications,
https://doi.org/10.1007/978-981-97-7909-3_13

Keywords Exploratory data analysis · Feature selection · Machine learning · Wind power prediction

13.1 Introduction

Energy plays a vital role in the rapid development of the economy as well as industry. Though fossil fuels occupy a major portion of the global energy resources, the usage of fossil fuels as a primary resource of energy contributes significantly to the global carbon dioxide (CO_2) emissions, thereby attributing to the threat of climate change and global warming (Ogbutor and Capstone 2020). Also, the tremendous utilization of fossil fuels over the last several decades led to a vast depletion of the fuels, thereby immensely increasing the energy costs. Therefore, owing to their various advantages such as being environmentally clean and freely available in nature, renewable energy sources are currently being exploited globally on a very large scale (Sachin Kumar et al. 2022). Among the various renewable sources, wind energy has gained increased attention due to its easier availability, cleaner production, higher energy density, and larger scope for large-scale generation aspects. However, the intermittent and unpredictable nature of wind poses major challenges to the safety and stability of the energy production grid when integrated at large scales. Thus, precise forecasting of the wind data can mitigate the effect of wind volatility and randomness on the stability of the grid (Feng et al. 2017).

The most widely used models for wind forecasting are physical models, statistical models, and machine learning (ML) models. The physical models develop a mathematical formulation using huge historical datasets to forecast wind power. They require large amounts of data and involve extensive time-expensive calculations making them unfit for short-term forecasting. Statistical models such as autoregressive exogenous input (ARX), autoregressive moving average (ARMA), and autoregressive integrated moving average (ARIMA) identify a linear relationship between the wind data and wind power to forecast the future power. However, they do not capture any nonlinear behavior present in the wind data and hence cannot be used for long-term forecasting. ML models, on the other hand, can effectively handle nonlinear relationships and do not involve any time-consuming mathematical calculations. Thus, ML models such as artificial neural networks (ANN), support vector machines (SVM), recurrent neural networks (RNN), long-short term memory networks (LSTM), convolutional neural networks (CNN), etc., are gaining tremendous momentum in wind forecasting. As the wind data is time-relevant, the traditional ANN models failed to capture the long-term dependencies present in the data and various deep learning methods such as RNN, LSTM, CNN, and transfer learning are currently being exploited.

Although the ML models are widely applicable because of their ability to predict quickly in both short-term and long-term forecasting, they highly rely on wind data (Xiao et al., n.d.; Singh and Rizwan 2022). The raw wind data consists of tremendous information such as wind speed, direction, temperature, pressure, various operating parameters of the turbine and generator, geographical information, etc., collected from various sensors. Since the data is obtained through sensors, there are high chances of the presence of abnormalities and irregularities in the data. Hence, many data preprocessing techniques, mainly different wavelet decomposition techniques, are being used to make the data reliable and useful (Delgado and Fahim 2021). Further, as the data contains various operational and geographical parameters, feature selection, and extraction techniques form a vital role before training the ML model, not only to reduce the computational expense of the model but also to remove the futile and redundant parameters present in the data, thereby improving the model's accuracy.

Though there are many studies reported in literature that performed feature selection and extraction techniques on wind data (Carta et al. 2015; Feng et al. 2017; Jiang and Huang 2017; Liu et al. 2020; Huang et al. 2021), there are very few studies that performed an exploratory data analysis (EDA) (Delgado and Fahim 2021; Kavun and Zamula 2023; Ponkumar et al. 2023), a technique used to get a detailed insight of the data and identify the patterns among various input parameters and the outputs (Morgenthaler, n.d.). The main work of this chapter is to perform EDA on ENGIE wind data, by considering various parameters such as wind speed, direction, pitch angle, various temperatures, convertor torque, speeds of the generator, rotor, nacelle angle, and active power. Further, after processing the data and removing any redundant and irrelevant parameters through EDA, the target is to identify the most contributing parameters to the wind power prediction using random forest regressor (RFR), extra trees regressor (ETR), and permutation feature importance (PFI) methods.

The rest of the chapter is organized as follows. In Sect. 13.2, a description of the dataset, a brief introduction to the EDA, and feature selection methods namely RFR, ETR, and PFI methods are presented. In Sect. 13.3, the results of EDA and feature selection techniques are discussed and finally, in Sect. 13.4, the conclusions of the work are collated.

13.2 Methodology

In this section, firstly the detailed description of the ENGIE dataset is presented. Then, the importance of EDA followed by the various techniques included in EDA is briefly introduced. Further, to identify the significant parameters in the prediction of wind power, the methodology of three feature selection techniques, viz. RFR, ETR, and PFI methods, is explained.

13.2.1 ENGIE Dataset Description

The raw data from ENGIE is of 10 min resolution and contains various operating conditions of the wind, wind turbine, generator, and converter along with the date-time stamp and is collected from four turbines over four years (i.e., 2013–2016). To avoid the large number of parameters, the average data of all four turbines is considered. These parameters of the average data include a total of 31 operating conditions, and Table 13.1 presents the parameter name along with its description.

The 10 min. data was converted to a 6 hr. resolution by averaging every 36 points. The total number of data points after this conversion was 5835 in number. Since after the 3000th data point, there were many missing values, only the data up to 3000 values is considered in this study. Any abnormal data were replaced with the average of the previous and the following time step values. Since the power generated by the turbine cannot have a negative value, any negative values under the wind power can be confidently claimed to be erroneous data, and hence, these negative values were replaced with zero. Further, all the missing values in the data were filled using the cubic spline interpolation technique. Since the parameters, grid frequency, and grid voltage are not of significant interest, they are eliminated. Also, only the absolute wind direction corrected, and nacelle angle corrected are included in the study. Furthermore, only the active power is considered, and the reactive, apparent power, and cos phi average are not considered as the outputs. Finally, this data, after these preprocessing techniques with 23 inputs and one output (active power), is used for further study, and the time-dependencies in this data are not considered in the study of this chapter.

13.2.2 Exploratory Data Analysis

To appropriately assess the predictions made by the ML model and make any further conclusions, a clear-cut understanding of the various information available in the dataset stands in the first place (Morgenthaler, n.d.). Real-time datasets contain huge information and observations, and hence, it becomes practically impossible to understand the insights provided by the data just by merely going through it. Also, developing an ML model with raw data, i.e., without processing and cleaning the data, could lead to large anomalies in the predictions. Thus, data mining plays a crucial role before the actual development of the ML model. EDA is a key solution and is usually the primary technique in data mining. The main aim of EDA is to reveal the underlying patterns and relationships among the various parameters in the data using various visualization techniques and correlation analysis, summarize the data, and spot outliers, and irregularities present in the data (Milo and Somech 2020; Patel et al. 2022). To outline, the primary target of EDA is to understand what the data can reveal before proceeding with the actual modeling. The widely used visualization techniques include box plots, bar plots, histograms, residual plots, mean plots,

Table 13.1 Description of all the 31 parameters of the raw ENGIE dataset

Parameter name	Description
Ba_avg	Pitch angle
Rt_avg	Hub temperature
DCs_avg	Generator converter speed
Cm_avg	Convertor torque
P_avg	Active power
Q_avg	Reactive power
S_avg	Apparent power
Cosphi_avg	Ratio of P_avg and S_avg
Ds_avg	Generator speed
Db1t_avg	Generator bearing 1 temperature
Db2t_avg	Generator bearing 2 temperature
Dst_avg	Generator stator temperature
Gb1t_avg	Gearbox bearing 1 temperature
Gb2t_avg	Gearbox bearing 2 temperature
Git_avg	Gearbox inlet temperature
Gost_avg	Gearbox oil sump temperature
Ya_avg	Nacelle angle
Yt_avg	Nacelle temperature
Ws1_avg	Wind speed 1
Ws2_avg	Wind speed 2
Ws_avg	Wind speed average
Wa_avg	Absolute wind direction
Va_avg	Vane position average
Ot_avg	Outdoor temperature
Nf_avg	Grid frequency
Nu_avg	Grid voltage
Rs_avg	Rotor speed
Rbt_avg	Rotor bearing temperature
Rm_avg	Torque
Wa_c_avg	Absolute wind direction corrected
Na_c_avg	Nacelle angle corrected

etc., for the univariate analysis and scatter plots, heatmaps, polar plots, etc., for the bivariate analysis. Some of the data summarization techniques include descriptive statistics, correlation statistics, summary tables, etc. (Morgenthaler, n.d.).

13.2.3 Feature Selection Methods

13.2.3.1 Random Forest Regressor

Decision tree (DT) built using the classification and regression tree (CART) method is a widely used ML model owing to its simple architecture and ease of interpretability. However, the major drawbacks of the decision tree are its tendency to over-fit the data and lack of robustness. To overcome these drawbacks, an RFR which is an ensemble of a large number of regression trees was developed (Breiman 2001). To reduce the variance and thereby the over-fitting, two approaches are followed in all the regression tress of the RFR: (i) Using bootstrapped data (i.e., random subset of the dataset sampled with replacement) instead of the entire dataset for training a DT; (ii) Using random subset (of fixed number) of the input features for the node-splitting in each tree. Once the random subset of features is generated, the best feature for node-splitting and its corresponding value are selected through impurity value. In a regression tree, impurity refers to the variance, i.e., $\frac{\sum_{i=1}^{N}(y_i-\mu)^2}{N}$ where N represents the total number of data points and y_i represents the output of the ith data point and μ represents the mean of the outputs of all the data points. The second approach (approach (ii)) is followed to avoid the possibility of all the trees being trained with the more significant input features leading to higher correlation among the predictions of all the DT (Grömping 2009). Thus, this randomness creates huge differences among the individual DT's predictions, and the overall prediction of the RFR is, therefore, the average of predictions of all the trees DTs calculated using Eq. (13.1).

$$\widehat{y_{\mathrm{RFR}}}(X) = \frac{\sum_{i=1}^{M} \widehat{y_i}(X_i')}{M} \tag{13.1}$$

where $\widehat{y_{\mathrm{RFR}}}$ is the final prediction of the RFR, X represents the entire input dataset with N features, X_i' represents the bootstrapped dataset with $K \in N$ features used for training the ith DT, $\widehat{y_i}$ represents the prediction of ith tree, and M is the total number of DTs in the RFR (Ahmad et al. 2018).

A striking feature of RFR is its capability to assess the relative importance of the features. The importance score of a particular feature is evaluated by measuring the decrease in the impurity value after the node-split at all those nodes which were split using the corresponding feature. Let a node-split, split the dataset of D observations into two nodes with D_{left} observations in the left child node and D_{right} observations in the right child node, and then the importance score (i.e., the information gain (IG)) for j^{th} feature is calculated using Eq. (13.2). The final importance score of a feature in RFR is then the average of all the information gain values calculated at all the nodes having that feature over all the DTs of the RFR. A higher importance score implies a higher importance of that corresponding feature in the prediction (Huang et al. 2016).

$$\mathrm{IG}_f = \mathrm{Impurity}(D) - \frac{D_{\mathrm{left}}}{D} \times \mathrm{Impurity}(D_{\mathrm{left}}) - \frac{D_{\mathrm{right}}}{D} \times \mathrm{Impurity}\left(D_{\mathrm{right}}\right) \quad (13.2)$$

13.2.3.2 Extra Trees Regressor

Like RFR, an extremely randomized trees (or extra trees) regressor is an ensemble of many DTs. It was developed as an extension to RFR by making it extremely randomized to further reduce the variance (Geurts et al. 2006). ETR is different from RFR in two main aspects. The first aspect is, that unlike RFR, the ETR does not generate bootstrapped data but uses the entire dataset for building the DTs. The second one is that in ETR, although a random subset of the features is chosen for building the DTs as in RFR, the node-splitting in each DT is done in a completely random manner, i.e., both the feature for node-split and its corresponding value are chosen randomly unlike in RFR where the feature and its value are chosen based on the impurity. These modifications in ETR were proven to make it less susceptible to over-fitting and improve its performance (Geurts et al. 2006; Ahmad et al. 2018). The feature importance assessment is performed in the same way as that of the RFR.

13.2.3.3 Permutation Feature Importance Technique

In the case of over-fitting, the feature ranking technique in tree-based ensemble methods is generally biased and favors numerical features over categorical or discrete ones. Also, since the feature scoring is done during the training process, features that might not be important on an unseen dataset could be given a higher score value. To avoid such erroneous conclusions, the PFI technique was developed which evaluates the importance scores based on unseen data. The PFI technique is model-independent, i.e., it can be applied to any trained ML model. The PFI technique evaluates the feature importance of a feature by measuring the increase in the model's prediction error after randomly permuting its values in the unseen data. The permutation of the feature's values disrupts the relationship between itself and the original output values. Then, it can be concluded that this permutation results in a decrease in the model's performance (i.e., an increase in the prediction error) and then that corresponding feature can be considered as important as it is affecting the model's prediction. On the contrary, if the permutation does not result in a big change in the model's performance, then that feature can be conveniently ignored as the model's performance is not affected by that feature (Otchere et al., n.d.; Huang et al. 2016). The algorithm for the calculation of the feature importance score is as follows.

Let M denote a trained model, N denote the total number of features, and MSE denote the MSE calculated using Eq. (13.3) on unseen dataset evaluated through the predictions of M then,

$$\mathrm{MSE} = \frac{\sum_{i=1}^{n} (y_i - \hat{y}_i)^2}{n} \tag{13.3}$$

where n is the total number of unseen samples, $y_{i,t}$, $\hat{y}_{i,t}$ represent the true value and the predicted value of the ith sample.

(a) For feature j:

(i) For n repetitions, the values of the feature j in the unseen dataset are randomly shuffled, while the other features are left unaltered.
(ii) The new outputs are predicted using the above n shuffled datasets and, in each case, MSE is evaluated. Let the average of these n MSEs be denoted as $(\text{MSE-shuffled})_j$.
(iii) The feature importance score is then calculated using Eq. (13.4)

$$\text{Score} = \text{MSE} - (\text{MSE-shuffled})_j \tag{13.4}$$

(b) The above step is repeated for all the N features.
(c) The higher the feature importance score implies more importance of that feature in the prediction.

Thus, the advantage of this technique is that it does not require the retraining of the model to find the best features. However, this technique does not talk about the intrinsic predictive importance of a feature but the importance of that feature in the prediction process (Otchere et al., n.d.).

13.2.4 *Complete Methodology*

The methodology followed to identify the most contributing factors in the wind power prediction using the ENGIE dataset is as follows, and Fig. 13.1 shows the corresponding flowsheet. After the required preprocessing techniques were performed on the ENGIE data, EDA was executed. The first step in EDA is to detect any outliers and eliminate them. After this step, to eliminate any redundant and irrelevant features, correlation analysis is performed. Further, after EDA, feature selection methods, viz. RFR, ETR, and PFI technique, using the trained RFR and ETR models are performed to assess the contribution of each parameter in the prediction of wind power.

13.3 Results and Discussions

All the techniques in EDA and feature selection methods were performed in PYTHON language. To generate the results, various packages, viz. *numpy*, *pandas*, and *sklearn*, were utilized, and for the visualization techniques, *matplotlib* and *seaborn* packages were used.

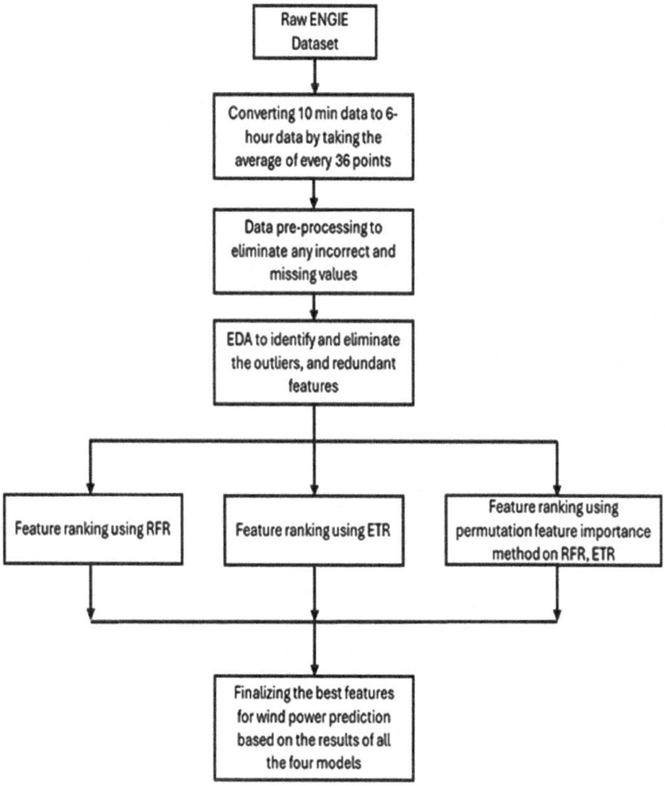

Fig. 13.1 Flowsheet of the methodology followed to identify the best features for wind power prediction

13.3.1 EDA

13.3.1.1 Outlier Detection

The primary step in EDA is the detection of outliers and their elimination, and this is done through the generation of box plots. Thus, the box plots for each feature were plotted using the *boxplot* command of *seaborn* package, and Fig. 13.2 shows these box plots. From the plots, it can be seen that there are outliers in the following parameters: Ba_avg, Cm_avg, Db1t_avg, Db2t_avg, Dst_avg, Gb1t_avg, Gb2t_avg, Yt_avg, Git_avg, Gost_avg, Ws1_avg, Ws2_avg, Ws_avg, Rm_avg, Rbt_avg, Va_avg, and P_avg. However, eliminating the outliers purely based on the data analysis might lead to the loss of important data in the domain aspect. Therefore, before eliminating the outliers, analysis of the points based on domain knowledge is considered as a good practice. Thus, analyzing the box plots based on the knowledge of wind power generation led to the following inferences:

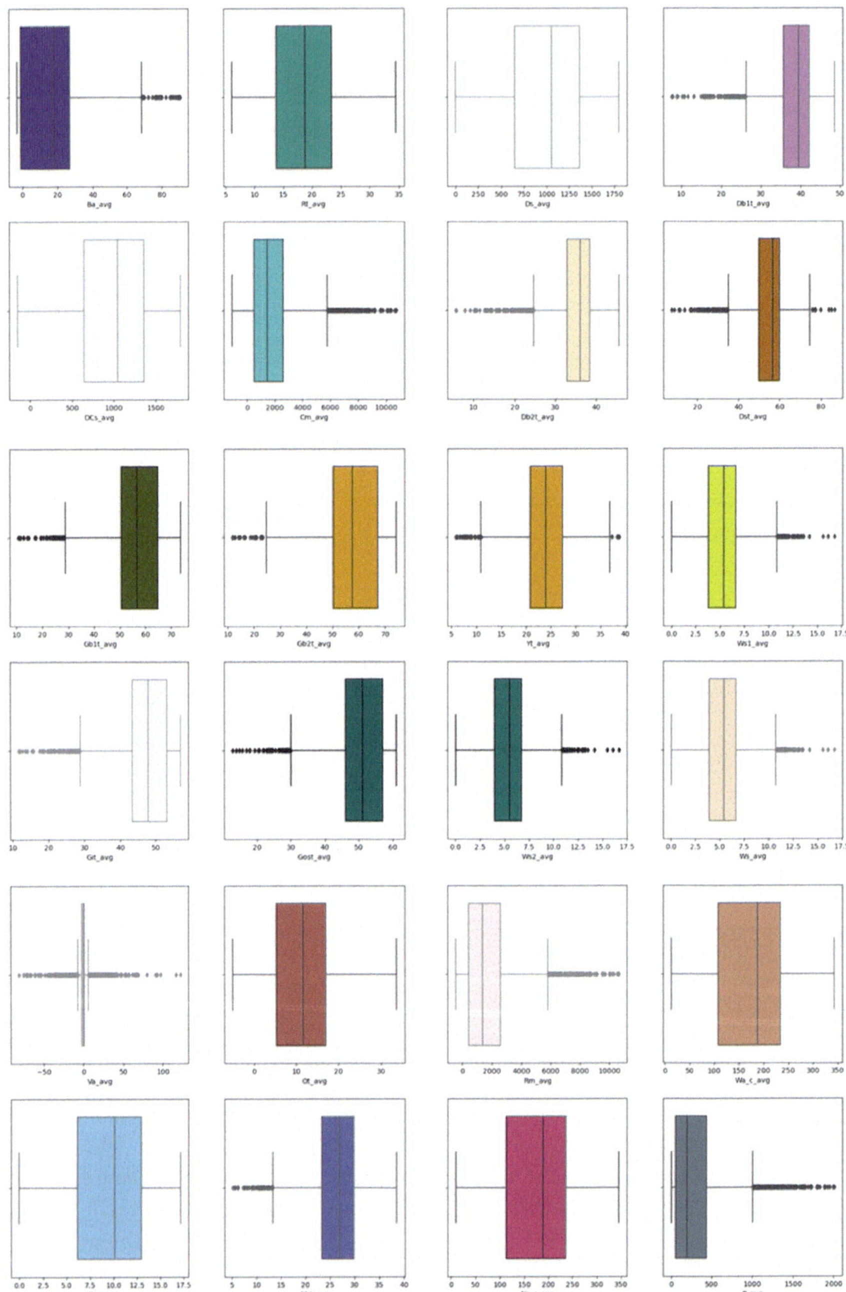

Fig. 13.2 Boxplots of 24 parameters of the ENGIE dataset

- Although the pitch angle ranges between 5 and 20° for maximum power production, in cases of low power generation, the pitch angle can be as high as 80° (Sudhamshu et al. 2016). Eliminating this information would lead to the model's incapability to capture the parameters in the cases of low power generation.
- For the various temperatures, viz. Db1t_avg, Db2t_avg, Dst_avg, Gb1t_avg, Gb2t_avg, Yt_avg, Git_avg, Gost_avg, and Rbt_avg, most of the data have temperatures ranging from 30 to 60 °C. However, it can be seen that the detected outlier points lie within 5–20 °C. For temperature, this difference can be conveniently ignored as the temperature might vary at random. Thus, this outlier elimination can also be avoided to enhance the model's capability to capture the randomness in the temperature.
- For the wind speed parameters, i.e., Ws1_avg, Ws2_avg, and Ws_avg, there might be rare cases of very high winds, and this increases the wind speed significantly higher than the normal range. To capture such rare cases, information of such high speeds is retained. Further, the torque (Rm_avg) and convertor torque (Cm_avg) are dependent on wind speed and direction, and hence, the extreme cases of torque are not ignored. Furthermore, although the normal power generated goes up to 500 kJ, in extreme cases of high winds, the power generated would also be sufficiently high. Hence, the data points with P_avg ranging from 1000 to 2000 are not eliminated.
- Finally, since the vane position of the turbine can be either negative or positive depending on the relative orientation with the wind speed, no points are eliminated.

Thus, based on the above inferences, it can be concluded that there are no notable outliers, and the data can be conveniently used for further analysis.

13.3.1.2 Correlation Analysis

To identify any insignificant and repetitive parameters, the correlation matrix was generated using the *corr*() function of the pandas package, and the corresponding heatmap was generated using the *seaborn* package, and Fig. 13.3 shows the heatmap of the correlation matrix. This function evaluates the Pearson correlation coefficient (R) value for all the possible pairs of features. A threshold value of 0.85 was fixed, i.e., for all the pair of features having a correlation coefficient value greater than 0.85 or less than − 0.85, either of two features is selected. Table 13.2 presents such pairs along with their threshold values.

Based on these results, after the elimination of linearly correlated features, eight features were finalized as inputs and active power (P_avg) as output. These are Ba_avg, Rt_avg, Db2t_avg, Git_avg, Ws_avg, Va_avg, Rbt_avg, and Wa_c_avg. To get brief information about the relationship between each of these inputs with the power, the pair plots are plotted, and Fig. 13.4 shows these pair plots.

From Fig. 13.4, it can be seen that there is high collinearity between wind speed (Ws_avg) and active power (P_avg), and this can also be confirmed by the correlation coefficient value of 0.903. However, the rest of all inputs are highly nonlinear in

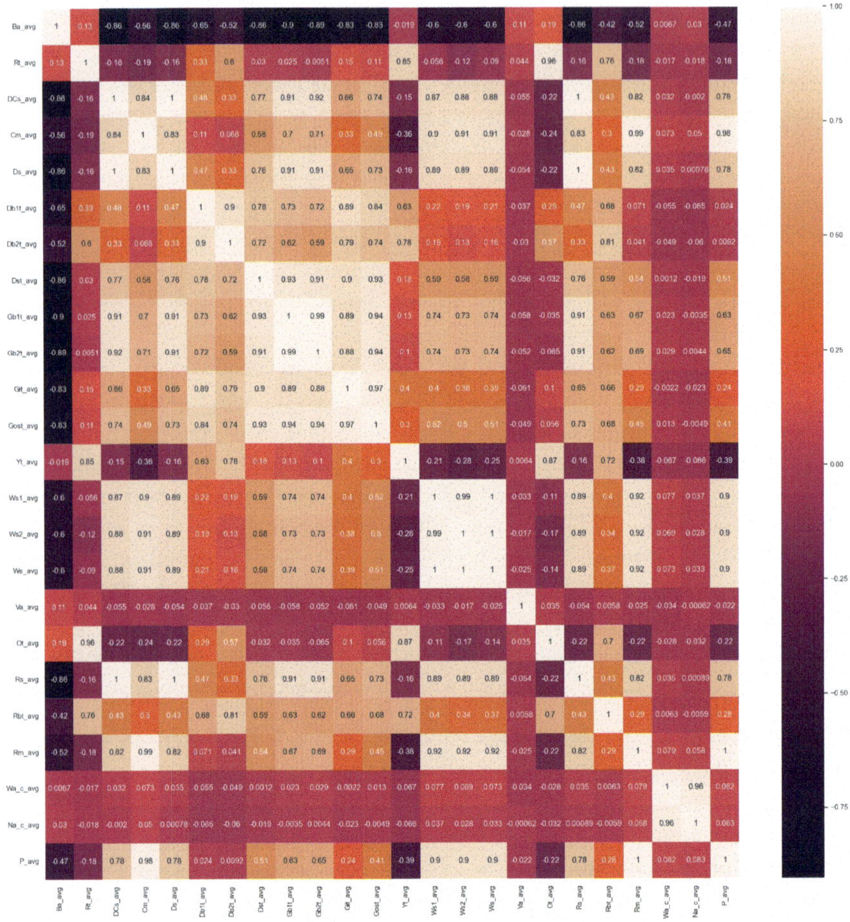

Fig. 13.3 Heatmap of the correlation matrix for the 24 parameters

nature, and hence, nonlinear modeling is necessary for the prediction of wind power using these features. Finally, to find the most significant features in the prediction of wind power, feature selection techniques were performed.

13.3.2 Feature Selection Methods

Before proceeding with the feature selection techniques (RFR, ETR, and PFI techniques), as a standard practice, the preprocessed data was normalized. Then, 80% of the normalized data was used for training (i.e., 2400 data points) and 20% for

Table 13.2 All the pairs of parameters whose correlation coefficient value is less than – 0.85 and greater than 0.85

Pair	*R*-value	Pair	*R*-value
Ba_avg and DCs_avg	– 0.862	Dst_avg and Gb1t_avg	0.928
Ba_avg and Ds_avg	– 0.856	Dst_avg and Gb2t_avg	0.914
Ba_avg and Dst_avg	– 0.860	Dst_avg and Git_avg	0.895
Ba_avg and Gb1t_avg	– 0.904	Dst_avg and Gost_avg	0.929
Ba_avg and Gb2t_avg	– 0.886	Gb1t_avg and Gb2t_avg	0.994
Ba_avg and Rs_avg	– 0.856	Gb1t_avg and Git_avg	0.888
Rt_avg and Yt_avg	0.852	Gb1t_avg and Gost_avg	0.937
Rt_avg and Ot_avg	0.962	Gb1t_avg and Rs_avg	0.910
DCs_avg and Ds_avg	0.997	Gb2t_avg and Git_avg	0.877
DCs_avg and Gb1t_avg	0.913	Gb2t_avg and Gost_avg	0.942
DCs_avg and Gb2t_avg	0.917	Gb2t_avg and Rs_avg	0.913
DCs_avg and Ws1_avg	0.874	Git_avg and Gost_avg	0.971
DCs_avg and Ws2_avg	0.875	Yt_avg and Ot_avg	0.866
DCs_avg and Ws_avg	0.877	Ws1_avg and Ws2_avg	0.990
DCs_avg and Rs_avg	0.997	Ws1_avg and Ws_avg	0.998
Cm_avg and Ws1_avg	0.901	Ws1_avg and Rs_avg	0.885
Cm_avg and Ws2_avg	0.906	Ws1_avg and Rm_avg	0.920
Cm_avg and Ws_avg	0.906	Ws1_avg and P_avg	0.900
Cm_avg and Rm_avg	0.986	Ws2_avg and Ws_avg	0.998
Cm_avg and P_avg	0.981	Ws2_avg and Rs_avg	0.886
Ds_avg and Gb1t_avg	0.910	Ws2_avg and Rm_avg	0.924
Ds_avg and Gb2t_avg	0.914	Ws2_avg and P_avg	0.903
Ds_avg and Ws1_avg	0.885	Ws_avg and Rs_avg	0.888
Ds_avg and Ws2_avg	0.886	Ws_avg and Rm_avg	0.924
Ds_avg and Ws_avg	0.887	Ws_avg and P_avg	0.904
Ds_avg and Rs_avg	1.000	Rm_avg and P_avg	0.997
Db1t_avg and Db2t_avg	0.898	Wa_c_avg and Na_c_avg	0.965
Db1t_avg and Git_avg	0.892		

testing (i.e., 600 data points). This data splitting was performed using the *traintest-split* function of the *sklearn* package. The *traintestsplit* function randomly splits the entire dataset into train data and test data for a specified test split percentage. Further, to ensure that the same train and test data are generated every time and thereby the reproducibility of the results, the random state of the *traintestsplit* function was set to 0.

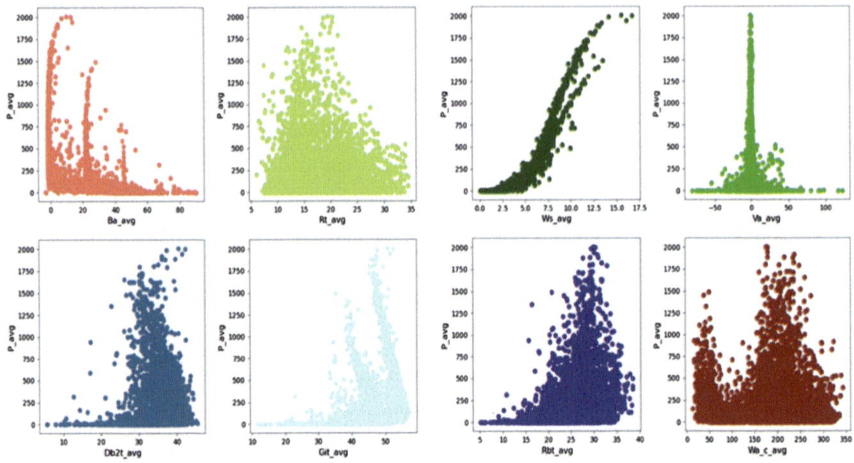

Fig. 13.4 Pair plots of pitch angle, rotor bearing temperature, wind speed, vane position, generator bearing temperature, gearbox inlet temperature, hub temperature, and the active power

13.3.2.1 RFR and ETR

The RFR and ETR models were built using the *sklearn.ensemble* package. The number of trees in both the models was fixed to 50, and to avoid very deep trees the depth of each tree was set as 5. Since, the bootstrapping and selection of a subset of features is performed randomly, to ensure reproducibility of the results, the random state was set to 0. The built models were trained using the training data and finally tested using the test dataset. Table 13.3 shows the $R2$ score for the training and testing of both models. From the table, since the $R2$ scores for both training and testing data are above 0.95, it can be concluded both the models' performance is substantial. Finally, using the *feature_importances* function, the feature importance scores were evaluated in both the cases, and Fig. 13.5 shows the feature ranking through RFR and ETR.

From the results, it is noticeable that wind speed has almost 90% of the contribution in the wind power prediction followed by pitch angle and gearbox inlet temperature. The contribution of the rest parameters is almost close to 0%.

Table 13.3 $R2$ scores of the RFR and ETR models for the training and testing datasets

$R2$ score	Training	Testing
RFR	0.991	0.986
ETR	0.972	0.966

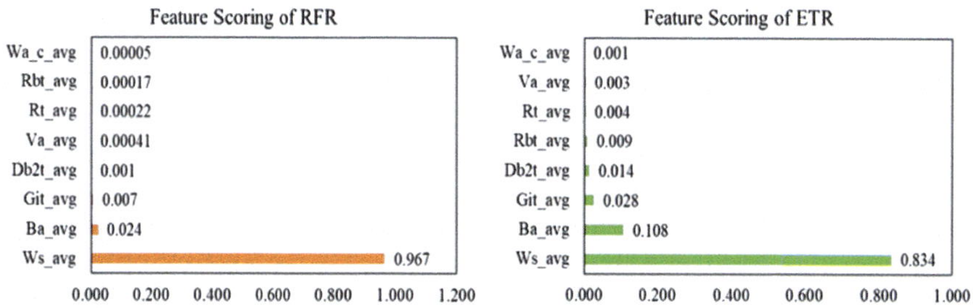

Fig. 13.5 Feature ranking of the eight selected inputs for the prediction of wind power using RFR and ETR models

13.3.2.2 Permutation Feature Selection

The PFI method was performed on the trained RFR and ETR models described in Sect. 13.3.2.1 using the *sklearn.inspection* package. For each of these models, the test data was randomly shuffled ten times, and the feature importance score was evaluated as mentioned in Sect. 13.2.4. To generate the same shuffled datasets and thereby ensure the reproducibility of the results, the random state was set to 0. Finally, the mean score of all the ten runs was evaluated, and Fig. 13.6 shows these results.

From these results, it can be seen that similar to the results of RFR and ETR, the wind speed is the most contributing feature in the wind power prediction followed by pitch angle while the importance of other factors is almost zero. Analyzing all the aforementioned results, it can be inferred that the wind power can be accurately predicted using wind speed alone, and this can be attributed to the fact that the wind power generated is proportional to the cube of wind speed.

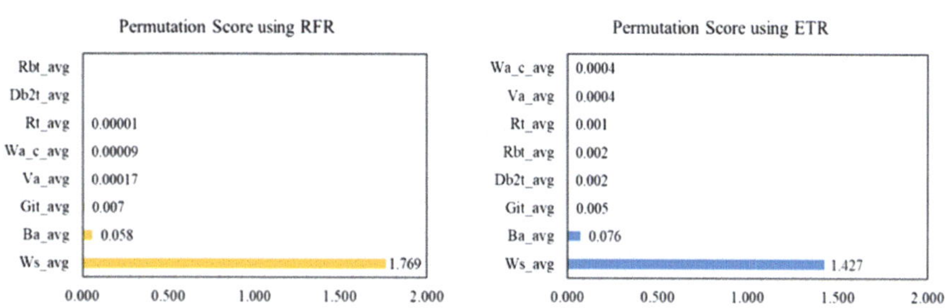

Fig. 13.6 Feature ranking of the eight selected inputs for the prediction of wind power using the PFI method for the RFR and ETR models

13.4 Conclusions

Wind energy is considered one of the primary renewable energy resources because of its sustainable and cleaner energy production and easier migration toward large-scale energy generation. However, the integration of the wind turbines with large-scale grids leads to the control and safety issues of the grids and thus requires accurate forecasting of the wind speed and power. Recently, ML models have been widely being used for this prediction because of their ability to easily predict in both short-term and long-term wind forecasting. However, the performance of these ML models greatly depends on the data used, and hence, data preprocessing and analysis of data is a crucial step in the training of ML models. In this study, EDA was performed on the ENGIE dataset to get clear insights from the wind data and eliminate the redundant parameters. The dataset initially consisted of 3000 data points with 23 parameters (wind speed, wind direction, torque, various operating temperatures, pitch angle, and nacelle angle) as inputs and the active power as the output. After the correlation analysis, 15 parameters were eliminated, and only the wind speed, pitch angle, gearbox inlet temperature, generator bearing temperature, hub temperature, rotor bearing temperature, vane position, and wind direction were retained. Further, different feature selection methods such as RFR, ETR, and PFI techniques were performed on this reduced dataset to identify the most significant features in the prediction of wind power. From the results, it can be concluded that wind speed is the major contributing parameter in the wind power prediction while the contributions of the other parameters are negligible. Although the ENGIE dataset was used in this study, this chapter highlights the need for data analysis in wind modeling and provides a basis for EDA and feature selection for wind data before proceeding to the actual wind forecasting techniques.

Acknowledgements We would like to acknowledge the support provided by the SPARC project (SPARC/2018-2019/P1084/SL) sponsored by the Ministry of Education (MoE), Government of India.

References

Ahmad MW, Reynolds J, Rezgui Y (2018) Predictive modelling for solar thermal energy systems: a comparison of support vector regression, random forest, extra trees, and regression trees. J Clean Prod 203:810–821. https://doi.org/10.1016/j.jclepro.2018.08.207

Breiman L (2001) Random forests

Carta JA, Cabrera P, Matías JM, Castellano F (2015) Comparison of feature selection methods using ANNs in MCP-wind speed methods. A case study. Appl Energy 158:490–507. https://doi.org/10.1016/j.apenergy.2015.08.102

Delgado I, Fahim M (2021) Wind turbine data analysis and LSTM-based prediction in SCADA system. Energies (Basel) 14. https://doi.org/10.3390/en14010125

Feng C, Cui M, Hodge BM, Zhang J (2017) A data-driven multi-model methodology with deep feature selection for short-term wind forecasting. Appl Energy 190:1245–1257. https://doi.org/10.1016/j.apenergy.2017.01.043

Geurts P, Ernst D, Wehenkel L (2006) Extremely randomized trees. Mach Learn 63:3–42. https://doi.org/10.1007/s10994-006-6226-1

Grömping U (2009) Variable importance assessment in regression: linear regression versus random forest. Am Stat 63:308–319. https://doi.org/10.1198/tast.2009.08199

Huang N, Lu G, Xu D (2016) A permutation importance-based feature selection method for short-term electricity load forecasting using random forest. Energies (Basel) 9. https://doi.org/10.3390/en9100767

Huang H, Jia R, Shi X et al (2021) Feature selection and hyper parameters optimization for short-term wind power forecast. Appl Intell 51:6752–6770. https://doi.org/10.1007/s10489-021-02191-y

Jiang Y, Huang G (2017) Short-term wind speed prediction: hybrid of ensemble empirical mode decomposition, feature selection and error correction. Energy Convers Manag 144:340–350. https://doi.org/10.1016/j.enconman.2017.04.064

Kavun S, Zamula A (2023) Exploratory data analysis in wind energy datasets. In: 2023 IEEE 4th KhPI week on advanced technology, KhPI week 2023—conference proceedings. Institute of Electrical and Electronics Engineers Inc.

Liu X, Zhang H, Kong X, Lee KY (2020) Wind speed forecasting using deep neural network with feature selection. Neurocomputing 397:393–403. https://doi.org/10.1016/j.neucom.2019.08.108

Milo T, Somech A (2020) Automating exploratory data analysis via machine learning: an overview. In: Proceedings of the ACM SIGMOD international conference on management of data. Association for Computing Machinery, pp 2617–2622

Morgenthaler S. Exploratory data analysis. https://doi.org/10.1002/wics.002

Ogbutor O, Capstone MA (2020) Multi-class root cause classification in wind turbine faults

Otchere DA, Aboagye M, Ayoub M et al. Enhancing drilling fluid lost-circulation prediction using model agnostic and supervised machine learning

Patel H, Guttula S, Mittal RS et al (2022) Advances in exploratory data analysis, visualisation and quality for data centric AI systems. In: Proceedings of the ACM SIGKDD international conference on knowledge discovery and data mining. Association for Computing Machinery, pp 4814–4815

Ponkumar G, Jayaprakash S, Kanagarathinam K (2023) Advanced machine learning techniques for accurate very-short-term wind power forecasting in wind energy systems using historical data analysis. Energies (Basel) 16. https://doi.org/10.3390/en16145459

Sachin Kumar R, Sakthiya Ram S, Arun Jayakar S, Senthil Kumar TK (2022) Failure prediction of turbines using machine learning algorithms. Mater Today Proc 66:1175–1182. https://doi.org/10.1016/j.matpr.2022.04.984

Singh U, Rizwan M (2022) SCADA system dataset exploration and machine learning based forecast for wind turbines. Results Eng 16. https://doi.org/10.1016/j.rineng.2022.100640

Sudhamshu AR, Pandey MC, Sunil N et al (2016) Numerical study of effect of pitch angle on performance characteristics of a HAWT. Eng Sci Technol Int J 19:632–641. https://doi.org/10.1016/j.jestch.2015.09.010

Xiao L, Meng A, Rong J et al. A dual-dimensionality reduction attention mechanism with fusion of high-dimensional features for wind power prediction

Chapter 14
Application of TimeSHAP, an eXplainable AI Tool, for Interpreting Time-Series Model of Wind Turbines

Vishnu Swaroopji Masampally, Rashi Verma, and Kishalay Mitra

Abstract The transition from traditional energy sources to renewable alternatives has become imperative in combating climate change. Wind power has emerged as a viable solution, driven by technological advancements and its role in addressing global energy needs. This study employs eXplainable AI (XAI), specifically TimeSHAP, to enhance the interpretability of a time-series model, providing insights into the contributions of each time-event and feature to the predictions. Utilizing a long short-term memory (LSTM) model for accurate wind power forecasting, the study leverages its ability to capture temporal dependencies in data. Through TimeSHAP's local explanations, the interpretability of the LSTM model in capturing nonlinear and dynamic wind energy patterns is showcased. This combination of LSTM model and TimeSHAP not only improves forecast accuracy but also enhances understanding of the factors influencing wind power predictions.

Keywords eXplainable AI · TimeSHAP · Long short-term memory · Wind power prediction

V. S. Masampally · R. Verma · K. Mitra (✉)
Department of Chemical Engineering, Indian Institute of Technology Hyderabad, Hyderabad, Telangana 502284, India
e-mail: kishalay@che.iith.ac.in

V. S. Masampally
e-mail: ch22resch14001@iith.ac.in

R. Verma
e-mail: ch22resch11004@iith.ac.in

V. S. Masampally
TCS Research, Tata Consultancy Services, Tata Research Development and Design Centre, Pune 411057, India

K. Mitra
Department of Climate Change, Indian Institute of Technology Hyderabad, Hyderabad, Telangana 502284, India

Department of Artificial Intelligence, Indian Institute of Technology Hyderabad, Hyderabad, Telangana 502284, India

K. Mitra et al. (eds.), *Optimization, Uncertainty and Machine Learning in Wind Energy Conversion Systems*, Engineering Optimization: Methods and Applications,
https://doi.org/10.1007/978-981-97-7909-3_14

14.1 Introduction

Using sustainable renewable energy sources is becoming vital in a world where fossil fuel supplies are running out and climate change related issues such as global warming have to be addressed. Reducing global warming to 1.5 °C needs the reduction of carbon dioxide emission by around 37 Gt. It will help in reaching net-zero emissions in the energy industry by 2050. According to IRENA's projections, the proportion of renewable energy in the world's energy mix would rise from 16% in 2020 to 77% by 2050 (IRENA 2023). Among the different renewable energy sources, wind power is one of them that stands out as a powerful force, providing a long-term and increasingly affordable alternative. The positive environmental impact of wind power is one of the main factors contributing to its increasing significance. Wind energy produces electricity without releasing any harmful pollutant or greenhouse gases, thus preventing climate change. However, to what extent is wind power impactful today?

The statistics are self-explanatory to show the impact of wind power. According to IRENA (2023) in the past few years between 2010 and 2022, the usage of wind power has increased by more than five times. According to IRENA (2023), global installed capacity for onshore wind power reached about 836 GW in 2022. This growth trajectory of wind power on a global scale reflects its importance in contemporary landscape. This has led to receiving a lot of attention these days and being encouraged by government, but there are major drawbacks when comes to its implementation. One of the main reasons is the nonlinear and unpredictable characteristics of wind speed, leading to inconsistency and uncertainty in energy output. We aim to design a model which will help us in predicting accurate wind power generation by incorporating the nonlinearity nature of wind speed such as instability and unpredictability. Wind power prediction holds immense importance due to multiple reasons impacting everything from grid stability to financial stability.

For example, accurate prediction of power helps in balancing the equilibrium between the production and demand of electricity by anticipating the variability of wind. Accurate prediction enables better planning and grid upgrades, resulting in minimizing wind curtailments and maximizing energy utilizations. Thus, precise forecasting is highly desirable for organizing dependable and efficient power system implementing and ensuring uninterrupted power supply. Also, power prediction helps in maximizing financial return by making informed investment decisions. By considering all these things, wind prediction acts as essential factor for ensuring grid stability, optimizing financial gain and reduce transmission and distribution losses and also advancing sustainability in the energy sector. In case of wind power, data is available in form of time series; there are two kind of prediction method: physics-based modeling and data-based modeling (Lange and Focken 2006).

There are several physics-based models that use weather data such as humidity, speed, and air pressure, for calculating thermodynamic equation to generate wind power (Foley et al. 2012). Numerical Weather Prediction (NWP) (Taylor and Buizza 2003) and Weather Researcher Forecasting (WRF) (Hu et al. 2020) are well-known physical models that use first principles and meteorological databases (Shahid et al. 2021), and hence, increases the complexity and computational time of modeling weather conditions to get accurate results. These models offer low accuracy in addition to its need for forecasted data to validate the models. To get around these challenges with physics-based models, the trend of prediction methods has shifted to data-based models.

Machine learning (ML) models, a type of data-based model, have become very popular due to its ability to extract useful information from time-series data. Time series, which are sequential data points gathered at regular intervals, are essential for comprehending and forecasting patterns, trends, and anomalies in a wide range of fields, including finance, healthcare, and climate research. When navigating through the complex web of temporal data, conventional method frequently proves inadequate due to long-term patterns, anomalies, and sudden changes of the data, whereas machine learning models have the good learning capability from data and perform forecasting based on those data. As with the advancement in area of data-based modeling, many deep learning-based models have been employed to forecasting wind speed/wind power such as gated recurrent units (GRU) (Niu et al. 2020), bidirectional GRU (BiGRU) (Neshat et al. 2021), convolution neural network (CNN) (Yu et al. 2020), long short-term memory (LSTM) (Liu et al. 2018), and deep belief network (DBN) (Wan et al. 2016). These deep learning-based models have gained more attention in wind power/wind speed forecasting because of their better ability to handle intricate nonlinear problems when compared to conventional statistical models. In this chapter, we are going to use long short-term memory (LSTM) model, a type of ML model that addresses the challenges presented by long-term temporal datasets. Using state-of-the-art techniques from statistical modeling, machine learning, and data processing, LSTM equips researcher with the knowledge and resources needed to fully understand the complexities of the time-series data. The objective of this chapter is to provide readers with a thorough understanding of long short-term trends and short-term irregularity detection by thoroughly examining various methodology. We also focused on eXplainable AI (XAI) that deals with "black box" problem of LSTM models. This "black box" nature makes the internal working of LSTM model opaque, hence making it difficult to understand how they arrive at their prediction causing lack of interpretability and potential biases in the model. Understanding the decision-making process of black-box models is crucial, especially in contexts involving safety-critical processes. Without explainability, deploying such models is often discouraged. In this scenario, an explainer should be model-agnostic and perturbation-based to adapt to various deployment scenarios, considering constraints such as access to the model solely through an inference API.

To solve this issue, we will employ the XAI technique to provide insight into these models and enhance their overall explainability by disclosing the process by which the models made their predictions. The contribution of each feature can also be explained via eXplainable AI, such as **SHAP** (**SH**apley **A**dditive ex**P**lanations) (Lundberg et al. 2020; Lundberg and Lee 2017), which aids in comprehending how various factors affect the model's output. XAI also helps in providing insights into the temporal relationship by determining which past time points the model focuses on most while making a prediction. We can take advantage of LSTMs' power for wind power prediction and obtain important insights into their decision-making processes by combining XAI. To ensure a more informed investment decision, an operator may utilize XAI to evaluate if the model fairly takes into account various geographical characteristics when estimating wind power output for a proposed wind farm. To sum up, XAI offers useful resources for realizing AI's full potential in wind power forecasting. It opens the door to a more efficient and sustainable energy future by dispelling misconceptions about the models and fostering confidence.

This chapter provides an exhaustive examination of data-based analysis for interpreting the wind power forecasts. The chapter is organized into four sections with Sect. 14.2 offering comprehensive insight into the used dataset and the different analyses conducted on the dataset. The objective is to identify important variables essential for constructing a time-series-based data model for predicting wind power. Section 14.2.1 delves into the correlation analysis, including Pearson coefficient to scrutinize the impact of changes in one variable on another, thereby isolating highly correlated variables. Section 14.2.2 expounds on LSTM and Sect. 14.2.3 on TimeSHAP. Moving to Sect. 14.3, an overview is presented on the outcomes derived from LSTM models and detailed examination of TimeSHAP results. This section encapsulates the results obtained from the implemented model, shedding light on their performance and the insight gained through the TimeSHAP analysis.

14.2 Methodology

14.2.1 Data Description and Analysis

A typical wind turbine consists of several key components such as tower, blades, rotor hub, generator, yaw system, nacelle, gearbox, and controller and monitoring system. The natural wind flows over the blades of the turbine, exerting force and making them rotate. The rotating blades turns the shaft that connects the turbine with the rotor hub. The torque at the main shaft is transferred to generator via gearbox that optimizes the speed of the generator. The generator converts the rotational energy to electrical energy. The nacelle is a large housing located at the top of the wind turbine tower consisting of key components such as the gearbox, generator, controller, and other electrical and mechanical systems. The yaw system orients the blades and the nacelle of the wind turbine with changing wind direction to capture maximum amount of wind energy.

Data from various sensors in and around the wind turbines are collected from four wind turbines installed in La Haute Borne, France by ENGIE, a French electricity utility company. These four wind turbines are placed in the corners of a rectangular farm of 9 km^2. The wind turbines are assumed to be far enough from each other for the wake effects to affect the power generation. The raw data is available at frequency of one observation per 10 min for over four years for different process and climate variables such as wind speed, wind direction, nacelle angle, hub temperature, etc.

The raw data of four wind turbines is averaged to obtain the average climate and wind turbine characteristics of the given wind farm. The data is further processed, as explained in Sect. 13.2.1, to remove noise and outliers. The data is averaged to change the frequency from one observation per 10 min to one observation per hour. This helps in reducing the dimensionality of the data. Although data is available for over four years, only 270 days of data is used for this analysis.

A total of 31 variables are present in the data. The output variables in the data are active power, reactive power, apparent power, and cosine of phase angle (also known as power factor, the ratio of active power to apparent power). Among these output variables, the amount of active power generated by the wind turbine is considered as a key performance indicator (KPI) for this study. Among variables related to the process and climate conditions, variables such as grid frequency, grid voltage, absolute wind direction, and nacelle angle are not considered for the study. Absolute wind direction and nacelle angle are highly correlated with corrected absolute wind direction and corrected nacelle angle, respectively, and therefore excluded from this study.

The variables considered in this study are summarized in Table 14.1 using statistical parameters such as mean, standard deviation (SD), 25 percentile, 50 percentile, 75 percentile, minimum, and maximum values. A further analysis is performed to determine the important variables that are considered to build models that predict active power generated by the wind turbine. Correlation analysis is a statistical technique that examines the relationship between two or more variables to understand the effect of changes in one variable on another variable. Some of the common correlation methods are Pearson correlation, Spearman's rank correlation, Kendall's tau correlation, point-biserial correlation, etc. (Akoğlu 2018). Among these, Pearson correlation is the simplest and commonly used method that measures the linear relationship between two continuous variables. Hence, Pearson correlation is used to filter out the variables that are highly correlated by retaining only one from each closely correlated group of variables. This approach aids in addressing multicollinearity concerns and reducing the dimensionality of the data. The details of correlation analysis and feature selection are discussed in Sect. 14.3.

14.2.2 Model Training with Long Short-Term Memory

LSTM is a type of recurrent neural system (Van Houdt et al. 2020) which is very popular model when comes to dealing with temporal data. Hochreiter and Schmidhuber (1997) proposed LSTM, especially designed to overcome the exploding vanishing gradient issues that frequently arise while learning long-term dependencies.

Table 14.1 Statistical summary of the variables in data

Variable	Mean	SD	Min.	25%	50%	75%	Max.
Pitch angle	18.85	19.45	− 1.01	− 0.94	22.25	22.98	90.78
Hub temperature	18.44	5.35	9.43	13.83	17.28	22.24	34.73
Generator converter speed	998.06	521.5	− 142.07	695.77	1051.02	1359.9	1799.99
Convertor torque	1915.4	2124.83	− 1090.1	350.46	1397.63	2918.93	10,801.5
Active power	348.99	395.85	0	47.43	208.13	518.6	2026.39
Generator speed	1030.17	510.31	0	751.95	1084.6	1361.56	1801.6
Generator bearing 1 temperature	38.05	4.88	15.75	35.48	38.63	41.72	50.93
Generator bearing 2 temperature	34.77	4.25	13.86	32.52	35.52	37.53	45.77
Generator stator temperature	53.77	8.27	15.64	49.53	52.76	61.03	87.56
Gearbox bearing 1 temperature	55.87	9.84	17.92	51.12	55.66	64.92	101.44
Gearbox bearing 2 temperature	56.79	10.85	18.4	50.42	56.5	67.24	74.69
Gearbox inlet temperature	46.62	6.64	18.24	43.01	46.69	52.14	57.92
Gearbox oil sump temperature	49.64	6.9	19.6	46.03	48.95	56.48	61.73
Nacelle temperature	23.41	4.48	10.14	20.57	23.34	26.15	38.48
Wind speed 1	5.59	2.47	0	4.06	5.58	6.97	17.7
Wind speed 2	5.7	2.45	0	4.19	5.69	7.07	17.58
Wind speed avg.	5.64	2.46	0	4.13	5.63	7.02	17.61
Vane position avg.	− 0.4	14.06	− 149.14	− 2.95	− 0.39	2.35	148.42
Outdoor temperature	10.96	6.64	− 0.77	5.38	9.71	15.46	33.21
Rotor speed	9.82	4.87	0	7.15	10.33	12.97	17.2
Rotor bearing temperature	25.74	4.46	10.71	23.44	26.49	28.57	38.36
Rotor torque	2036.9	2079.72	− 269.43	425.5	1488.14	2978.21	10,740.3
Absolute wind direction corrected	178.26	79.1	5.52	126.92	184.31	234.91	352.83
Nacelle angle corrected	180.71	80.24	3.62	131.86	187.54	238.53	357.13

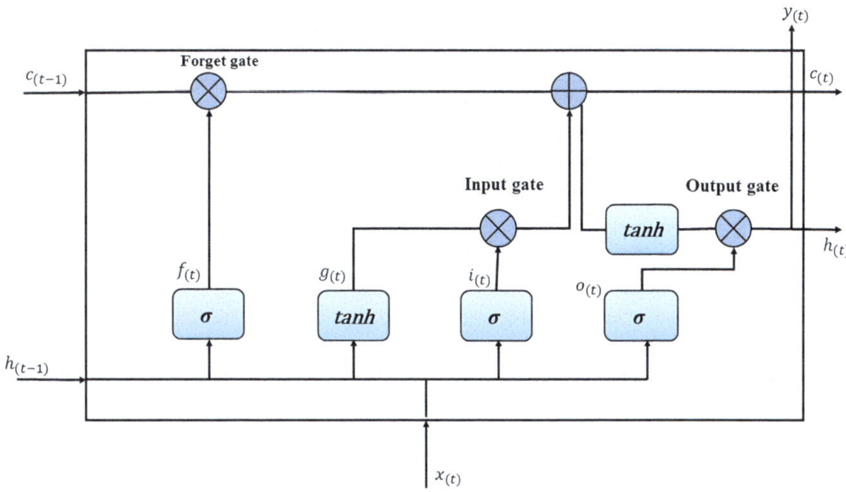

Fig. 14.1 Schematic diagram of LSTM unit

The fundamental assumption governing the LSTM (Yu et al. 2019) principle is that LSTM cells have the potential to notice significant inputs and retain them for as long as they are required. This explains why these cells have demonstrated remarkable success in identifying long-term patterns in audio, lengthy texts, time series, and other media. As shown in Fig. 14.1, LSTM unit architecture consists of three gates: output gate, an input gate, and a forget gate to track the sequence information.

Each gate performs a particular set of tasks as described below:

- *Forget Gate*: Controls which segment of information should be eliminated.
- *Input Gate*: Determines which segment of long-term state information should be included.
- *Output Gate*: Decides which long-term state information should be read and gives outputs at a given time step.

The computation of the cell's long-term state, short-term state, and outputs at each time step for a single instance is represented in Eq. (14.1), where $x_{(t)}$ is current input and $h_{(t-1)}$ is the previous short-term state and $c_{(t-1)}$ is the long-term state from traverse across the network. In Eq. (14.1), the weight matrices for four-layers connected to input and short term state are denoted as (W_{xi}, W_{xf}, W_{xo}, W_{xg}) and (W_{hi}, W_{hf}, W_{ho}, W_{hg}), respectively, whereas (b_i, b_f, b_o) are bias terms for the four connected layers.

$$
\begin{aligned}
i_{(t)} &= \sigma\left(W_{x\mathrm{i}} x_{(t)} + W_{h\mathrm{i}} h_{(t-1)} + b_{\mathrm{i}}\right) \\
f_{(t)} &= \sigma\left(W_{x\mathrm{f}} x_{(t)} + W_{h\mathrm{f}} h_{(t-1)} + b_{\mathrm{f}}\right) \\
o_{(t)} &= \sigma\left(W_{x\mathrm{o}} x_{(t)} + W_{h\mathrm{o}} h_{(t-1)} + b_{\mathrm{o}}\right) \\
g_{(t)} &= \tanh\left(W_{x\mathrm{g}} x_{(t)} + W_{h\mathrm{g}} h_{(t-1)} + b_{\mathrm{g}}\right) \\
c_{(t)} &= f_{(t)} \otimes C_{(t-1)} + i_{(t)} \otimes g_{(t)} \\
y_{(t)} &= h_{(t)} = o_{(t)} \otimes \tanh\left(c_{(t)}\right)
\end{aligned}
\tag{14.1}
$$

Given the sequential nature of the data employed in wind power forecasting, the LSTM model emerges as a suitable model, owing to its adeptness at capturing and retaining information over longer temporal intervals. In preparation for the application of the LSTM model, meticulous data preprocessing is imperative. Before feeding the raw data to a LSTM model, it is crucial to do thorough data preprocessing that involves handling missing data, outliers making it in a suitable format for input into the LSTM network. The significance of this preprocessing step lies in ensuring that the LSTM model can effectively learn and generalize pattern from the input data, thereby enhancing its predictive accuracy in the context of wind power forecasting. The LSTM model is then trained using 70% of the data, which is fed into the network and adjusting the model's parameters through back propagation and optimizing the model to minimize the prediction error. Parameters such as learning rate, number of layers, and cell units impact the model's ability to generalize and adapt to different pattern within the data. Once trained, the LSTM model needs to be rigorously evaluated to assess its performance using evaluation metrics such as mean absolute error (MAE), root mean square error (RMSE), and correlation coefficient.

14.2.3 TimeSHAP

TimeSHAP is an explainer that's model-agnostic and post-hoc, specifically tailored for sequential or time-series models (Bento et al. 2021). TimeSHAP leverages KernelSHAP that consolidates various explanation methods to a family of "additive feature attribution methods" (Lundberg and Lee 2017). It builds upon the SHAP (SHapley Additive exPlanations) method, which is a feature attribution method approximating the local behavior of a model with a linear model using a subset of features (Lundberg and Lee 2017). To grasp the concept of TimeSHAP effectively, it is essential to introduce and explain several topics, including model-specific explainers, model-agnostic explainers, and KernelSHAP. In this section, a concise overview of these topics is presented before delving into a comprehensive explanation of TimeSHAP.

14.2.3.1 Model-Specific Explainers

Model-specific explainers are used in machine learning to interpret the predictions of a specific type or class of models. These are tailored to the particular characteristics and behaviors of a specific model architecture leveraging the internal structure, parameters, and assumptions of the target model to provide insights into model predictions (Ribeiro et al. 2016). Model-specific explainers include techniques such as gradient-based methods, decision tree interpreters, rule-based methods, saliency maps, and attention mechanisms (Ismail et al. 2019; Sha and Wang 2017; Zhang et al. 2018; Choi et al. 2016). The major shortcomings of model specific explainers are that they are not generalized and they often rely on assumptions regarding the model's architecture. Therefore, when assumptions are violated or inaccurate, the explanations provided by the explainer may be inaccurate.

14.2.3.2 Model-Agnostic Explainers

Model-agnostic explainers are used in machine learning to interpret the predictions of predictive models of any model architecture. Model-agnostic explainers rely on their ability to perturb the input data and observe the changes in the model's predictions (Guidotti et al. 2018). Perturbation, h_x, of an input vector $x \in X^m$ refers to the process of intentionally altering or modifying the values of features in the input sample as explained by Eq. (14.2).

$$h_x(z) = x \odot z + b \odot (1 - z) \tag{14.2}$$

Here, $\odot$ denotes the component-wise product, and the vector $\mathbf{b} \in X^m$ embodies an uninformative input sample or background vector. This vector is typically chosen as the zero vector, $\mathbf{b} = \mathbf{0}$, or it can be constructed from the average feature values in the input dataset, such that $b_i = x_i$. The **z** vector, $z \in \{0, 1\}^m$, enables the coalition of features such that $z_i = 1$ signifies that feature i retains its original value x_i and $z_i = 0$ indicates that feature i adopts an uninformative background value b_i, signifying its exclusion. Therefore, perturbation function $h_{\mathbf{x}}$: $\{0, 1\}^m \rightarrow R_{m\times 1}$ maps a coalition vector $z \in \{0, 1\}^m$ to the original input space $R_{m\times 1}$.

Lundberg and Lee (2017) consolidated various explainers, encompassing both model-agnostic and model-specific techniques, into a unified family known as "additive feature attribution methods." They further proved that there exists a unique solution within this class of methods that satisfies local accuracy, ensuring that the explanation model closely mirrors the complex model at a local level; missingness, ensuring that the absence of features has no influence on predictions; and consistency, meaning that if a feature's contribution increases, its attributed importance remains unchanged. Furthermore, Lundberg and Lee (2017) proposed KernelSHAP, a model-agnostic explainer that satisfies properties such as local accuracy, handling missing data, and ensuring consistency.

14.2.3.3 KernelSHAP

KernelSHAP, a model-agnostic explainer, achieves three critical properties in explaining model predictions. It approximates the local behavior of a complex prediction model f with a linear model of feature importance g, ensuring $g(\mathbf{z}) \approx f(\boldsymbol{h_x}(\mathbf{z}))$. The training of $g(\mathbf{z})$ task is framed as a cooperative game, distributing the score or prediction of the complex model, $f(\mathbf{x})$ fairly among the features $i \in \{1, \ldots, m\}$ using Shapley values formulation. Since exact Shapley values for all features are computationally infeasible due to exponential scaling with feature count m, KernelSHAP approximates them via random feature coalition sampling. Additionally, a single coalition weighing kernel $\pi_x(\mathbf{z})$ and loss metric $L(f, g, \pi_x)$, as described in Eqs. (14.3) and (14.4), are shown to yield optimal Shapley value approximations. The $|z|$ in Eq. (14.3) signifies the number of nonzero elements of z.

$$\pi_x(z) = \frac{(m-1)}{\binom{m}{|z|}|z|(m-|z|)} \tag{14.3}$$

$$L(f, g, \pi_x) = \sum_{z \in \{0,1\}^m} \left[f(h_x(z)) - g(z)\right]^2 \cdot \pi_x(z) \tag{14.4}$$

14.2.3.4 TimeSHAP

KernelSHAP is popularly known for its effectiveness in elucidating models with single input instance features. TimeSHAP is an extension of the KernelSHAP method specifically designed for time-series data, allowing for the explanation of important features and time steps contributing to the predictions. It maintains the key desirable properties associated with importance attribution from Shapley values, including local accuracy, handling missing data, and ensuring consistency (Shapley 1953). The explanation is enabled by assigning an importance value to each feature or event in the input sequence of the features, indicating the extent to which that particular feature or event influenced the model prediction. By fitting a linear explainer g to approximate the local behavior of a complex prediction model f, TimeSHAP provides explanations for sequential inputs, treating events as features in the temporal dimension. The process for explaining both features $x \in R^{1\times l}$ and events $x \in R^{d\times 1}$ is conceptually same, and the term "feature" is used interchangeably for both rows and columns of the input matrix. TimeSHAP utilizes the loss function described in Eq. (14.4) to approximate explainer g to prediction model f and the expression for the modified explainer g is as given in Eq. (14.5).

$$f(h_x(z)) \approx g(z) = w_0 + \sum_{i=1}^{m} w_i z_i \tag{14.5}$$

The term $h_x(0)$ represents no perturbation in the input vector and produces an uninformative input sample or background matrix, $\mathbf{B} \in R^{d \times l}$. The bias term $w_0 = f(h_x(0) = \boldsymbol{B})$ in the above equation represents the model's output without the perturbation to input vector, known as the base score. The weights w_i denotes the importance of each feature, $i \in \{1, \ldots, d\}$ or each event, $i \in \{1, \ldots, l\}$. The perturbation function h_x is defined similar to Eq. (14.2) and is explained in detail subsequently. It can observed from Eq. (14.5) that the sum of all feature importances leads the explainer g's output from the base score of $f(h_x(0))$ to $f(h_x(z))$, signifying the additive contribution of each feature or event on the final output of the prediction model f.

TimeSHAP distinguishes itself from KernelSHAP by its approach to perturbing input data. While KernelSHAP leverages Eq. (14.2) that utilizes a background vector solely focusing on features for perturbing the input vector, TimeSHAP employs a background matrix that encompasses both features and time-events. Authors of the TimeSHAP initially described that background matrix $B \in R^{l \times d}$ is calculated using average feature values in the dataset. They implemented this by calculating the median for numerical features and the mode for categorical features present in dataset and repeating it N times to form the background sequence. The background matrix B is mathematically expressed as given in Eq. (14.6).

$$\begin{bmatrix} \overline{x_1} & \cdots & \overline{x_1} \\ \overline{x_2} & \cdots & \overline{x_2} \\ \vdots & \ddots & \vdots \\ \overline{x_l} & \cdots & \overline{x_l} \end{bmatrix} \tag{14.6}$$

A coalition vector, either $z \in \{0, 1\}^d$ or $z \in \{0, 1\}^l$, is calculated similar to KernelSHAP depending whether a feature or an event is being explained. While explaining the features, $z_i = 0$ ensures that the feature i is set to the background value for all the events in the sequence. Similarly, while explaining the events, $z_j = 0$ ensures that all features of event j are set to background value. Therefore, the perturbation function for explaining the features, $h^f{}_x$ and the perturbation function for explaining the events, $h^e{}_x$, are defined as given below.

$$h_X^f(Z) = D_z X + (I - D_Z)B \tag{14.7}$$

$$h_X^e(Z) = X D_z + B(I - D_Z) \tag{14.8}$$

Here, D_z is constructed as a diagonal matrix from coalition vector z, D_z = diagonal (z). It is interesting to note at $z = 0$, $h^f{}_x(0) = h^e{}_x(0) = B$, the perturbation functions in Eqs. (14.7) and (14.8) will transform to the background matrix B. Similarly, at $z = \mathbf{1}$, $h^f{}_x(1) = h^e{}_x(1) = X$, the perturbation functions in Eqs. (14.7) and (14.8) will transform to the input data X.

Temporal Explanation and Pruning

Obtaining Shapley values for all features entails generating every conceivable coalition of the input, denoted by $z \in \{0, 1\}^m$, a task that grows exponentially with the number of features, m in the model. However, this computational task is impractical due to its exponential complexity. KernelSHAP addresses this challenge by resorting to approximating the exact values by randomly sampling the feature coalitions. Similarly, TimeSHAP grapples with increased complexity due to the exponential rise in event coalitions with sequence length, compounding existing challenges. TimeSHAP utilizes a novel temporal coalition pruning algorithm, namely pruning algorithm that mitigates scalability challenges in temporal domain.

The pruning method entails consolidating less significant older events into a single coalition, reducing computational complexity by a factor of 2^{i-1}, where i is the number of grouped events. This enhances runtime speed and precision for crucial events in explanation while sacrificing granularity on older events. However, all pruned events are still considered in the explanation process, with their Shapley values summed up to a single value.

The pruning method involves dividing the input sequence $X \in R^{d\times l}$ into two sub-sequences $X_{:,1:i}$ and $X_{:,i+1:l}$ for $i \in \{1, \ldots, l-1\}$, with $X_{:,l}$ representing the most recent event. True Shapley values are computed for each sub-sequence, requiring consideration of four coalitions ($Z \leftarrow \{[0, 0], [0, 1], [1, 0], [1, 1]\}$ for each i). TimeSHAP identifies the largest i where the importance value for pruned events, $X_{:,1:i}$, is less than a predefined threshold, η. This algorithm is explained in detail in *Algorithm 1 Temporal Coalition Pruning* in TimeSHAP (Bento et al. 2021).

Cell-Level Explanations

TimeSHAP extends its explanation capabilities to include cell-wise attributions, akin to feature-level and event-level explanations, by perturbing cells of the input matrix. Since, the number of cell-level coalitions scale exponentially, $O(2^{ld})$, with the product of number of events and number of features, achieving reliable cell-level explanations necessitates either a drastic reduction in considered cells or the implementation of a pruning/grouping strategy. TimeSHAP utilizes cell grouping strategy that aggregates cells into single coalition, enabling a range of explanation granularity while mitigating computational complexity and ensuring the reliability of explanations. TimeSHAP considers a feature or an event as significant if its absolute attribution exceeds a user-defined threshold θ.

TimeSHAP constructs five different groups that represent all possible coalitions of the given cells. TimeSHAP first prunes older events (P) and then computes event-level and feature-level explanations to identify the most relevant columns and rows. The intersection ($|C^*|$) of these most relevant columns and rows yields the most important cells. TimeSHAP then forms groups of cells belonging to relevant events but not features (E'), and vice versa (F'), while the remaining cells form a single coalition (C'). Therefore, the coalitions to be perturbed are determined by the groups defined as $U = P \cup C^* \cup E' \cup F' \cup C'$, and this approach drastically reduces the number of coalitions to be perturbed from $O(2^{ld})$ to 2^5, resulting in a significant reduction in computational complexity. The perturbation function in this case, $h^c{}_X$, is detailed in Eq. (14.9).

$$h_X^c(Z) = X \odot Z + B \odot (J - Z) \tag{14.9}$$

In Eq. (14.9), coalition matrix $Z_{r,c}$ is defined as $Z_{r,c} = z_i \ \forall (r, c) \in m[z_i] \ \forall z_i \in z$. Further, the matrix $J \in \{1\}^{l \times d}$, and the background matrix $B \in R^{d \times l}$ are similar to the matrices explained in Eqs. (14.7) and (14.8).

14.3 Results and Discussion

14.3.1 Data Analysis

The active power generated by the turbine depends on different process and climate variables and varies as shown in Fig. 14.2. It is observed that active power varies from 0 to 2026.39 kW with a mean value of 348.99 kW and a standard deviation of 395.84 kW. It is also evident from Fig. 14.2 that a significant number of observations (~ 73.77%) have active power values less than 500 kW and 11.05% of the total observations have zero active power. This is because, the turbine only starts to turn with a wind speed of 3 m/s and there are 14.63% observations with wind speed < 3 m/s. Although the turbines are shut down when the wind speed exceeds 25 m/s, there are no observations with wind speeds exceeding 25 m/s and the maximum average wind speed is 17.61 m/s as mentioned in Table 14.1.

The multicollinearity concerns and dimensionality reduction of the data are performed using the Pearson correlation. The Pearson correlation coefficient, r, quantifies the linear relationship between two continuous variables and r ranges from − 1

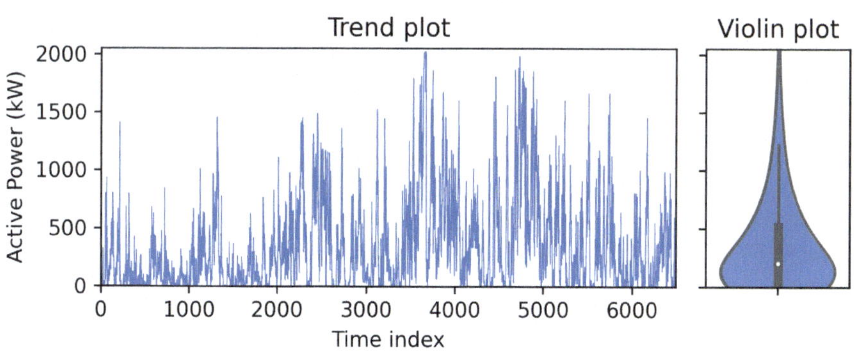

Fig. 14.2 Trend plot of active power (kW) and the violin plot of active power (kW)

to 1 while $r = 1$ indicates a perfect positive linear relationship and $r = -1$ indicates a perfect negative linear relationship. However, $r = 0$ indicates no linear relationship between the variables. Pearson correlation combined with hierarchical clustering of the dataset helps identifying the clusters of variables that exhibit similar patterns or relationships. Hierarchical clustering is a method in which a hierarchy of clusters are built iteratively by merging highly correlated variables, using r, into smaller clusters and then subsequently merging the smaller clusters into larger ones. Figure 14.3 displays a hierarchically clustered heatmap of the given dataset, and eight important clusters are inferred from the same. The important clusters are detailed in Table 14.2. One feature from each important cluster is identified for building a predictive model for active power and is highlighted in Table 14.2. A heatmap of the Pearson correlation factors of the selected variables is as shown in Fig. 14.4. Among these variables, as shown in Fig. 14.4, pitch angle has a correlation factor − 0.781 with gearbox inlet temperature. Generator bearing 2 temperature has a correlation factor of 0.736 and 0.814 with gearbox inlet temperature and rotor bearing temperature, respectively. Correlation factors of the remaining variables are less than 0.7 and greater than − 0.7. The time-series trend plots of important variables are plotted in Fig. 14.5.

14.3.2 Predictions of Active Power

The preprocessed data from Sect. 14.2 is used to build LSTM model for the prediction of active power. The data is at hourly frequency, and 6 h of the past data is considered to estimate the active power at the current timestamp. The given data is converted to sequential data and is explained in Fig. 14.6. The data of 6500 observations is hence converted to 6495 sequences as the first five observations cannot have past 6 h data. The entire dataset is normalized to ensure that the scale of the variables does not affect the model training procedure. This data is randomly split into train (4547 sequences) and test (1948 sequences) sequences in the ratio of 70:30.

The LSTM model architecture is designed such that it has one input layer with 6 neurons for each input variable, 2 hidden layers with 64 and 16 LSTM cells, respectively, and one output layer with one neuron for the output variable (active power). Tanh activation function and Adam optimizer are chosen for this model. This architecture has a total of 23,889 parameters. 4547 training sequences are used for training the LSTM model till convergence. The prediction accuracy of the LSTM model with train and test sequences were evaluated by coefficient of regression (R^2), mean absolute error (MAE), mean square error (MSE), and root mean square error (RMSE). The accuracy metrics evaluated for train and test sequences are presented in Table 14.3. The R^2 value for the train and test sequences are 0.997 and 0.996, respectively. The predictions of the LSTM model with train and test sequences are shown in Figs. 14.7 and 14.8. It is observed from Figs. 14.7 and 14.8 that the LSTM model captured the time-series behavior of the active power accurately with neither over-fitting nor under-fitting.

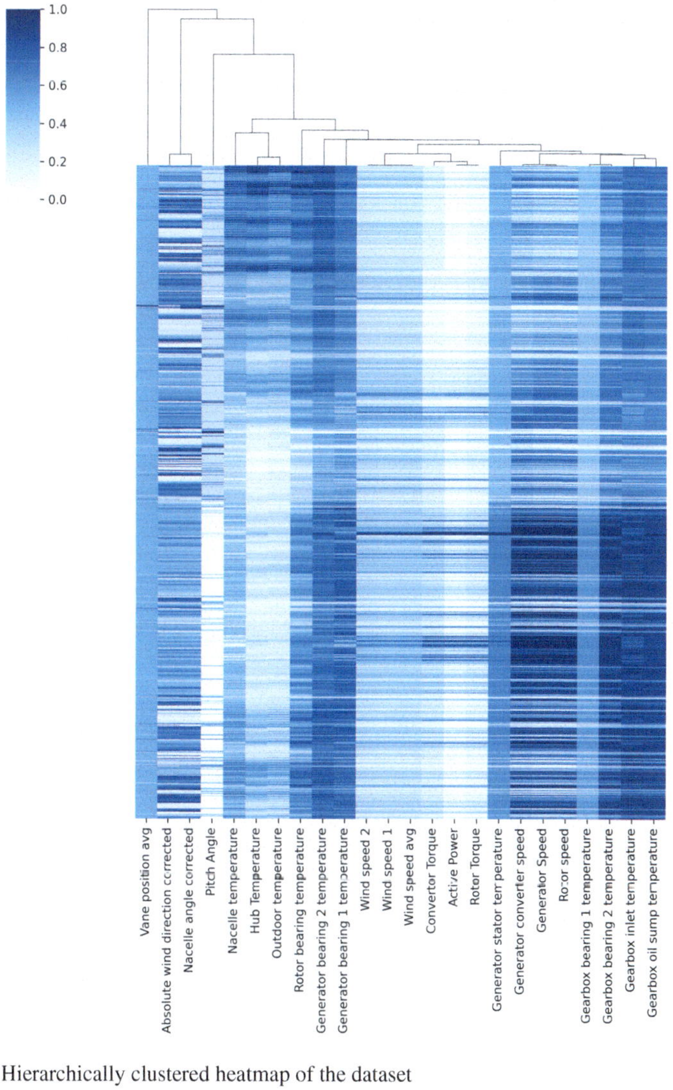

Fig. 14.3 Hierarchically clustered heatmap of the dataset

14.3.3 *TimeSHAP*

TimeSHAP is used to explain the predictions of active power from LSTM model that considers six events from the past of six variables, namely pitch angle, hub temperature, generator bearing 2 temperature, gearbox inlet temperature, average wind speed, average vane position, rotor bearing temperature, and corrected absolute wind direction. It was reported that the maximum number of coalitions of 32,000

Table 14.2 Clusters and their corresponding variables

Clusters	Variables
Cluster 1	'*Average vane position*'
Cluster 2	'*Corrected absolute wind direction*' and corrected nacelle angle
Cluster 3	'*Pitch angle*'
Cluster 4	Nacelle temperature, '*hub temperature*' and outdoor temperature
Cluster 5	'*Rotor bearing temperature*'
Cluster 6	Wind speed 1, wind speed 2, '*average wind speed*', convertor torque, rotor torque and active power
Cluster 7	Generator stator temperature, generator convertor speed, generator speed, rotor speed, gearbox bearing 1 temperature, gearbox bearing 2 temperature, '*gearbox inlet temperature*', and gearbox oil sump temperature
Cluster 8	Generator bearing 1 temperature and '*generator bearing 2 temperature*'

The highlighted variable of each cluster is considered as representative variable of that cluster

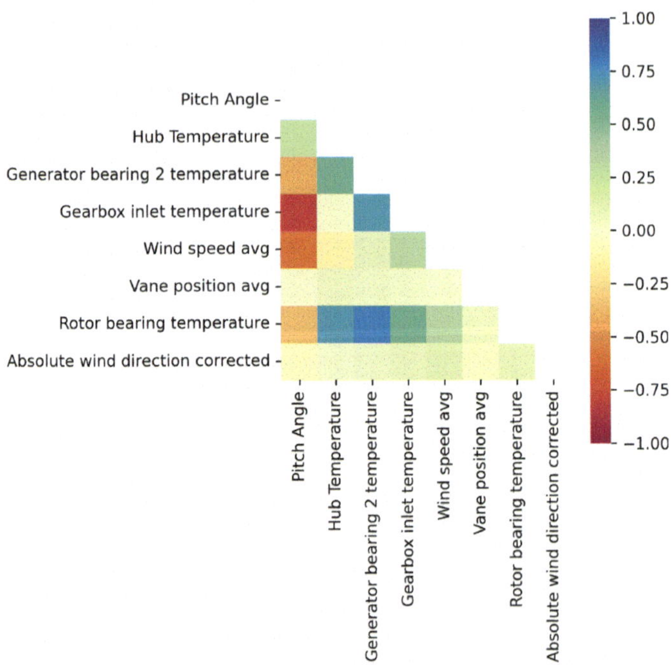

Fig. 14.4 Correlation heatmap of the variables that are considered for training of LSTM model

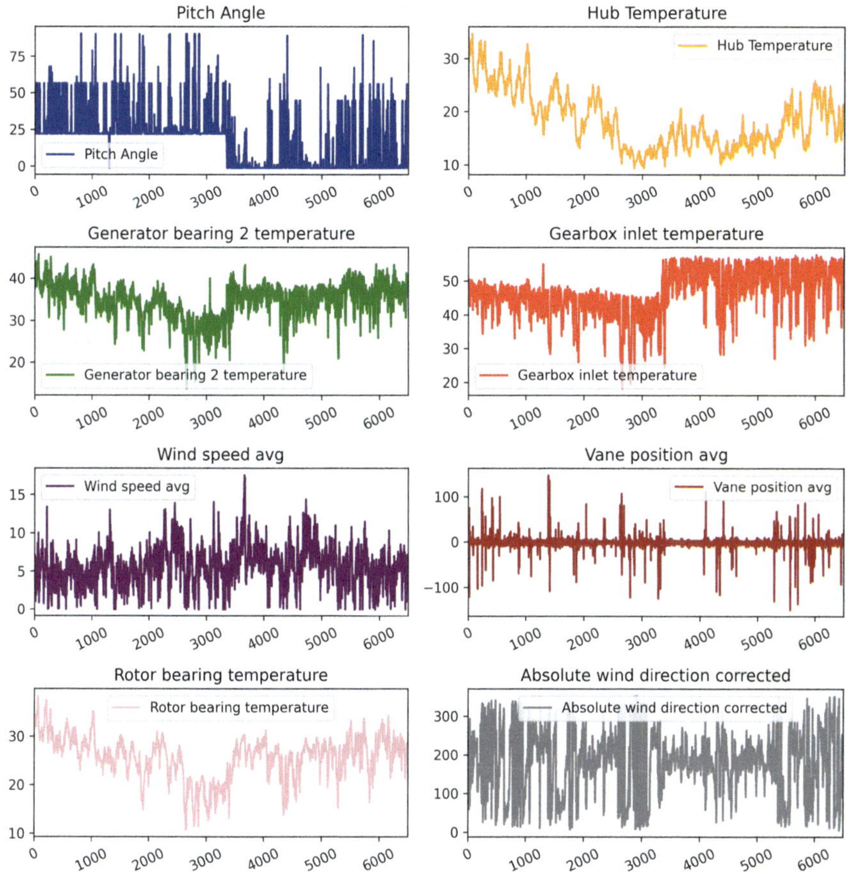

Fig. 14.5 Trend plots of the variables that are considered for training of LSTM model

(a)

Time	Input Values	Active Power
1		
2		
3		
4		
5		
6		
7		
8		
9		
10		

(b)

Time	Input Values	Active Power
1		
2		
3		
4		
5		
6		
7		
8		
9		
10		

(c)

Time	Input Values	Active Power
1		
2		
3		
4		
5		
6		
7		
8		
9		
10		

Fig. 14.6 Explanation of extraction of sequences, for example, **a** sequence-1, **b** sequence-2, and **c** sequence-3 from the time-series data. A total of six time-events, including the current time step, is used to estimate the active power at the current time-event

Table 14.3 Performance metrics of LSTM model with train and test sequences

Metric	Train	Test
R^2	0.997	0.996
MAE	15.279	16.562
RMSE	21.326	23.712
MSE	454.784	562.269

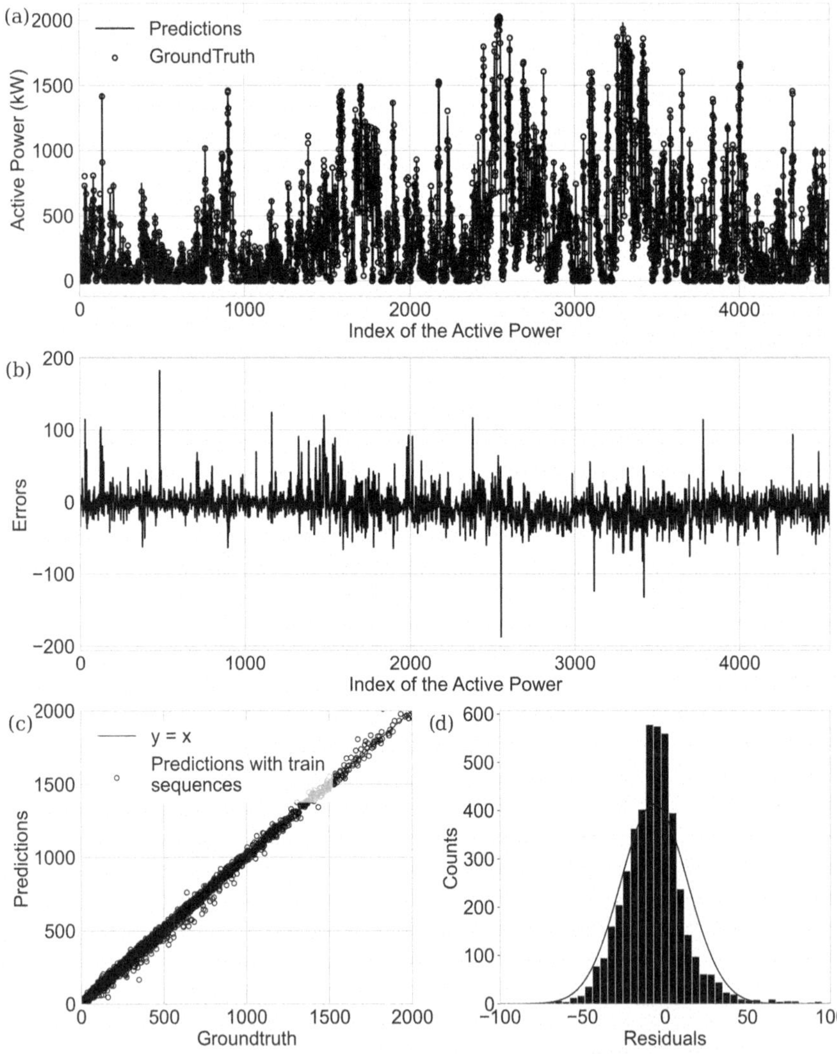

Fig. 14.7 Comparative analysis of model predictions of train sequences with actual data. **a** Trend plot of the actual and predicted active power, **b** trend plot of the residuals, **c** parity plot comparing the actual and predicted active power, and **d** density plot of the residuals of the model predictions

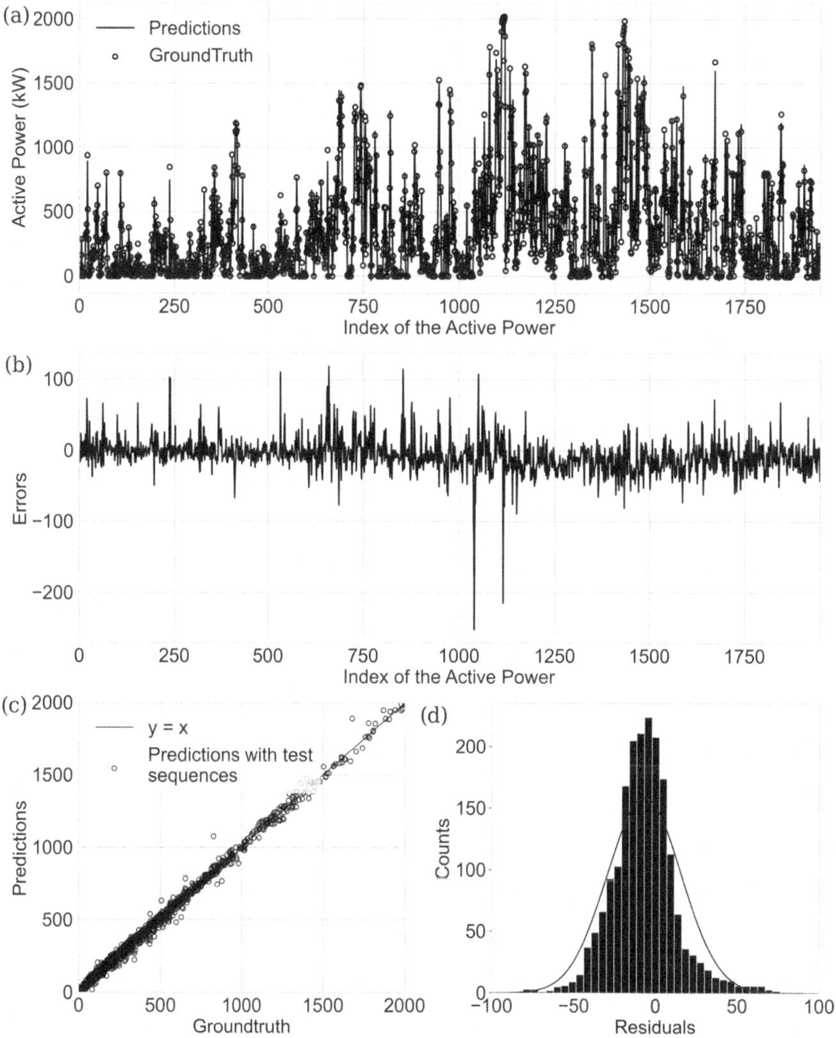

Fig. 14.8 Comparative analysis of model predictions of test sequences with actual data. **a** Trend plot of the actual and predicted active power, **b** trend plot of the residuals, **c** parity plot comparing the actual and predicted active power, and **d** density plot of the residuals of the model predictions

provides an optimal trade-off between TimeSHAP's computational cost and reliability, and hence, it is set to 32,000. For cell-level explanations, a threshold $\theta = 0.1$ is set for a feature or an event to be considered as significant.

Table 14.4 Statistical summary of pruning results for various tolerances

Tolerance	Mean	Standard deviation
0.01	1.446	1.022
0.025	0.910	0.436
0.05	0.760	0.459
0.075	0.616	0.495
0.1	0.384	0.487
0.25	0.152	0.359
No pruning	6.000	0.000

14.3.3.1 Pruning Results

To obtain the pruning results, we applied the temporal coalition pruning algorithm of TimeSHAP to a subset of 500 sequences randomly selected from the total 1948 test sequences. In this analysis, TimeSHAP focuses solely on the event-level explanations, as the pruning does not impact feature-level explanations.

Table 14.4 outlines the mean and standard deviation of the number of events for unpruned sequences across various pruning tolerances (η) ranging from 0.01 to 0.25, with an additional category for sequences with no pruning applied. We observe that lower pruning tolerances result in more detailed event-level explanations, albeit with greater variability. A clear inverse relationship is evident between pruning tolerance and relative standard deviation (RSD) values. For instance, evaluating pruning algorithm with the lowest tolerance, $\eta = 0.01$, yields notably high variance (RSD = 1.022), while the highest pruning tolerance, $\eta = 0.25$, exhibits least variance (RSD = 0.359). A well-balanced trade-off between detailed event-level explanations and RSD is achieved at $\eta = 0.025$ with a mean = 0.910 and an RSD = 0.436. These results indicate that the most recent event of the sequence has the highest impact on the prediction of active power and the effect of subsequent time-events is negligible for tolerances higher than 0.025.

14.3.3.2 Local Explanations

Local explanations elucidate the model's decision-making process for a particular instance, serving various purposes such as troubleshooting, etc. In this work, we present a use case for which the active power was 802.107 kW.

In Fig. 14.9a, the importance or the Shapley values of grouped events ($X_{:,:t}$) calculated using pruning algorithm are depicted as a function of time-event t. The cumulative importance of events declines sharply with increasing distance from the current time-event, t. This observed trend validates the assumption and underscores the effectiveness of the coalition pruning algorithm. At a coalition pruning tolerance of $\eta = 0.025$, the sequence considered is pruned to only one event, amalgamating the five older events.

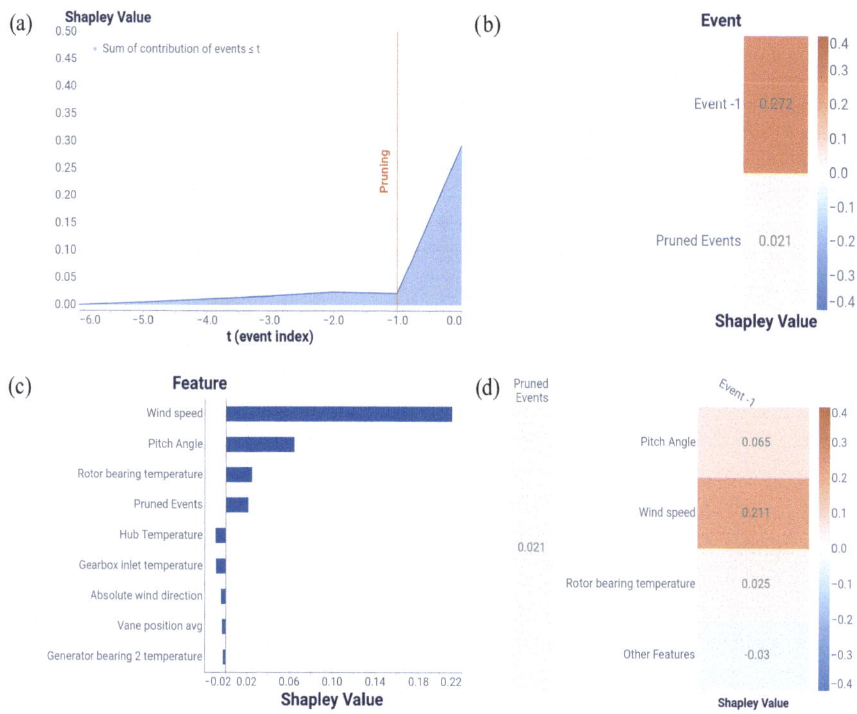

Fig. 14.9 Local level explanation of LSTM model **a** visualization of pruning statistics, **b** event-level local explanation, **c** feature-wise local explanation, and **d** cell-level local explanation

Event-wise explanations for this sequence are depicted in Fig. 14.9b, while its feature-wise explanations are illustrated in Fig. 14.9c. This analysis reveals that only one event, the event being explained ($t = -1$), holds the highest importance for the model's prediction, with a Shapley value of 0.272 contributing 92.83% of the sequence score. The subsequent events are pruned, resulting in an aggregate Shapley value of 0.021 for the pruned events contributing just 7.17% of the sequence score. A similar, but an overall conclusion of event-level and feature-wise explanation can be obtained from cell-level explanations using Fig. 14.9d.

In terms of feature importances, we note that the most significant features, ranked in descending order of importance, are the average wind speed, the pitch angle, and the rotor bearing temperature, aligning with conventional understanding of the wind turbine operation. Furthermore, the Shapley value attributed to the average wind speed is notably high at 0.211, underscoring its crucial role as a predictor of active power, given its direct correlation with active power, as depicted in Fig. 14.10. Additionally, the pitch angle contributes to active power prediction with a Shapley value of 0.065.

Fig. 14.10 Effect of average wind speed on active power with a strong linear correlation between an average wind speed of 5 and 12 m/s

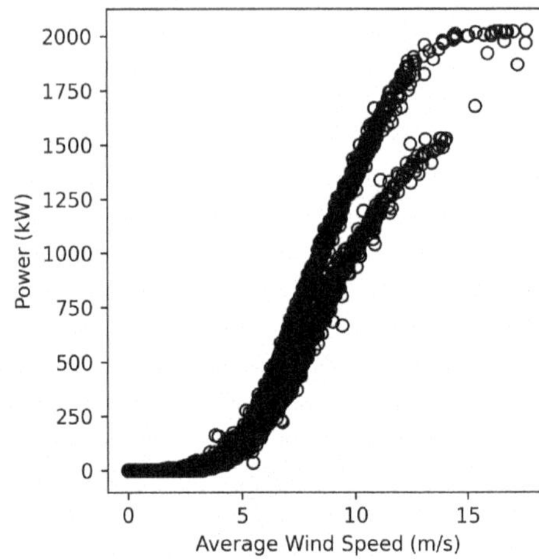

Examining the cell-level explanations in Fig. 14.9d for this sequence with $\theta = 0.1$, the earlier observation indicating that the average wind speed measurement at $t = -1$ holds the highest importance which is validated. Remarkably, this single cell accounts for 0.211 of the score, demonstrating its exceptional importance within a sequence consisting of six cells. Furthermore, these cell-level explanations reveal that pertinent cells are situated within relevant events, while cells associated with non-relevant events collectively receive an importance score of 0.021.

14.4 Conclusion

In the realm of artificial intelligence for the prediction of active power at wind turbines, LSTM-based models are often preferred for their effectiveness. However, their adoption in the real-world is hindered by concerns about their lack of transparency.

In response, we utilized TimeSHAP, a model-agnostic and post-hoc recurrent explainer designed to elucidate the predictions of the LSTM model. Leveraging only the features of each instance and an inference API, TimeSHAP offers three levels of explanations: event, feature, and cell attributions, computed through customized perturbation functions tailored for sequences.

Using TimeSHAP, we provide insights into the decision-making processes of an LSTM-based model for the wind turbine. Our findings underscore the model's reliability, as the features and events identified as significant by TimeSHAP align with domain experts' knowledge, particularly in detecting the effect of average wind speed on active power. We observe that the first event typically accounts for 92.83%

of the sequence's score, while preceding events contribute the remaining 7.17%. Furthermore, our analysis reveals the importance of the average wind speed and the pitch angle in the model's predictions.

Acknowledgements We would like to acknowledge the support provided by the SPARC project (SPARC/2018–2019/P1084/SL) sponsored by the Ministry of Education (MoE), Government of India.

References

Akoğlu H (2018) User's guide to correlation coefficients. Turk J Emerg Med 18:91–93. https://doi.org/10.1016/j.tjem.2018.08.001

Bento J et al (2021) TimeSHAP: explaining recurrent models through sequence perturbations. In: Proceedings of the 27th ACM SIGKDD conference on knowledge discovery & data mining, pp 2565–2573. https://doi.org/10.1145/3447548.3467166

Choi E et al (2016) RETAIN: an interpretable predictive model for healthcare using reverse time attention mechanism. Neural Inf Process Syst 29:3504–3512. https://papers.nips.cc/paper/6321-retain-an-interpretable-predictive-model-for-healthcare-using-reverse-time-attention-mechanism.pdf

Foley A et al (2012) Current methods and advances in forecasting of wind power generation. Renew Energy 37:1–8. https://doi.org/10.1016/j.renene.2011.05.033

Guidotti R et al (2018) A survey of methods for explaining black box models. ACM Comput Surv 51:1–42. https://doi.org/10.1145/3236009

Hochreiter S, Schmidhuber J (1997) Long short-term memory. Neural Comput 9:1735–1780. https://doi.org/10.1162/neco.1997.9.8.1735

Hu J et al (2020) Deterministic and probabilistic wind speed forecasting with de-noising-reconstruction strategy and quantile regression based algorithm. Renew Energy 162:1208–1226. https://doi.org/10.1016/j.renene.2020.08.077

IRENA (2023) World energy transitions outlook 2023: 1.5°C pathway. International Renewable Energy Agency. https://mc-cd8320d4-36a1-40ac-83cc-3389-cdn-endpoint.azureedge.net/-/media/Files/IRENA/Agency/Publication/2023/Jun/IRENA_World_energy_transitions_outlook_2023.pdf?rev=db3ca01ecb4a4ef8accb31d017934e97

Ismail AA et al (2019) Input-cell attention reduces vanishing saliency of recurrent neural networks. Cornell University. arXiv preprint https://arxiv.org/pdf/1910.12370

Lange M, Focken U (2006) Physical approach to short-term wind power prediction. Springer eBooks. https://doi.org/10.1007/3-540-31106-8

Liu H, Mi X, Li Y (2018) Smart multi-step deep learning model for wind speed forecasting based on variational mode decomposition, singular spectrum analysis, LSTM network and ELM. Energy Convers Manage 159:54–64. https://doi.org/10.1016/j.enconman.2018.01.010

Lundberg S, Lee S-I (2017) A unified approach to interpreting model predictions. Neural Inf Process Syst 30:4768–4777. https://papers.nips.cc/paper/7062-a-unified-approach-to-interpreting-model-predictions.pdf

Lundberg S et al (2020) From local explanations to global understanding with explainable AI for trees. Nat Mach Intell 2:56–67. https://doi.org/10.1038/s42256-019-0138-9

Neshat M et al (2021) A deep learning-based evolutionary model for short-term wind speed forecasting: a case study of the Lillgrund offshore wind farm. Energy Convers Manage 236:114002. https://doi.org/10.1016/j.enconman.2021.114002

Niu Z et al (2020) Wind power forecasting using attention-based gated recurrent unit network. Energy 196:117081. https://doi.org/10.1016/j.energy.2020.117081

Ribeiro MT, Singh S, Guestrin C (2016) Model-agnostic interpretability of machine learning. Cornell University. arXiv preprint https://arxiv.org/pdf/1606.05386

Sha Y, Wang MD (2017) Interpretable predictions of clinical outcomes with an attention-based recurrent neural network. In: ACM-BCB'17: proceedings of the 8th ACM international conference on bioinformatics, computational biology, and health informatics, pp 233–240. https://doi.org/10.1145/3107411.3107445

Shahid F, Zameer A, Muneeb M (2021) A novel genetic LSTM model for wind power forecast. Energy 223:120069. https://doi.org/10.1016/j.energy.2021.120069

Shapley LS (1953) A value for n-person games. In: Contributions to the theory of games, 2nd edn, pp 307–317

Taylor JW, Buizza R (2003) Using weather ensemble predictions in electricity demand forecasting. Int J Forecast 19:57–70. https://doi.org/10.1016/s0169-2070(01)00123-6

Van Houdt G, Mosquera C, Nápoles G (2020) A review on the long short-term memory model. Artif Intell Rev 53:5929–5955. https://doi.org/10.1007/s10462-020-09838-1

Wan J et al (2016) Day-ahead prediction of wind speed with deep feature learning. Int J Pattern Recognit Artif Intell 30:1650011. https://doi.org/10.1142/s0218001416500117

Yu R et al (2019) LSTM-EFG for wind power forecasting based on sequential correlation features. Future Gener Comput Syst 93:33–42. https://doi.org/10.1016/j.future.2018.09.054

Yu Y et al (2020) Probabilistic prediction of regional wind power based on spatiotemporal quantile regression. IEEE Trans Ind Appl 56:6117–6127. https://doi.org/10.1109/tia.2020.2992945

Zhang J et al (2018) Patient2VEC: a personalized interpretable deep representation of the longitudinal electronic health record. IEEE Access 6:65333–65346. https://doi.org/10.1109/access.2018.2875677

The manufacturer's authorised representative in the EU is Springer Nature Customer Service Centre GmbH, Europaplatz 3, 69115 Heidelberg, Germany. If you have any concerns regarding our products, please contact ProductSafety@springernature.com

Printed and bound by CPI Group (UK) Ltd, Croydon, CR0 4YY
30/09/2026
02225302-0007